VOLUME ONE HUNDRED

ADVANCES IN GENETICS

Fungal Phylogenetics and Phylogenomics

ADVANCES IN GENETICS, VOLUME 100

VOLUME ONE HUNDRED

ADVANCES IN GENETICS

Fungal Phylogenetics and Phylogenomics

Edited by

JEFFREY P. TOWNSEND

Departments of Biostatistics and Ecology and Evoilutionary Biology, Yale University, New Haven, CT, United States

ZHENG WANG

Department of Biostatistics, Yale University, New Haven, CT, United States

ELSEVIER

ACADEMIC PRESS

An imprint of Elsevier

Academic Press is an imprint of Elsevier
50 Hampshire Street, 5th Floor, Cambridge, MA 02139, United States
525 B Street, Suite 1800, San Diego, CA 92101–4495, United States
The Boulevard, Langford Lane, Kidlington, Oxford OX5 1GB, United Kingdom
125 London Wall, London, EC2Y 5AS, United Kingdom

First edition 2017

Notices
Knowledge and best practice in this field are constantly changing. As new research and experience broaden our understanding, changes in research methods, professional practices, or medical treatment may become necessary.

Practitioners and researchers must always rely on their own experience and knowledge in evaluating and using any information, methods, compounds, or experiments described herein. In using such information or methods they should be mindful of their own safety and the safety of others, including parties for whom they have a professional responsibility.

To the fullest extent of the law, neither the Publisher nor the authors, contributors, or editors, assume any liability for any injury and/or damage to persons or property as a matter of products liability, negligence or otherwise, or from any use or operation of any methods, products, instructions, or ideas contained in the material herein.

ISBN: 978-0-12-813261-6
ISSN: 0065-2660

For information on all Academic Press publications
visit our website at https://www.elsevier.com/books-and-journals

Publisher: Zoe Kruze
Acquisition Editor: Zoe Kruze
Editorial Project Manager: Alina Cleju
Production Project Manager: James Selvam
Cover Designer: Miles Hitchen

Typeset by SPi Global, India

CONTENTS

CONTRIBUTORS

M. Catherine Aime
Purdue University, West Lafayette, IN, United States

Debashish Bhattacharya
Rutgers University, New Brunswick, NJ, United States

Alex Dornburg
North Carolina Museum of Natural Sciences, Raleigh, NC, United States

Sébastien Duplessis
IUnité Mixte de Recherche INRA/Université de Lorraine, Champenoux, France

Rhys A. Farrer
Imperial College London, London, United Kingdom

Matthew C. Fisher
Imperial College London, London, United Kingdom

David A. Fitzpatrick
Maynooth University, Maynooth, County Kildare, Ireland

Jing Luo
Rutgers University, New Brunswick, NJ, United States

Li-Jun Ma
University of Massachusetts Amherst, Amherst, MA, United States

Charley G.P. McCarthy
Maynooth University, Maynooth, County Kildare, Ireland

Alistair R. McTaggart
University of Pretoria, Pretoria, South Africa

Stephen J. Mondo
US Department of Energy Joint Genome Institute, Walnut Creek, CA, United States

László G. Nagy
Hungarian Academy of Sciences, Szeged, Hungary

Jason C. Slot
The Ohio State University, Columbus, OH, United States

Gergely Szöllősi
Eötvös Loránd University, Budapest, Hungary

Jeffrey P. Townsend
Yale University, New Haven, CT, United States

Zheng Wang
Yale University, New Haven, CT, United States

Ning Zhang
Rutgers University, New Brunswick, NJ, United States

Yong Zhang
University of Massachusetts Amherst, Amherst, MA, United States

PREFACE

Advances in the rapid and voluminous acquisition of gene sequence data have been dramatically reshaping the study of fungal systematics and evolution for more than two decades. The recent increase in availability of genomic, metagenomic, and transcriptomic tools—extending from model to nonmodel fungal species—has brought fungal phylogenetics and systematics research into a new age of "big data," presenting new challenges in data analysis and enabling tremendous opportunities to reveal new science via creative approaches. This Thematic Volume—the 100th in the *Advances in Genetics* series—contains eight comprehensive reviews laying out recent progress and future challenges in fungal phylogenetics and phylogenomics. Chapter 1 reviews the broad importance of fungal phylogenetics and phylogenomics. It highlights the new role of quantification of phylogenetic power in the resolution of diverse questions in phylogenomics, particularly in addressing the challenge of curating big data when it cannot be handled manually. Chapter 2 reviews the important consequences of genome-scale datasets applied to fungal phylogenetic inferences. Chapter 3 focuses on the important application of genomic and epigenomic data toward understanding evolution and distribution of fungal pathogens. Chapter 4 focuses on the diversity of fungal gene clusters, which are important to fungal pathogenesis and to secondary metabolic biochemistry, reviewing recent discoveries regarding how gene clusters evolve along the fungal phylogeny. Chapter 5 focuses further, revealing recent breakthroughs and prospects in research on the pathogen model *Fusarium oxysporum*, covering evolutionary analyses of phylogenomic and pathogenicity data. Chapter 6 provides a benchmarking comparison of popular approaches in phylogenomics, and capitalizing on that comparison, presents a reconstructed phylogenetic backbone for the Fungal Kingdom based on available fungal genome data. Chapter 7 reviews recent advances of the phylogenetics and phylogenomics of rust fungi, one of largest and most systematically problematic fungal groups, that contains pathogenic species that target diverse plants. Chapter 8 then reviews how recent developments in fungal phylogenomics have impacted systematic practice for fungi.

We are most grateful to the authors for their excellent contributions, and urge readers—whether fungal biologists, phylogeneticists of any stripe, or otherwise—to engage in thoughtful contemplation of these excellent reviews. It will be time well spent.

Jeffrey P. Townsend
Zheng Wang
New Haven, USA

CHAPTER ONE

Maximizing Power in Phylogenetics and Phylogenomics: A Perspective Illuminated by Fungal Big Data

Alex Dornburg*, Jeffrey P. Townsend†, Zheng Wang†,[1]
*North Carolina Museum of Natural Sciences, Raleigh, NC, United States
†Yale University, New Haven, CT, United States
[1]Corresponding author: e-mail address: Wang.Zheng@Yale.edu

Contents

Abstract

Since its original inception over 150 years ago by Darwin, we have made tremendous progress toward the reconstruction of the Tree of Life. In particular, the transition from analyzing datasets comprised of small numbers of loci to those comprised of hundreds of loci, if not entire genomes, has aided in resolving some of the most vexing of evolutionary problems while giving us a new perspective on biodiversity. Correspondingly, phylogenetic trees have taken a central role in fields that span ecology, conservation, and medicine. However, the rise of big data has also presented phylogenomicists with a

Advances in Genetics, Volume 100
ISSN 0065-2660
https://doi.org/10.1016/bs.adgen.2017.09.007

new set of challenges to experimental design, quantitative analyses, and computation. The sequencing of a number of very first genomes presented significant challenges to phylogenetic inference, leading fungal phylogenomicists to begin addressing pitfalls and postulating solutions to the issues that arise from genome-scale analyses relevant to any lineage across the Tree of Life. Here we highlight insights from fungal phylogenomics for topics including systematics and species delimitation, ecological and phenotypic diversification, and biogeography while providing an overview of progress made on the reconstruction of the fungal Tree of Life. Finally, we provide a review of considerations to phylogenomic experimental design for robust tree inference. We hope that this special issue of *Advances in Genetics* not only excites the continued progress of fungal evolutionary biology but also motivates the interdisciplinary development of new theory and methods designed to maximize the power of genomic scale data in phylogenetic analyses.

1. INTRODUCTION

Phylogenetic studies of genomic scale data have yielded new insights into topics spanning the relationships of major organismal lineages (Dunn et al., 2008; Hejnol et al., 2009; McCormack et al., 2012; Qiu et al., 2006) to the tempo and mode of cancer evolution (Gerlinger et al., 2014; Sun et al., 2017; Zhao et al., 2016). Now, more than 150 years after its inception (Darwin, 1859), the Tree of Life is coming into view. As we enter a time of global change unparalleled with any other in human history (Bellard, Bertelsmeier, Leadley, Thuiller, & Courchamp, 2012; Hansen et al., 2013; Hooper et al., 2012; Seto, Güneralp, & Hutyra, 2012), achieving a resolved Tree of Life is more critical than ever (Buckley & Kingsolver, 2012). We need an understanding of the evolutionary history that gave rise to the extant and historical biodiversity of our planet if we are to accurately forecast how lineages or communities will respond to shifts in climate or habitat (Williams, Henry, & Sinclair, 2015; Willis, Ruhfel, Primack, Miller-Rushing, & Davis, 2008). We need an understanding of what catalyzes the rise and diversification of virulent pathogens or pest organisms if we are to safeguard our crops, products, and citizenry (Biek, Pybus, Lloyd-Smith, & Didelot, 2015; Boldin & Kisdi, 2012; Grenfell et al., 2004; Ploch et al., 2011; Suzán et al., 2015). We need an understanding of the evolutionary pathways that pathogens take to develop effective therapies (Dean et al., 2012; Howlett, Lowe, Marcroft, & Wouw, 2015; Hu et al., 2014; Inoue et al., 2017; Moran, Coleman, & Sullivan, 2010; Shang et al., 2016; Sillo, Garbelotto, Friedman, & Gonthier, 2015). The ways in which phylogenetic trees can illuminate biology and inform decision-making for conservation, economic, or medical purposes are extensive and rapidly growing. Nevertheless, resolving the Tree of Life is far from trivial.

Despite major breakthroughs in sequencing technology and the development of analytical tools, the empirical resolution of some evolutionary relationships continues to defy resolution (Brandley et al., 2015; Eytan et al., 2015; Federman et al., 2015; Hibbett et al., 2007; Krüger, Krüger, Walker, Stockinger, & Schüßler, 2011; Misof et al., 2014; Pöggeler & Wöstemeyer, 2011; Schoch et al., 2009; Spatafora et al., 2016; Spriggs et al., 2015). While it is true that genome-scale sequence data contain tremendous amount of information, their analysis retains numerous analytical challenges and pitfalls that can impede or mislead evolutionary tree inferences (Gatesy, DeSalle, & Wahlberg, 2007; Jeffroy, Brinkmann, Delsuc, & Philippe, 2006; Kumar, Filipski, Battistuzzi, Pond, & Tamura, 2012; Posada, 2016; Salichos, Stamatakis, & Rokas, 2014). We are seeing increasingly frequent instances of strongly supported topological incongruence between research studies, despite analyses of thousands or even millions of molecular characters (Chakrabarty et al., 2017; Fernández et al., 2014; Reddy et al., 2017; Sharma et al., 2014; Shen, Hittinger, & Rokas, 2017; Simion et al., 2017; Simon, Narechania, Desalle, & Hadrys, 2012). Fortunately, the continued development of new theory, models, and analytical tools support the optimistic view that such challenges may not be insurmountable.

Since the extraction of phylogenetic information from some of the very first genomes sequenced (Dujon, 2010; Galagan, Henn, Ma, Cuomo, & Birren, 2005; Levy, 1994; Muers, 2011), fungal phylogeneticists have long been at the forefront of overcoming challenges to phylogenomic inference. In this special volume of *Advances in Genetics*, we review recent achievements and continued challenges in fungal phylogenetics and phylogenomics. We begin by introducing fungi and providing an overview of select topics in evolutionary biology where fungal phylogenetics have made exciting breakthroughs, while highlighting contributed works to this volume. We then provide an overview of key insights into evolutionary relationships that have been gained in phylogenetic/omic analyses, and then move on to focus on how to consider phylogenomic experimental design to refine future tree inferences. Our hope is that this volume not only excites more work in fungal evolutionary biology but also motivates consideration and the development of new theory and methods geared toward maximizing the amount of phylogenetic information extracted from fungal genomes while minimizing error.

2. PHYLOGENETICS AND PHYLOGENOMICS: THE CORE OF BIODIVERSITY SCIENCE

2.1 Why Fungi?

The Kingdom Fungi represents an extraordinary amount of eukaryotic morphological and ecological variation (Stajich et al., 2009). Fungal species richness is estimated to be anywhere between 1.5 and 7.1 million species, with the list of living species continually expanding as new fungal species are identified from around the world (Blackwell, 2011; Hawksworth, 2001; Hibbett, 2007; Martin, Gazis, Skaltsas, Chaverri, & Hibbett, 2015; Schoch et al., 2012; Tedersoo, Bahram, Rasmus, Henrik Nilsson, & James, 2017). While incredible, this staggering richness has provided a monumental challenge to fungal systematics and classification (Hibbett & Taylor, 2013; Wang, Henrik Nilsson, James, Dai, & Townsend, 2016). The simple morphology of many fungi and lack of a detailed fossil record have caused difficulties in fungal classification and systematics (Bard, 2008; Botstein, 1997; Davis, 2000; Hibbett, 2007; Lutzoni et al., 2004; Money, 2016; Stajich et al., 2010). While the simple and easily manipulated bauplan of many fungi has been highly advantageous for studies of developmental or cell biology, fungal phylogenetics and subsequent phylogenomic initiatives have made tremendous strides toward a robust fungal Tree of Life.

Given their staggering biodiversity, ecological importance, and long history of both positive and negative interactions with humans and wildlife (Werner & Kiers, 2012), resolution of the fungal Tree of Life is far from an esoteric objective, but instead of pivotal importance to emerging conservation, management, and human health concerns over the next hundred years. Global changes in climate have begun to drive the movement of entire ecological communities, expanding or collapsing the ranges of individual species (Barnosky et al., 2017; Last et al., 2010). As reviewed in chapters "Describing Genomic and Epigenomic Traits Underpinning Emerging Fungal Pathogens" by Farrer and Fisher, "Deciphering Pathogenicity of Fusarium Oxysporum From a Phylogenomics Perspective" by Zhang and Ma, as well as "Phylogenetics and Phylogenomics of Rust Fungi" by Catherine Aime et al. genomics and epigenomics are becoming key tools in the fight against fungal pathogens. Over the past decade, we have already seen how the rapid spread of fungal pathogens like *Geomyces* or *Batrachochytrium* (chytrid) can devastate wildlife populations (Blehert et al., 2009; Lorch et al., 2011; Skerratt et al., 2007; Warnecke et al., 2012).

As we continue to modify our planet at historically unparalleled speeds, an evolutionary foundation from which to accurately forecast how fungi will respond to change is of critical importance if we are to predict the movement of botanical communities that rely on fungal interactions, the spread of crop pests, or the spread and evolution of virulent fungal pathogens.

2.2 Systematics, Classification, and Species Delimitation

> *The time will come I believe ... when we shall have fairly true genealogical trees of each great kingdom of nature.*
>
> ***Darwin (1897)***

We may look back at the early 21st century as a period of systematic renaissance. Genomic data have fundamentally restructured our understanding of higher-level relationships across all parts of the Tree of Life (Abbott, 2012; Crandall, 2004; Hao, Qi, & Wang, 2003; Harish & Kurland, 2017; Hug et al., 2016; Jones, 2015; Ludmir & Enquist, 2009; Oliverio & Katz, 2014; Parfrey, Lahr, & Katz, 2008; Pennisi, 2008; Prum et al., 2015; Simonson et al., 2005; Walker, 2014; Wang, Xu, Gao, & Hao, 2009; Williams et al., 2017). This pace of progress is unprecedented, and systematic revisions of major lineages remain a common occurrence. A dialogue has been opened as to whether (or perhaps the correct term should be *when*) systematists should update classic Linnaean taxonomy and make the ultimate move to an entirely rank-free taxonomic framework (Casiraghi, Galimberti, Sandionigi, Bruno, & Labra, 2016; Kraichak, Crespo, Divakar, Leavitt, & Thorsten Lumbsch, 2017; Money, 2013). Concomitantly, molecular species delimitation has blossomed into an established practice that is revolutionizing our understanding of biodiversity as well as our ability to validate or refine existing species boundaries (Crous, Hawksworth, & Wingfield, 2015; De Queiroz, 2007; Jackson, Carstens, Morales, & O'Meara, 2016; Petit & Excoffier, 2009; Yang & Rannala, 2010). Examples abound of molecular data validating contentious species boundaries (Dornburg et al., 2015), revealing clusters of "cryptic" species (Crespo & Thorsten Lumbsch, 2010; Sato, Yumoto, & Murakami, 2007), identifying species with clear genetic differentiation but little to no morphological differentiation (Cai et al., 2011; Dornburg, Federman, Eytan, & Near, 2016; Nguyen, Vellinga, Bruns, & Kennedy, 2016; Perkins, 2000; Smith, Harmon, Shoo, & Melville, 2011; Taylor, Turner, Townsend, Dettman, & Jacobson, 2006), or lumping multiple species that had been delimited based on morphology into one taxon (Singh & Gupta, 2017; Taylor et al., 2006).

Biological information commonly used for the creation of a fungal classification system is an amalgamation of morphology, physiology, biochemistry, and ecological traits (Alexopoulos, 2007; Bessey, 1942, 1950; Blackwell, 2009; Frisvad, 1998; Lutzoni et al., 2004; Martin, 1951; Wang et al., 2016). Early synthesis of this information yielded a systematic framework that has remained generally stable (reviewed in chapter "Advances in Fungal Phylogenomics and Its Impact on Fungal Systematics" by Zhang et al. as well as in chapter "Multiple Approaches to Phylogenomic Reconstruction of the Fungal Kingdom" by McCarthy and Fitzpatrick), although controversies over some lineage-specific characters have occurred within all taxonomic ranks of fungal classifications (Singh & Gupta, 2017). In particular, morphology is contentious. While fungi are one of the major groups in Eukaryotes, they have been considered similar to microbes with regard to their comparatively small and simple morphology. This lack of phenotypic variation is problematic, especially for pathogenic species with extremely simple appearances (reviewed in chapter "Describing Genomic and Epigenomic Traits Underpinning Emerging Fungal Pathogens" by Farrer and Fisher). Many originally described morphological species have been disclosed as complexes of morphologically cryptic lineages (Howard, 2014). Lack of genetic information or developmental understanding of the characters used for fungal taxonomy and systematics challenges our ability to scrutinize the taxonomic or phylogenetic value of morphological characters (Taylor et al., 2006; figure delimitation). Compared with morphological characters, the wealth of information in molecular markers has greatly enhanced our ability to classify fungi (Fig. 1). Correspondingly, there have been numerous recent initiatives to reclassify on the basis of molecular data (Hibbett et al., 2011; Li, 2016; McLaughlin & Spatafora, 2015; Wang et al., 2016, 2014; Young & Peter, 2012).

Using molecular markers has not only facilitated several major taxonomic revisions (reviewed in chapter "Advances in Fungal Phylogenomics and Its Impact on Fungal Systematics" by Zhang et al.) but has also contributed to the discovery of new fungal lineages highly divergent from previously defined fungal groups. In particular, discoveries of new taxa from diverse environmental samples that are either morphologically similar or cannot be cultured under laboratory conditions have given us access to another dimension of fungal biodiversity (Hibbett et al., 2007; James & Berbee, 2012; James et al., 2006; Jones et al., 2011; Rosling et al., 2011). Methods have also been developed that help to identify novel single-copy genes appropriate for low-level fungal taxonomy (Feau, Decourcelle,

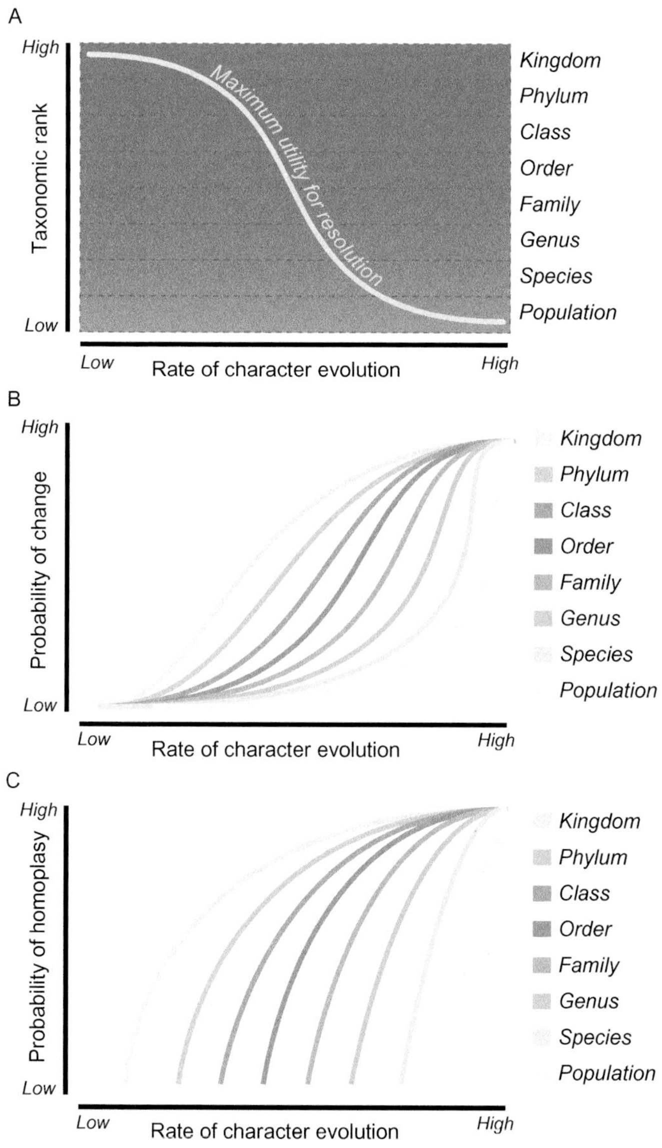

Fig. 1 Relationship between the rate of character evolution and degree of phylogenetic resolution at different taxonomic levels. (A) Higher rates of character change are useful for resolving relationships at finer taxonomic scales. (B) Illustrative diagram of how the probability of observing character change between taxa at different taxonomic levels changes based on the rate of character change. (C) Graphical representation of the probability of accruing homoplasious site patterns at different taxonomic levels. Taxonomic levels are used here to reflect an illustration of widely recognizable evolutionary temporal depth, not to subscribe or endorse any rank-based taxonomic framework.

Husson, Desprez-Loustau, & Dutech, 2011; Walker, Castlebury, Rossman, & White, 2012). However, there are pitfalls to using genomic data for species delimitation in fungi. For example, delimitation methods can be misled by complex patterns of population structure (Sukumaran & Knowles, 2017) or by lack of care taken during the initial assignment of individuals (Edwards & Knowles, 2014. Further, species delimitation methods or frameworks may disagree on what constitutes species boundaries (Carstens, Pelletier, Reid, & Satler, 2013), as in a case of lichen taxonomy where both biological and phylogenetic species concepts have failed to robustly delimit a species (Wei et al., 2016). While comparing results from analyses conditioned on coalescent or genealogical concordance phylogenetic species recognition methods may provide some insights into the number of independent evolutionary lineages (Dettman, Jacobson, Turner, Pringle, & Taylor, 2003; Liu, Wang, Damm, Crous, & Cai, 2016; Parnmen et al., 2012), the continued growth of analytical tools for molecular species delimitation will be critical to understanding fungal diversity. The continual advancement of theory and methods is particularly important for the numerous fungal pathogens and endophytes, such as species complex in *Fusarium* and plant endophytes, that possess little morphological variation, yet high levels of molecular divergence (Cai et al., 2011; Kepler et al., 2013; Oono et al., 2014; Taylor, 2006; U'Ren et al., 2009; Wingfield et al., 2011). As we move further into the era of fungal genomics, it is our hope that the need for molecular data in fungal species delimitation will drive the development of new strategies for utilizing phylogenomic data to inform species discovery.

2.3 Ecological Diversification

Understanding how ecological communities and organismal niches evolve are fundamental goals of ecology and evolution that are critical for forecasting how biodiversity will respond to continued global change (Catullo, Ferrier, & Hoffmann, 2015; Federman et al., 2016; Ikeda et al., 2017; Thuiller, Lavorel, & Araujo, 2005). Fortunately, the rapid proliferation of genomic data and associated phylogenomic and phylogeographic studies are driving unparalleled insights into tempo and mode of ecological diversification at both local and global scales (Forrestel, Donoghue, & Smith, 2014, 2015; Losos, 2008; Mahler, Ingram, Revell, & Losos, 2013). For example, integration of population genetic/omic analyses allows for fine-scale consideration of how the interaction between ecotypes, geography, and genetic covariance within or between populations might drive heterogeneous responses to climatic change within a species (Ikeda et al., 2017). Further, the availability of species-level phylogenetic trees offers a chance to disentangle historical processes such as the geography of speciation (e.g., allopatry or colonization dynamics) from ecological processes, thereby refining estimates of an organism's niche (Warren, Cardillo, Rosauer, & Bolnick, 2014). Finally, for lineages whose evolutionary relationships are difficult to resolve based on morphology, a genomic perspective also enables a more accurate estimate of how often and under what conditions certain ecomorphs originate or proliferate (Capella-Gutiérrez, Marcet-Houben, & Gabaldón, 2012; Chang et al., 2015; Jiang, Xiang, & Liu, 2017; Marcet-Houben, Marceddu, & Gabaldón, 2009; Spatafora & Bushley, 2015).

For groups such as fungi that are both critically important from the perspective of human health and for the continued functionality of ecological communities, integration of genomics into ecological studies is critical. Prior to the use of DNA sequence data, our understanding of fungal ecological diversity had been largely restricted to fungal pathogens and symbionts (Alexopoulos, 2007). However, genomic tools now allow for detection of fungal species inhabiting environmentally sampled habitats, providing a rapid way to assess the global distribution or movement of fungal biodiversity (Clemmensen, Ihrmark, Durling, & Lindahl, 2016; Gherbawy & Voigt, 2010; Hibbett et al., 2011; Lindahl et al., 2013; Nilsson, Abarenkov, Larsson, & Kõljalg, 2011). However, we often lack basic information on fungal ecology and life history. This problem is not unique to fungi, as the sheer diversity of life on earth has left the natural history of numerous lineages across the Tree of Life understudied. As a result, our understanding of the natural world is filled with both gaps (Bland, Collen, Orme, & Bielby, 2015; Ribeiro, Teixido, Barbosa, & Silveira, 2016) and taxonomically or geographically biased perspectives (Dornburg, Townsend, et al., 2017; Loiselle et al., 2008; Reddy & Dávalos, 2003). Although this lack of information has the potential to mislead investigations concerning the conservation of the planet's biodiversity, numerous methodological developments have been, and continue to be, devised to overcome such potential pitfalls (Boria, Olson, Goodman, & Anderson, 2014; Qiao, Townsend Peterson, Ji, & Junhua, 2017; Varela, Anderson, García-Valdés, & Fernández-González, 2014; Warren, Wright, Seifert, & Bradley Shaffer, 2013). Despite their limitations, phylogenetic analyses have been highly valuable in facilitating the formation of new hypotheses and insights into fungal ecology (e.g., Bärlocher, 2006; Hwang, Qi, Yang, Wang, & Townsend, 2015; James et al., 2013; Lopez-Llorca, Jansson, Vicente, & Salinas, 2006; Richards, Jones, Leonard, & Bass, 2012; Unterseher & Schnittler, 2010; Wang et al., 2011; Wang, Johnston, Yang, & Townsend, 2009). For example, ITS sequences in rRNA regions were used to probe fungal components within 40 soil samples collected from 365 sites worldwide to understand the soil fungal diversity distribution across the fungal Tree of Life (Tedersoo et al., 2014). Such a perspective on fungal ecological diversity allows for testing of how convergent or divergent evolution shaped the evolution of fungal ecomorphs and is a critical first step to understanding how fungal ecological communities assemble and shift through time.

2.4 The Evolution of Phenotypic Disparity

The uneven distribution of morphological diversity across the Tree of Life is one of the most striking patterns in evolutionary biology. What drives some lineages to experience rapid bursts of phenotypic diversification vs morphological stasis? How do morphological novelties evolve? What is the relationship between phenotypic disparity and lineage diversification? Over the last decade, the rapid development of phylogenetic comparative methods (Garamszegi, 2014; Harmon, Weir, Brock, Glor, & Challenger, 2008; Paradis, Claude, & Strimmer, 2004; Revell, 2011; Revell & Graham Reynolds, 2012) has spurred the writing of thousands of manuscript pages devoted to answering these questions about morphological diversity investigating lineages spanning the entire Tree of Life (Adams, Berns, Kozak, & Wiens, 2009; Blackwell, Hibbett, Taylor, & Spatafora, 2006; Harish & Kurland, 2017; Lanier & Williams, 2017; Medina, Jones, & Fitzpatrick, 2011; Near et al., 2014; Ragan, 2015; Wägele & Bartolomaeus, 2014; Zeigler, 2014). The answers reveal staggering complexity. We have repeatedly seen evidence for putative key innovations either coupled or decoupled from diversification (Hulsey, García de León, & Rodiles-Hernández, 2006; Near et al., 2012; Sánchez-García & Matheny, 2016). Likewise, phenotypic and lineage diversification may or may not be correlated (e.g., James et al., 2006; Looney, Ryberg, Hampe, Sánchez-García, & Matheny, 2016; Rabosky & Adams, 2012; Raghukumar, 2017; Seena & Monroy, 2016; Tanabe, Watanabe, & Sugiyama, 2005; Torruella et al., 2015). In some cases, traits that were thought to represent single evolutionary origins have in fact been independently evolved, such as wood decay types in mushroom-forming fungi (Floudas et al., 2015; Hibbett & Donoghue, 2001; Hori et al.,

2013; Nagy et al., 2017; Riley et al., 2014). In other cases, the presence of an unrelated species may hinder diversification despite the presence of open niches (Fukami, Beaumont, Zhang, & Rainey, 2007; Knope, Forde, & Fukami, 2011). At a glance, such heterogeneity may seem frustratingly difficult to reconcile into any sort of principles that govern biological diversity. However, despite this complexity, we are seeing consistency between molecular phylogeny and previous hypotheses of trait evolution based on morphological data in various fungal groups as well as the emergence of new phenotypic paradigms (Blackwell et al., 2006; Ebersberger et al., 2012; Hibbett et al., 2011; Pacheco-Arjona & Ramirez-Prado, 2014; Riley et al., 2014). Further, the amalgamations of case studies demonstrating exceptions to expectations are guiding the creation of novel nuanced lineage- or habitat-specific hypotheses. Recent insights into the rise and diversification of fungal phenotypes are no exception to these trends (e.g., Hwang et al., 2015; Nagy et al., 2014; Pena et al., 2017; Raghukumar, 2017; Trail, Wang, Stefanko, Cubba, & Townsend, 2017; Wang et al., 2009).

Understanding the origin and evolution of fungal morphological diversity has long been a major objective of fungal phylogeneticists (Bruns et al., 1992; Hibbett, Fukumasa-Nakai, Tsuneda, & Donoghue, 1995; Saenz, Taylor, & Gargas, 1994), but one that has been fraught with challenges. Recent insights into fungal evolutionary relationships, especially those gained through the effort of the Fungal Tree of Life project, have greatly improved our understanding of evolution of fungal morphology, from body plans to ultrastructures (James et al., 2013; Liu & Hall, 2004; Luo et al., 2017; Lutzoni et al., 2004; McLaughlin & Spatafora, 2015; Nagy, 2017; Nagy et al., 2014; Schoch et al., 2009; Wang et al., 2009; Yang, Yang, An, & Liu, 2007). For example, mapping traits onto a phylogenetic framework of all major fungal lineages revealed that early-diverging lineages consisted of various zoosporic fungi, implying that earliest fungi were not terrestrial but instead produced flagellated spores in aquatic environments (James et al., 2006). Similarly, ancestors of the largest fungal group, Ascomycota, have been suggested as filamentous (Liu & Hall, 2004); however, this conclusion has been challenged by phylogenies using different data (Gabaldón & Marcet-Houben, 2014). Phylogenetic and genomic data suggest that multicellular fungi, such as one basal ascomyceteous lineage of *Neolecta*, can have small genomes that resemble those of unicellular yeasts (Nagy, 2017; Nguyen et al., 2017). Further morphological and genomic investigations of ascomycetes led to the identification of a rapidly evolving gene that may play a key role in mitigating fungal morphological development (Wang et al., 2016).

Molecular phylogenetics has also made pivotal contributions to the identification of instances of both plesiomorphic traits and convergence in character state. Ultrastructures, such as septal pores in fungal hyphae or ascus dehiscence as a spore release mechanism in ascomycetes, have been regarded as conserved traits for fungal classification. However, molecular phylogenies suggest that the current descriptions of septal structure probably include plesiomorphic traits (Lutzoni et al., 2004; Schoch et al., 2009). Further environmental factors have promoted trait convergence within the Leotiomycetes (Wang et al., 2009), a group for which inference of evolutionary relationships has been problematic (Peterson & Pfister, 2010; Wang et al., 2006; Zhang & Wang, 2015). Robust phylogenetic histories of major fungal lineages also make it possible to perform "phylostratigraphy" of elements in fungal genomes, i.e., categorizing genes for their minimal ages in a genome along a given phylogeny. Based on phylostratigraphy, hourglass gene expression patterns have been identified in plants and animals, where upregulated expression of "old" genes can be observed during the conserved phylotypic stages of embryogenesis (Drost, Gabel, Grosse, & Quint, 2015; Kalinka et al., 2010; Quint et al., 2012). Such an hourglass expression pattern was also identified for the first time in a mushroom species (Cheng, Hui, Lee, Wan Law, & Kwan, 2015). These case studies represent just a very small snapshot of the amazing diversity of case studies conducted on various fungal groups over the last decade. Clearly, this is a very exciting time to be a fungal comparative biologist.

2.5 Historical Biogeography and the Geographic Movement of Fungi in the Anthropocene

As human influence becomes the predominant force shaping both the environment and climate globally, we are witnessing a massive biotic response. Both locally and globally, we are witnessing a steady reorganization of the world's biomes. Locally, the creation of novel local ecosystems, such as urban centers, has simultaneously driven extirpation of some species while promoting rapid adaptation in others (Littleford-Colquhoun, Clemente, Whiting, Ortiz-Barrientos, & Frère, 2017). Increasingly, dispersal between urban centers or other habitats is better modeled by the level of human commerce than classic biogeographic species–area models for some taxa (Helmus, Luke Mahler, & Losos, 2014). Globally, our ability to move species and modify habitats has resulted in alien species displacing native species to some degree in virtually all of the world's biomes (Capinha, Essl, Seebens, Moser, & Pereira, 2015). Further, rapid patterns of climatic change and habitat alteration are driving global shifts in species ranges that we are only beginning to understand (Barnosky et al., 2017; Last et al., 2010). To effectively conserve and manage biodiversity, we must draw inferences from a historical perspective of how lineages and ecological communities have persisted through evolutionary time despite changes analogous to those forecasted to occur over the next hundred years (Barnosky et al., 2017; Dornburg, Federman, Lamb, Jones, & Near, 2017; Hellmann & Pfrender, 2011; Kennicutt et al., 2014).

Given their tremendous ecological diversity, species richness, and global distributions, fungi have tremendous potential to advance our understanding of terrestrial biogeography of not only entire communities but also of major drivers of biogeographic processes (Davis, Phillips, Wright, Linde, & Dixon, 2015; Feurtey et al., 2016; Liu, Wang, Gao, Bartlam, & Wang, 2015; Richards et al., 2015; Talbot et al., 2014; Tedersoo et al., 2014). Fungal distributional patterns often mirror those found in terrestrial plants and animals. For example, a recent study of poisonous amanitas mushrooms found that the disjunct distribution of closely related lethal amanitas in East Asia and eastern North America reflects a loss of connectivity due to climate change in the Oligocene (Cai et al., 2014; Cavalier-Smith et al., 2014; Nagy et al., 2014; Schoch et al., 2009). This biogeographic break is among the most well-known patterns in global plant distributions (Liu, Wen, & Yi, 2017; Wen, Nie, & Ickert-Bond, 2016). Taken together, findings of climate-driven biogeography between plants and fungi illustrate the growing concern that current climate trends will shift or replace entire biological communities over the next few centuries (Barnosky et al., 2017; Last et al., 2010). Considering the global movements of fungi raises an important question: How do we expect fungi that impact the health of humans or wildlife to move in response to shifting climates?

Just like microbial pathogens, many extant fungal pathogens were introduced to human populations thousands of years ago and have become geographically widespread as a result of our activities. Often, potential fungal pathogens are maintained in healthy human populations asymptomatically. However, humanity's increased connectivity and the increased prevalence of immunocompromised health states, coupled with the often large distributions of potentially lethal pathogenic fungi, raise the risk of disseminating either rare strains or newly evolved and highly aggressive strains. From this perspective, the study of fungal biogeography and phylogeography is fundamental to understanding the spatial component of how fungal pathogens have evolved virulence. Zhang and Ma in the chapter "Deciphering Pathogenicity of *Fusarium oxysporum* From a Phylogenomics Perspective" illustrate the need for population-level assessments of the pathogenetic fungi *Fusarium oxysporum*, while chapter "Describing Genomic and Epigenomic Traits Underpinning Emerging Fungal Pathogens" by Farrer and Fisher provide an overview of using genomics in the fight against fungal pathogens. Additionally, a recent, global phylogenomic study of the fungal pathogen *Cryptococcus neoformans* revealed extensive loss of genetic diversity in one African strain, suggestive of a history of population bottlenecks (Desjardins et al., 2017). *C. neoformans* causes approximately 625,000 deaths per year from nervous system infections, and this study highlighted the complex evolutionary interplay between adaptation to natural environments and opportunistic infections for fungal pathogens. Opportunities for fungal infections may increase over the next century, as dispersal capability is no longer restricted to natural means. Continuing to develop an accurate understanding of the origin and distribution of reservoirs of fungal pathogen diversity—as well as past and current migration routes of pathogenic components of the mycobiome—is crucial if we are to predict and manage the human and wildlife health needs of the Anthropocene.

3. ADVANCES IN FUNGAL PHYLOGENETICS AND PHYLOGENOMICS

The use of molecular phylogenetics to infer the evolutionary relationships of fungal groups began in early 1990s with the use of rRNA markers (Saenz et al., 1994; Swann & Taylor, 1993). Some of these markers are still widely used. The ITS regions, in particular, are being adopted as a primary universal marker for species identification and delimitation (Bruns et al., 1992; Schoch et al., 2012). Over the subsequent 30 years, more genetic markers were tested and optimized. Successful experiences with several markers led to their elevation to standard use as part of the practice established by the Assembling the Fungal Tree of Life (AFTOL) project (Blackwell et al., 2006). However, single individual markers cannot act as silver bullets that resolve all possible evolutionary problems. To answer some of the most vexing questions in fungal evolutionary research, studies have increasingly incorporated many loci derived from fungal genomes to yield power (Gabaldón & Marcet-Houben, 2014; Ren et al., 2016; Schoch et al., 2009; Shelest & Voigt, 2014). From the very first sequenced fungal genome of *Saccharomyces cerevisiae* (Goffeau, 2000) to the recently launched 1000 fungal genomes project at the Joint Genome Institute and the projects proposed by the Fungal Genome Initiative (Grigoriev, 2011; Stukenbrock & Croll, 2014), genome-scale data representative of all major fungal groups are increasingly entering the realm of public data. Here we briefly review advances and changes in phylogenetics and phylogenomics in our efforts to assemble the fungal Tree of Life.

3.1 Redefining Multilocus: Transitioning From Phylogenetics to Phylogenomics

The AFTOL project tested and proposed a set of candidate loci for resolving relationships among fungal lineages (Binder et al., 2013; Lutzoni et al., 2004; Schoch et al., 2009). These loci included genes encoding the rRNA subunits, elongation factors, subunits of DNA-directed RNA polymerase, and subunit 6 of ATP synthase, all of which have since been widely used in fungal phylogenetics in all major fungal clades (Goffeau, 2000; James et al., 2006; Spatafora, Hughes, & Blackwell, 2006). From the humble origins of using single genes to infer fungal relationships (Saenz et al., 1994; Swann & Taylor, 1993), these markers offered a dramatic increase in phylogenetic information content and corresponding power for the robust resolution of a set of evolutionary events. For example, six of these AFTOL markers were used to infer the phylogeny of Ascomycota, the largest phylum of Fungi (Schoch et al., 2009). By sampling 420 taxa, this study was the first to confirm that Ascomycetes sporocarp characters are consistent with two independent origins of multicellular sexual reproductive structures—in common ancestors of Pezizomycotina and in common ancestors of Taphrinomycotina (Schoch et al., 2009). From morphological and biogeographic problematic groups in lichens (Leavitt, Esslinger, Divakar, Crespo, & Thorsten Lumbsch, 2016; Leavitt, Esslinger, Spribille, Divakar, & Thorsten Lumbsch, 2013; Mark et al., 2016; Saag, Mark, Saag, & Randlane, 2014) to disjunct distributed fungal species that have evolved various associations with plants (Ge et al., 2014; Song & Cui, 2017; Wilson, Hosaka, & Mueller, 2017) to environmental species relying on multilocus barcoding for species recognition (Roe, Rice, Bromilow, Cooke, & Sperling, 2010), the use of a modest number of genes has transformed our understanding of fungal biodiversity.

The rise of public genome data availability for fungal model organisms and other fungi has catalyzed multilocus phylogenetic initiatives to continue to harvest increasing numbers of orthologous genes (e.g., Traeger et al.,

2013; Waterhouse, Tegenfeldt, Li, Zdobnov, & Kriventseva, 2013). Phylogenetic analyses using 173–192 genes from 73 to 122 eukaryote-wide taxa suggested that animals, fungi, choanozoans, and Amoebozoa shared common protozoan ancestors (Cavalier-Smith et al., 2014). To understand the evolution of yeast forms, 594 orthologous genes were used to reconstruct phylogeny of 59 species, demonstrating that yeast forms could have originated early in fungal evolution then became dominant independently in different clades via parallel diversification of Zn-cluster transcription factors (Nagy et al., 2014).

With so many fungal genome sequences available and many more in the sequencing pipeline, mycologists have reason to be optimistic that major questions in fungal phylogeny will be resolved within our lifetimes. We have already seen genomics drive a robust resolution of early-diverging fungal lineages whose inferred relationships had previously been incongruent based on different datasets (Capella-Gutiérrez et al., 2012; Gabaldón & Marcet-Houben, 2014; James et al., 2006). Further, analyses of newly sequenced genomes from early-diverging fungal basal lineages, including *Gonapodya prolifera* and many other flagellated, predominantly aquatic fungi, suggested that ancient aquatic fungi had evolved the ability to use plant cell walls as a nutrient resource (Chang et al., 2015). Additionally, a maximum age for the divergence of aquatic Chytridiomycota from other terrestrial fungal lineages at 750 million years—300 million years before the appearance of land plants (Wellman, Osterloff, & Mohiuddin, 2003)—corresponds with the estimated maximum age of the origin of the pectin-containing streptophytes (Chang et al., 2015). Continued investigations of this timescale will be critical in developing a robust understanding of the early evolution of fungi and the origins of both fungi–plant associations and terrestrial ecosystem colonization. Despite the limited number of taxa with genomic sequence data currently available, genomic data have already been used to resolve some long lasting questions in fungal evolution, such as horizontal gene transfer, gene duplication and loss, origin and evolution of fungal characters, and lifestyles across taxonomic scales (Chang et al., 2015; Gabaldón, Naranjo-Ortíz, & Marcet-Houben, 2016; Gladieux et al., 2014; Ješovnik, González, & Schultz, 2016; Luo et al., 2015; Marcet-Houben et al., 2009; Medina et al., 2011; Mixão & Gabaldón, 2017; Ropars et al., 2015; Spatafora & Bushley, 2015; Torruella et al., 2015). Phylogenomics analyses have proven particularly useful in aiding in the resolution of evolutionary patterns for gene families with complicated evolutionary histories. For example, integration of the study of the evolution of secondary metabolism, reviewed in chapter "Fungal Gene Cluster Diversity and Evolution" by Slot, into phylogenomic analyses, mitigates the misleading impacts of horizontal gene transfer in plant-associated fungi (Spatafora & Bushley, 2015), increasing the accuracy of nutrient and infection models in fungal pathogens of plants (Luo et al., 2015). These examples of systematics illustrate some of the exciting advances made by fungal evolutionary biologists, with a more detailed overview provided by Zhang et al. in the chapter "Advances in Fungal Phylogenomics and Its Impact on Fungal Systematics."

So what will the future look like? It is clear that genomic data hold tremendous power. For fungal groups with no closely related reference genomes, sequencing techniques such as phylotranscriptomics, the use of sequences of mRNA to infer phylogeny, could be a short-term promising approach for effectively gathering large datasets that have proven effective in nonfungal organisms (Bazinet et al., 2016; Breinholt & Kawahara, 2013; Janouškovec et al., 2017; Oakley, Wolfe, Lindgren, & Zaharoff, 2013; Suvorov et al., 2017). While challenges to sequence data acquisition still remain on the immediate horizon, a larger set of challenges has cast a shadow over the enterprise of phylogenomics. How do we disentangle sources of error from sources of signal in genomic data? How do we handle the computational complexity of big data? McCarthy and Fitzpatrick in the chapter "Multiple Approaches to Phylogenomic Reconstruction of the Fungal Kingdom" evaluate the consistency of phylogenomic inference under different models of tree inference, addressing the question of how robust our inferences are to changes in analytical frameworks. Such questions are not unique to fungal phylogenomics (Brinkmann, Giezen, Zhou, Raucourt, & Philippe, 2005; Gouy, Baurain, & Philippe, 2015; Jeffroy et al., 2006; Philippe et al., 2011) and are reviewed by Nagy et al. in the chapter "Fungal Phylogeny in the Age of Genomics: Insights Into Phylogenetic Inference From Genome-Scale Datasets" in addition to being discussed in the next section. Answering such challenges is critical if we are to unlock and harness the true power of genomic data and confidently resolve a stable Fungal Tree of Life.

4. EXPERIMENTAL DESIGN: EFFECTIVE HARNESSING OF PHYLOGENOMIC POWER

The continuous development of more efficient and inexpensive sequencing technology holds the promise of a comprehensive resolution of the Fungal Tree of Life by the end of the 21st century. However, simply sequencing more genes will not lead to confident resolution. Just as in other parts of the Tree of Life, some nodes in Fungal phylogeny continue to defy resolution, while others are resolved yet incongruent between studies (Hibbett et al., 2007; Krüger et al., 2011; Pöggeler & Wöstemeyer, 2011; Schoch et al., 2009; Spatafora et al., 2016). For over a decade, phylogenomicists have been striving toward developing new approaches to investigate causes of incongruence or lack of support in genomic scale datasets (Eytan et al., 2015; Liu, Wu, & Yu, 2015; Philippe et al., 2011; Romiguier, Ranwez, Delsuc, Galtier, & Douzery, 2013). While missing data due to sequencing type and gene-tree conflicts have been implicated in driving some cases of incongruence (Kuo, Wares, & Kissinger, 2008; Song, Liang, Edwards, & Shaoyuan, 2012; Watanabe et al., 2011), it is increasingly recognized that convergence or parallelisms in character state in only a handful of genes can drive strong but erroneous support for a given node (Salichos & Rokas, 2013; Shen et al., 2017). Likewise, low levels of homoplasy can positively mislead divergence time estimates across an entire tree topology. This issue of positively misleading sequences points out the utility of investigating how one should select the right gene sequences for resolving a specific phylogenetic question.

4.1 Experimental Design, Marker Scrutiny, and Topological Incongruence

It has long been known that, for a given rate of nucleotide change, time will eventually erode the signal of evolutionary history (Graybeal, 1993). This observation presents an informatic challenge of how to select markers based on expectations of character change. This challenge has not been ignored and has driven the development of a diverse array of both theory and tools with which to predict the phylogenetic utility of large multilocus and phylogenomic datasets (Goremykin, Nikiforova, & Bininda-Emonds, 2010; Pisani, Feuda, Peterson, & Smith, 2012; Xia, 2013). While highly useful, many of these methods only account for the temporal depth of a given phylogenetic problem and do not simultaneously account for the time between speciation events (internode distance). Quantification of the predicted probability of correctly resolving an ultrametric four-taxon tree demonstrates the necessity of accounting for the interaction of evolutionary depth, internode length, and the rate of character change when determining the phylogenetic utility of a marker (Townsend, Su, & Tekle, 2012). Theory has established expectations based on time-reversible substitution models and a Poisson model of molecular evolution that enable (1) direct quantification of the probability that a locus will contribute toward correctly resolving a given phylogenetic quartet; (2) the probability of a quartet internode reconstructed incorrectly due to homoplasy; and (3) the result of a polytomy due to low power.

Consideration of experimental design often focuses on evolutionary rates. However, tree structure exerts significant influence on which rates will be useful for phylogenetic resolution of a specific node. A marker composed of characters evolving at a given rate can be of high utility for given tree depths when internodes are long, yet positively misleading when internodes are short. Given a set of five genes used in fungal phylogenetic studies also in the dataset of Nagy et al. (2014), we can quantify the probability of correct topological resolution for each node (Fig. 2). This visualization clearly demonstrates the nonlinear relationship between time and probability of resolution, as well as demonstrating the heterogeneity in the utility of different markers. Nodes representing evolution occurring in similar temporal periods have both low and high probabilities of correct resolution, reflecting variation in internode length (Fig. 2). The rate of molecular evolution of the individual characters making up a marker is similarly important. For example, markers featuring rapidly evolving sites such as RPA3 (DNA-directed RNA polymerase I and III polypeptide) are of high utility for resolving early, rapid divergences, yet decay in utility even for large internode distances at deeper temporal depths (Fig. 2). Clearly, both the rate of evolution and the lengths of internodes matter for phylogenetic experimental design, as do unequal subtending branch lengths (Su & Townsend, 2015) and, to a lesser extent, the model of molecular evolution—ranging from Jukes-Cantor to the General Time Reversible (Rodríguez, Oliver, Marín, & Medina, 1990; Tavaré, 1986) model of nucleotide substitution (Su, Wang, López-Giráldez, & Townsend, 2014).

While this expanding analytical framework allows for more realistic modeling of substitution processes, there are still several key aspects missing from this approach to experimental design. First, current theory for experimental design does not go beyond time-reversible Markov chain assumptions to account for convergent mutation and selection processes. This assumption of process stationarity is problematic, as evolutionarily biased changes between synonymous codon variants have been repeatedly implicated as drivers of topological incongruence (Betancur-R, Li, Munroe, Ballesteros, & Ortí, 2013; Dornburg, Townsend, et al., 2017; Liu, Cox, Wang, & Goffinet, 2014; Reddy et al., 2017; Romiguier et al., 2013). Extending the approaches to incorporate the expected effects of

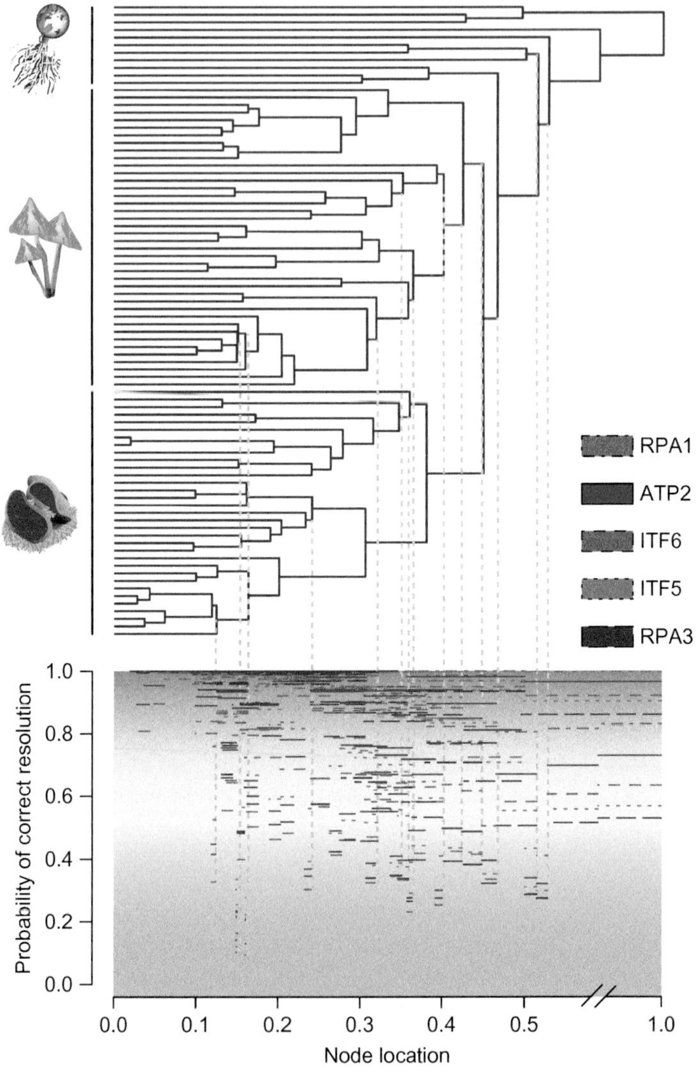

Fig. 2 Predicted probabilities (*y*-axis) of correct topological resolution quantified for six loci used in fungal phylogenetics for every node in the Nagy et al. (2014) dataset (tree and *x*-axis). *Horizontal lines* correspond with the results from the protein coding genes: DNA-directed RNA polymerase I subunit (RPA1), vacuolar membrane ATPase 2 (ATP2), eukaryotic translation initiation factors 5 (IFT5) and 6 (ITF6), and DNA-directed RNA polymerase I and III polypeptide (RPA3) for each node, respectively. *Vertical lines* link nodes across the phylogeny with computed low probabilities, demonstrating how internode distances can impact utility of different loci at any temporal scale.

convergences in nucleotide states, such as elevated GC content of third codon positions (GC3), represents an important next step in predicting the utility of genomic sequence data. Additionally, these models currently do not fully incorporate the impact of different taxon sampling strategies, despite theoretical expectations concerning how taxon sampling can impact inference (Townsend & Lopez-Giraldez, 2010). It is also important to point out that these approaches assume that orthologs have already been identified correctly (Narechania et al., 2016) and aligned appropriately. Further, they do not consider other potential sources of error in analyses, such as model performance or fit, that are also key to successful evolutionary inferences (Brown, 2014; Lanfear, Calcott, Kainer, Mayer, & Stamatakis, 2014). The degree to which resolution of focal nodes is robust to these or other considerations will be heterogeneous. However, if we are to confidently resolve the vexing problems of the Tree of Life, the continued application and development of methods that scrutinize sequence data for phylogenetic utility of specific phylogenetic problems will be critical.

4.2 Designing an Effective Taxon Sampling Strategy to Maximize Power

"Should more taxa or more genes be sequenced to help resolve this node?" is a classic question in phylogenetics (Pollock, Zwickl, McGuire, & Hillis, 2002; Rannala, 1998; Rokas, 2005; Zwickl & Hillis, 2002). Although we have begun to change the narrative of this question from *genes* to *genomes*, the problem remains the same. Its supposed controversiality belies extensive understanding of the impact of taxon sampling on experimental design. Dense taxon sampling has repeatedly been shown to increase phylogenetic accuracy by reducing the effects of long-branch attraction (Heath, Zwickl, Kim, & Hillis, 2008). However, it is also known that increased taxon sampling increases the complexity of phylogenetic inference (Felsenstein, 2004). This increase in complexity demands more information from the same loci—information that must address superexponentially increasing numbers of hypothetical relationships at finer and finer resolution. Choosing to add certain taxa can add long branches that can compromise accuracy, introduce new rate heterogeneities, and incorporate dominant model violating branches (Poe & Swofford, 1999). Given that personnel, computational, and funding resources are not infinite, finding a taxon gene sampling strategy that minimizes the probability of errors such as long-branch attraction—while maximizing the potential of resolution—represents an ideal design strategy.

Theory of phylogenetic experimental design has recently advanced to provide significant guidance as to when increased taxon or gene sampling would be cost-effective. For reasonable rates of sequence evolution, selection of taxa proximate to a focal internode is a good rule of thumb—if the goal is to resolve a specific node (Fig. 3). This finding has been invoked in the experimental design of phylogenomic studies characterized by sparse taxon but high gene sampling. By concentrating taxon sampling to taxa that diverge close to a focal node, there should be a considerable increase in power to resolve previously insoluble, rapidly radiating lineages (Chakrabarty et al., 2017; Prum et al., 2015). However, even when there are taxa readily available for sequencing, increasing taxon sampling is not always the optimal solution to resolving a node of interest (Reddy et al., 2017). The rate of character change factors into the choice of gene vs taxon sampling (Fig. 3). When rates are slow, adding more characters is almost always preferred. This relationship is intuitive: slow rates will yield few changes within the clade of interest and, therefore, more sites are required for power. Furthermore, sampling additional taxa when characters are evolving slowly will be unlikely to reveal novel variation within the clade

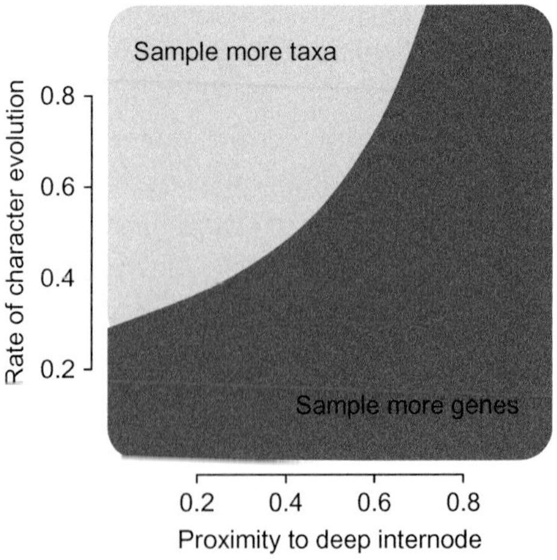

Fig. 3 Contour plot depicting when it is more cost-effective to sample additional characters or additional taxa for aiding in the resolution of a specific phylogenetic internode. At higher rates of character evolution, it is more cost-effective to sample taxa close to the node of interest. In contrast, at low rates of character evolution—or when no taxa that diverge close to the node of interest are available—it is more cost-effective to sample additional molecular characters. *Modified from Townsend, J. P., & Lopez-Giraldez, F. (2010). Optimal selection of gene and ingroup taxon sampling for resolving phylogenetic relationships.* Systematic Biology, *59(4): 446–457.*

that is useful to phylogenetic inference. Assessing characters evolving at rapid rates of evolution within the clade of interest will yield more changes and, therefore, information that presumably should be of use to phylogenetics. However, if the rates are fast enough, that information can become subject to noise and bias, which can continue to accumulate as additional characters are assessed. Both noise and bias can be counteracted by additional taxon sampling. Although additional taxon sampling can break up long branches, driving better identification of ancestral states and leading to more robust inference of phylogeny, a broader understanding of how noise, bias, and complex taxon sampling patterns effect topological inference is needed.

Rates of character change are integral to effectively choosing whether more genes or more taxa should be added to a study (Fig. 3). The results of Townsend and Lopez-Giraldez (2010) incorporate the loss of signal in characters due to their fast rate of evolution. However, they do not specifically account for the increased probability of convergence or parallelism (homoplasy) in character state that occurs at faster rates of character evolution. It is not clear how the corresponding changes in phylogenetic information content through time effect taxon sampling strategies. It has been demonstrated that failure to account for homoplasy when adding additional taxa can mislead investigators into erroneously concluding that taxon sampling strategies have driven topological incongruence between studies (Reddy et al., 2017). Homoplasy can be particularly problematic for short internodes (Townsend et al., 2012) and careful scrutiny of markers is therefore critical, even when adding only a few new sequences to a phylogenomic study (Shen et al., 2017). Theory by Townsend and Lopez-Giraldez (2010) has not quantified potential utility of assessing characters of taxa outside of the clade defined by the quartet of interest. If proximity to an internode can aid in supplying information to accurately model sequence evolution, what is the utility of adding outgroup taxa? How does this utility change as time and predicted levels of information content are taken into account? Answers to these and related questions represent exciting avenues of research that can greatly aid in efficiently resolving key nodes in the fungal Tree of Life.

4.3 Estimating a Time-Calibrated Fungal Tree of Life Requires Careful Marker Scrutiny

A time-calibrated fungal Tree of Life holds tremendous potential for understanding the early origins of the planet's terrestrial biodiversity (Berbee & Taylor, 2010; Heckman et al., 2001). However, age estimates concerning the origins of early fungi have varied dramatically, in some cases predating the Cambrian explosion by hundreds of millions of years (reviewed in Taylor & Berbee, 2006). This incongruence between divergence time studies is certainly not restricted to fungi and has been observed in lineages spanning mammals (Reis et al., 2012; Springer et al., 2017), fishes (Dornburg, Friedman, & Near, 2015; Dornburg et al., 2011; Near et al., 2012; Santini, Sorenson, & Alfaro, 2013), birds (Prum et al., 2015), and plants (Herendeen, Friis, Pedersen, & Crane, 2017). Investigations of what drives incongruence in age estimates have led to the recognition of numerous pitfalls that can mislead molecular age estimates, including the violation of clock model assumptions (Dornburg, Brandley, McGowen, & Near, 2012), conflict or modeling of prior age calibrations (Dornburg, Beaulieu, Oliver, & Near, 2011; Rannala, 2016; Warnock, Parham, Joyce, Lyson, & Donoghue, 2015), or branch length priors (Heath, Huelsenbeck, & Stadler, 2014). Additionally, phylogenetic experimental design (e.g., careful selection of loci with levels of convergence that are low enough not to mislead inference of the true evolutionary history) has also become recognized as a critical component of divergence time studies (Dornburg, Townsend, Friedman, & Near, 2014; Phillips, 2009; Wilke, Schultheiß, & Albrecht, 2009).

Simulations by Phillips (2009) illustrate how hidden substitutions in character state (noise) can bias the branch length distribution of an entire topology. Consider dating a tree that is calibrated at its deepest split, inferred based on a rapidly evolving marker that has accrued a high level of noise before subsequent branching events (Fig. 4A). If sites high in noise have converged in a pattern that mistakenly fits a slow rate of substitution relative to the real rates near the calibration, then inferred deep branches will extend and drive older age estimates (Fig. 4A). Under this scenario, rates for recent divergences are estimated with higher accuracy as a result of higher taxon sampling and greater recency toward the tips, therefore promoting these ages to appear "older." If a calibration is alternatively placed at the tips for the very same fast-evolving marker (Fig. 4B), then the accurately estimated faster rates at the tips will lead to dating the root to be very young due to the noise-driven slower-than-actual apparent amount of deep evolution. Consequently, this will compress the estimated branch length distribution of nontipward lineages in the topology (Fig. 4B). Alternatively, this pattern of convergence can be inverted. For example, if sites high in noise have converged in a pattern that appears fast relative to the real rates and a tipward calibration is used, then the deep branches of the tree will extend. Conversely, using a deep calibration in this same scenario of noise and real rates

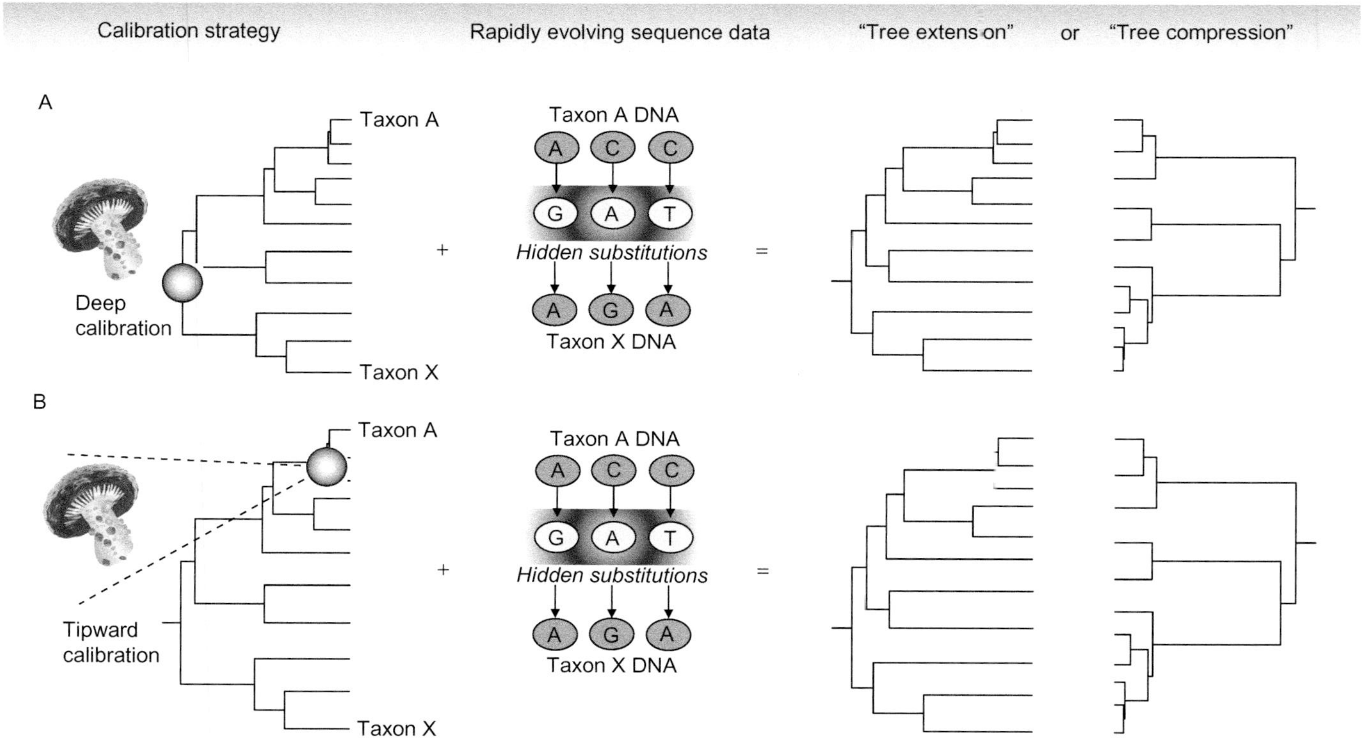

Fig. 4 The effect of calibration strategy and hidden substitutions on divergence time estimates. Utilizing sequence data characterized by high levels of hidden substitutions can often result in erroneous divergence time estimates regardless if calibrations were placed at a deep (A) or a tipward (B) node. In either case, failure to account for hidden changes results in a mismatch between rates of change estimated between tipward and deep splits. Depending on the direction of the rate mismatch, this effect of calibration strategy can lead to global patterns of tree extension (older ages) or tree compression (younger ages).

will cause tipward branches to compress (Fig. 4A and B). This example may seem overly simplistic. Dating a tree based on only a single calibration and a single rapidly evolving marker is an unlikely design for a 21st century phylogenomic study. However, utilizing a combination of loci characterized by high or low noise and a differential representation of deep or shallow calibrations for divergence time studies is subject to the same principles. Brandley et al. (2011) demonstrated a loss of accuracy in age estimates of lizards when they used a combination of noisy and conserved markers. These marker types yielded nearly mutually exclusive distributions of ages when analyzed independently. However, when marker types were combined for analysis, they yielded age estimates that were largely absent from both distributions, demonstrating an averaging effect that was driving a wholesale misestimation of ages (Brandley et al., 2011). While this study serves as an ominous warning for multilocus studies of divergence times, tools developed for phylogenetic experimental design can be utilized as a predictive framework to prevent these sorts of errors.

The predicted utility of a locus can be plotted across the temporal history of a focal group (Dornburg, Fisk, Tamagnan, & Townsend, 2016; López-Giráldez & Townsend, 2011; Townsend, 2007). The resulting phylogenetic informativeness (PI) profile provides a useful roadmap of information content through time, though it should be noted here that the height of the profile does not provide a direct indication of increased node support or that one locus with higher PI values will necessarily perform better than another (Townsend & Leuenberger, 2011). Instead, the shape of the resulting PI profile should be examined (Fig. 5). Moving from the tips to the apex of a profile represents a rise of phylogenetic information. Declines in informativeness following the apex of the primary peak reflect a steady decline or loss in phylogenetic information toward the root of a given focal tree (Fig. 5). Dubbed a "rainshadow of noise," such a change in profile shape is the signature of the expected loss of phylogenetic information, giving rise to homoplasious site patterns (Townsend & Leuenberger, 2011). Dornburg et al. (2014) demonstrated the utility of PI profile shapes as an experimental design tool to guide locus selection for divergence time studies. Retaining loci with high losses in PI along the timescale of interest greatly increases the risk of either tree compression or expansion (Fig. 5). In contrast, retaining only loci with low losses in PI increases the accuracy of divergence time estimates (Fig. 5). Empirically, Dornburg et al. (2014) found that retaining loci with high PI loss resulted in age estimates that were nearly twice as old as those estimated with low loss loci under similar conditions.

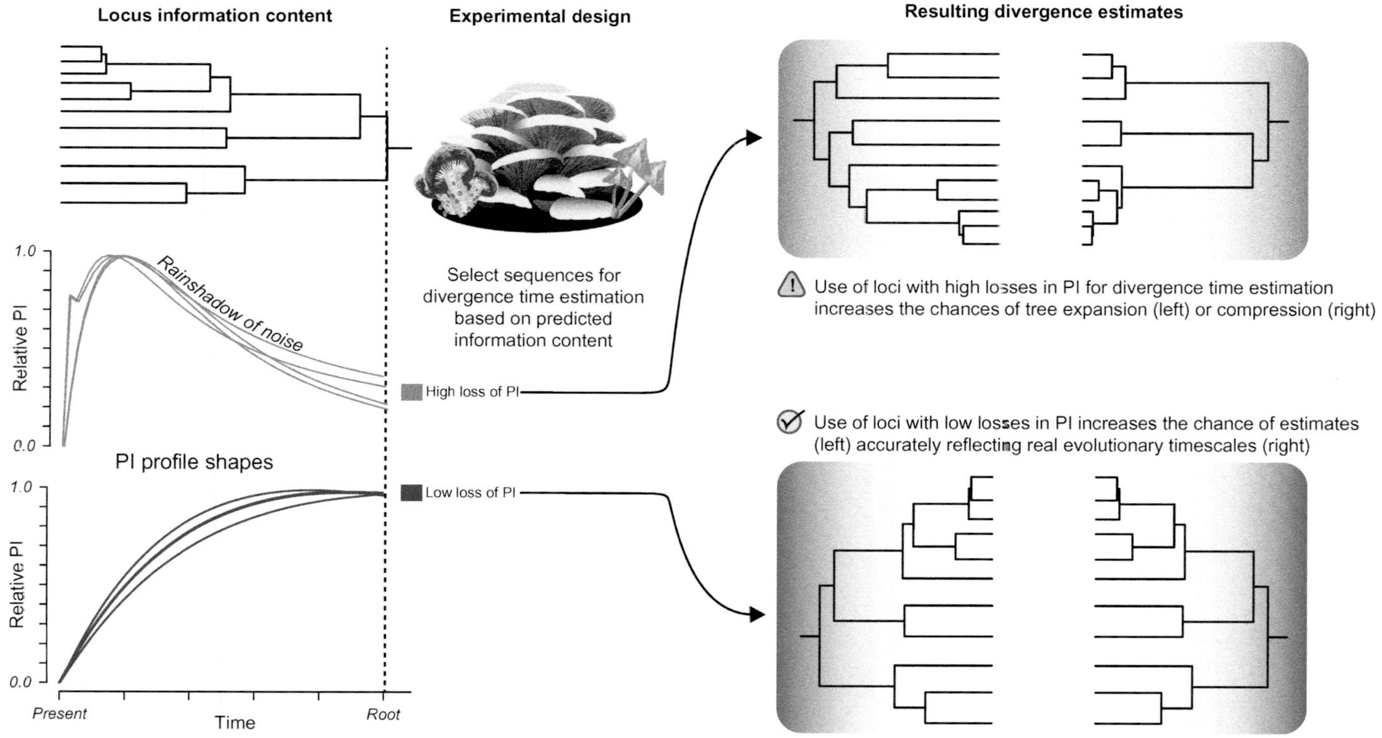

Fig. 5 Utility of phylogenetic informativeness (PI) profiles in the experimental design of divergence time studies. By calculating the PI of a set of loci for a given tree (*left*), PI profile shapes can be compared and utilized as a marker selection tool for experimental design (*middle*). Experimental design choice will determine risk of resulting divergence times subject to tree compression/extension (*right*).

Approaches such as the quantification of PI loss through time as a criteria for locus selection can easily be added to any genomic pipeline and represent a fundamental first step for assessing the validity of age estimates. Although a useful heuristic, it is also important to point out limitations and desirable extensions of this approach. In particular, PI profiles make no direct calculation of noise (Townsend & Leuenberger, 2011). As such, it is possible that a highly informative locus can also contain a high frequency of saturated sites and high noise. For estimating divergence times such a situation would clearly be problematic. Additionally, information content may not be distributed evenly across all taxa. PI approaches assume that substitutions across all taxa reflect a Poisson process or superposition of a number of Poisson processes via a Markov model of molecular evolution (Su et al., 2014). However, lineage-specific molecular rates are common across the Tree of Life (Dornburg et al., 2012; Hirt et al., 2017; Liu, Medina, & Goffinet, 2014; Soltis, Soltis, Savolainen, Crane, & Barraclough, 2002; Villarreal et al., 2015) and violate this assumption. A better understanding of rates of molecular change coupled with further development of approaches to phylogenetic experimental design that accommodate lineage-specific rate heterogeneity represents an exciting frontier for future phylogenetic experimental design.

5. EXPANDING THE PHYLOGENOMIC FRONTIER TO INCLUDE MORE EXPERIMENTAL DESIGN

In this special volume of *Advances in Genetics*, various aspects of new directions and challenges to fungal phylogenetics and phylogenomics have been reviewed. With our increasing ability to collect more and more genetic and genomic data, it is our view that the time is right for us to think of how to efficiently harness this incredible resource. One critical aspect is the continued development of software that enables scrutiny of collected data and/or efficient design of preselected regions for targeted genomic capture. Effective experimental design will greatly decrease cases of topological incongruence among research studies, aiding comparative genetic or genomic studies that critically depend on robust phylogenies. Recent research has laid out the theoretical framework for modeling the relationship between rates of character change and phylogenetic information. This framework provides a baseline assessment of which portions of which genome could be targeted to successfully resolve recalcitrant nodes. Extant theory can be used to aid the analysis of existing data and can help to optimize the selection of target regions in genomic capture. However, current algorithms and applications implementing these methods provide limited functionality compared to their potential and existing theory does not account for prior knowledge of the tree topology beyond the canonical quartet, the effects of heterogeneity of information between taxa, or the potential for changes in molecular character state that deviate from a random process—three critical gaps that are especially problematic in large datasets. Here we urge the society to develop new theories to account for taxon sampling and nonrandom patterns of convergence in sequence data and to implement these approaches in software that will aid in detecting, harvesting, and visualizing regions of a given dataset that are appropriate for phylogenomic analyses.

ACKNOWLEDGMENTS

We thank A. Bogan for helpful suggestions to earlier drafts of this manuscript, L. Nagy for sharing the fungal multilocus data for phylogenetic analysis, U. Ugwuowo for help with data analysis, and K. Zapfe for help with graphical design elements.

This research was supported by the National Science Foundation grant IOS-1457044 to J.P.T.

REFERENCES

Abbott, A. (2012). The genomes of the giants: A walk through the forest of tree genomes. *Tree Genetics & Genomes*, *8*(3), 443.

Adams, D. C., Berns, C. M., Kozak, K. H., & Wiens, J. J. (2009). Are rates of species diversification correlated with rates of morphological evolution? *Proceedings of the Royal Society B: Biological Sciences*, *276*(1668), 2729–2738.

Alexopoulos, C. J. (2007). *Introductory mycology* (4th ed.). Hoboken, NJ, USA: John Wiley & Sons.

Bard, J. (2008). Anatomical ontologies for model organisms: The fungi and animals. In A. Burger, D. Davidson, & R. Baldock (Eds.), *Anatomy ontologies for bioinformatics Computational biology*: *6*. London: Springer.

Bärlocher, F. (2006). Fungal endophytes in submerged roots. In B. J. E. Schulz, C. J. C. Boyle, & T. N. Sieber (Eds.), *Microbial root endophytes Soil biology*: *9*. Berlin, Heidelberg: Springer.

Barnosky, A. D., Hadly, E. A., Gonzalez, P., Head, J., David, P., Polly, A. M. L., et al. (2017). Merging paleobiology with conservation biology to guide the future of terrestrial ecosystems. *Science*, *355*(6325). https://doi.org/10.1126/science.aah4787, pii: eaah4787.

Bazinet, A. L., Mitter, K. T., Davis, D. R., Nieukerken, E. J. V., Cummings, M. P., & Mitter, C. (2016). Phylotranscriptomics resolves ancient divergences in the *Lepidoptera*. *Systematic Entomology*, *42*(2), 305–316.

Bellard, C., Bertelsmeier, C., Leadley, P., Thuiller, W., & Courchamp, F. (2012). Impacts of climate change on the future of biodiversity. *Ecology Letters*, *15*(4), 365–377.

Berbee, M. L., & Taylor, J. W. (2010). Dating the molecular clock in fungi—How close are we? *Fungal Biology Reviews*, *24*(1-2), 1–16.

Bessey, E. A. (1942). Some problems in fungus phylogeny. *Mycologia*, *34*(4), 355.

Bessey, E. A. (1950). *Morphology and taxonomy of fungi*. Philadelphia, USA.

Betancur-R, R., Li, C., Munroe, T. A., Ballesteros, J. A., & Ortí, G. (2013). Addressing gene tree discordance and non-stationarity to resolve a multi-locus phylogeny of the flatfishes (Teleostei: Pleuronectiformes). *Systematic Biology*, *62*(5), 763–785.

Biek, R., Pybus, O. G., Lloyd-Smith, J. O., & Didelot, X. (2015). Measurably evolving pathogens in the genomic era. *Trends in Ecology & Evolution*, *30*(6), 306–313.

Binder, M., Justo, A., Riley, R., Salamov, A., Lopez-Giraldez, F., Sjökvist, E., et al. (2013). Phylogenetic and phylogenomic overview of the polyporales. *Mycologia*, *105*(6), 1350–1373.

Blackwell, M. (2009). Fungal evolution and taxonomy. In H. E. Roy, F. E. Vega, M. S. Goettel, D. Chandler, J. K. Pell, & E. Wajnberg (Eds.), *The ecology of fungal entomopathogens*. (pp. 7–16). Netherlands: Springer.

Blackwell, M. (2011). The fungi: 1, 2, 3 ... 5.1 million species? *American Journal of Botany*, *98*(3), 426–438.

Blackwell, M., Hibbett, D. S., Taylor, J. W., & Spatafora, J. W. (2006). Research coordination networks: A phylogeny for kingdom fungi (Deep Hypha). *Mycologia*, *98*(6), 829–837.

Bland, L. M., Collen, B., Orme, C. D. L., & Bielby, J. (2015). Predicting the conservation status of data-deficient species. *Conservation Biology*, *29*(1), 250–259.

Blehert, D. S., Hicks, A. C., Behr, M., Meteyer, C. U., Berlowski-Zier, B. M., Buckles, E. L., et al. (2009). Bat white-nose syndrome: An emerging fungal pathogen? *Science*, *323*(5911), 227.

Boldin, B., & Kisdi, É. (2012). On the evolutionary dynamics of pathogens with direct and environmental transmission. *Evolution*, *66*(8), 2514–2527.

Boria, R. A., Olson, L. E., Goodman, S. M., & Anderson, R. P. (2014). Spatial filtering to reduce sampling bias can improve the performance of ecological niche models. *Ecological Modelling*, *275*, 73–77.

Botstein, D. (1997). Genetics: Yeast as a model organism. *Science*, *277*(5330), 1259–1260.

Brandley, M. C., Bragg, J. G., Singhal, S., Chapple, D. G., Jennings, C. K., Lemmon, A. R., et al. (2015). Evaluating the performance of anchored hybrid enrichment at the tips of the tree of life: A phylogenetic analysis of Australian *Eugongylus* group scincid lizards. *BMC Evolutionary Biology*, *15*, 62.

Brandley, M. C., Wang, Y., Guo, X., Oca, A. N. M. d., Fería-Ortíz, M., Hikida, T., et al. (2011). Accommodating heterogenous rates of evolution in molecular divergence dating methods: An example using intercontinental dispersal of Plestiodon (Eumeces) lizards. *Systematic Biology*, *60*(1), 3–15.

Breinholt, J. W., & Kawahara, A. Y. (2013). Phylotranscriptomics: Saturated third codon positions radically influence the estimation of trees based on next-gen data. *Genome Biology and Evolution*, *5*(11), 2082–2092.

Brinkmann, H., Giezen, M. v. d., Zhou, Y., Raucourt, G. P. d., & Philippe, H. (2005). An empirical assessment of long-branch attraction artefacts in deep eukaryotic phylogenomics. *Systematic Biology*, *54*(5), 743–757.

Brown, J. M. (2014). Detection of implausible phylogenetic inferences using posterior predictive assessment of model fit. *Systematic Biology*, *63*(3), 334–348.

Bruns, T. D., Vilgalys, R., Barns, S. M., Gonzalez, D., Hibbett, D. S., Lane, D. J., et al. (1992). Evolutionary relationships within the fungi: Analyses of nuclear small subunit rRNA sequences. *Molecular Phylogenetics and Evolution*, *1*(3), 231–241.

Buckley, L. B., & Kingsolver, J. G. (2012). Functional and phylogenetic approaches to forecasting species' responses to climate change. *Annual Review of Ecology, Evolution, and Systematics*, *43*(1), 205–226.

Cai, L., Giraud, T., Zhang, N., Begerow, D., Cai, G., & Shivas, R. G. (2011). The evolution of species concepts and species recognition criteria in plant pathogenic fungi. *Fungal Diversity*, *50*(1), 121–133.

Cai, Q., Tulloss, R. E., Tang, L. P., Tolgor, B., Zhang, P., Chen, Z. H., et al. (2014). Multilocus phylogeny of lethal amanitas: Implications for species diversity and historical biogeography. *BMC Evolutionary Biology*, *14*, 143.

Capella-Gutiérrez, S., Marcet-Houben, M., & Gabaldón, T. (2012). Phylogenomics supports microsporidia as the earliest diverging clade of sequenced fungi. *BMC Biology*, *10*, 47.

Capinha, C., Essl, F., Seebens, H., Moser, D., & Pereira, H. M. (2015). Biogeography. The dispersal of alien species redefines biogeography in the anthropocene. *Science*, *348*(6240), 1248–1251.

Carstens, B. C., Pelletier, T. A., Reid, N. M., & Satler, J. D. (2013). How to fail at species delimitation. *Molecular Ecology*, *22*(17), 4369–4383.

Casiraghi, M., Galimberti, A., Sandionigi, A., Bruno, A., & Labra, M. (2016). Life with or without names. *Evolutionary Biology*, *43*(4), 582–595.

Catullo, R. A., Ferrier, S., & Hoffmann, A. A. (2015). Extending spatial modelling of climate change responses beyond the realized niche: Estimating, and accommodating, physiological limits and adaptive evolution. *Global Ecology and Biogeography: A Journal of Macroecology*, *24*(10), 1192–1202.

Cavalier-Smith, T., Chao, E. E., Snell, E. A., Berney, C., Fiore-Donno, A. M., & Lewis, R. (2014). Multigene eukaryote phylogeny reveals the likely protozoan ancestors of opisthokonts (animals, fungi, choanozoans) and Amoebozoa. *Molecular Phylogenetics and Evolution*, *81*, 71–85.

Chakrabarty, P., Faircloth, B. C., Alda, F., Ludt, W. B., McMahan, C. D., Near, T. J., et al. (2017). Phylogenomic systematics of ostariophysan fishes: Ultraconserved elements support the surprising non-monophyly of characiformes. *Systematic Biology*. (p. 15). https://doi.org/10.1093/sysbio/syx038.

Chang, Y., Wang, S., Sekimoto, S., Aerts, A. L., Choi, C., Clum, A., et al. (2015). Phylogenomic analyses indicate that early fungi evolved digesting cell walls of algal ancestors of land plants. *Genome Biology and Evolution*, 7(6), 1590–1601.

Cheng, X., Hui, J. H., Lee, Y. Y., Wan Law, P. T., & Kwan, H. S. (2015). A 'developmental hourglass' in fungi. *Molecular Biology and Evolution*, *32*(6), 1556–1566.

Clemmensen, K. E., Ihrmark, K., Durling, M. B., & Lindahl, B. D. (2016). Sample preparation for fungal community analysis by high-throughput sequencing of barcode amplicons. In F. Martin & S. Uroz (Eds.), *Microbial environmental genomics (MEG). Methods in molecular biology*: *vol. 1399*. New York, NY: Humana Press.

Crandall, K. A. (2004). Evolution: Genomic databases and the tree of life. *Science*, *306*(5699), 1144–1145.

Crespo, A., & Thorsten Lumbsch, H. (2010). Cryptic species in lichen-forming fungi. *IMA Fungus*, *1*(2), 167–170.

Crous, P. W., Hawksworth, D. L., & Wingfield, M. J. (2015). Identifying and naming plant-pathogenic fungi: Past, present, and future. *Annual Review of Phytopathology*, *53*, 247–267.

Darwin, C. (1859). *On the origin of species by means of natural selection, or the preservation of favoured races in the struggle for life*. London, UK: Clowes and Sons, Ltd.

Darwin, F. (1897). *The life and letters of Charles Darwin: Including an autobiographical chapter*. London, UK: William Clowes and Sons, Ltd.

Davis, R. H. (2000). *Neurospora: Contributions of a model organism*. Oxford, UK: Oxford University Press.

Davis, B. J., Phillips, R. D., Wright, M., Linde, C. C., & Dixon, K. W. (2015). Continent-wide distribution in mycorrhizal fungi: Implications for the biogeography of specialized orchids. *Annals of Botany*, *116*(3), 413–421.

Dean, R., Van Kan, J. A., Pretorius, Z. A., Hammond-Kosack, K. E., Di Pietro, A., Spanu, P. D., et al. (2012). The top 10 fungal pathogens in molecular plant pathology. *Molecular Plant Pathology*, *13*(4), 414–430.

De Queiroz, K. (2007). Species concepts and species delimitation. *Systematic Biology*, *56*(6), 879–886.

Desjardins, C. A., Giamberardino, C., Sykes, S. M., Chen-Hsin, Y., Tenor, J. L., Chen, Y., et al. (2017). Population genomics and the evolution of virulence in the fungal pathogen *Cryptococcus neoformans*. *Genome Research*, *27*(7), 1207–1219. https://doi.org/10.1101/gr.218727.116.

Dettman, J. R., Jacobson, D. J., Turner, E., Pringle, A., & Taylor, J. W. (2003). Reproductive isolation and phylogenetic divergence in *Neurospora*: Comparing methods of species recognition in a model eukaryote. *Evolution*, *57*(12), 2721–2741.

Dornburg, A., Beaulieu, J. M., Oliver, J. C., & Near, T. J. (2011). Integrating fossil preservation biases in the selection of calibrations for molecular divergence time estimation. *Systematic Biology*, *60*(4), 519–527.

Dornburg, A., Brandley, M. C., McGowen, M. R., & Near, T. J. (2012). Relaxed clocks and inferences of heterogeneous patterns of nucleotide substitution and divergence time estimates across whales and dolphins (Mammalia: Cetacea). *Molecular Biology and Evolution*, *29*(2), 721–736.

Dornburg, A., Eytan, R. I., Federman, S., Pennington, J. N., Stewart, A. L., Jones, C. D., et al. (2015). Molecular data support the existence of two species of the Antarctic fish genus *Cryodraco* (Channichthyidae). *Polar Biology*, *39*(8), 1369–1379.

Dornburg, A., Federman, S., Eytan, R. I., & Near, T. J. (2016). Cryptic species diversity in sub-antarctic islands: A case study of lepidonotothen. *Molecular Phylogenetics and Evolution*, *104*, 32–43.

Dornburg, A., Federman, S., Lamb, A. D., Jones, C. D., & Near, T. J. (2017). Cradles and museums of Antarctic teleost biodiversity. *Nature Ecology & Evolution*, *1*(9), 1379.

Dornburg, A., Fisk, J. N., Tamagnan, J., & Townsend, J. P. (2016). PhyInformR: Phylogenetic experimental design and phylogenomic data exploration in R. *BMC Evolutionary Biology*, *16*(1), 262.

Dornburg, A., Forrestel, E., Moore, J., Iglesias, T., Jones, A., Rao, L., et al. (2017). An assessment of sampling biases across studies of diel activity patterns in marine ray-finned fishes (Actinopterygii). *Bulletin of Marine Science*, *93*(2), 611–639.

Dornburg, A., Friedman, M., & Near, T. J. (2015). Phylogenetic analysis of molecular and morphological data highlights uncertainty in the relationships of fossil and living species of *Elopomorpha* (Actinopterygii: Teleostei). *Molecular Phylogenetics and Evolution*, *89*, 205–218.

Dornburg, A., Sidlauskas, B., Santini, F., Sorenson, L., Near, T. J., & Alfaro, M. E. (2011). The influence of an innovative locomotor strategy on the phenotypic diversification of triggerfish (Family: Balistidae). *Evolution*, *65*(7), 1912–1926.

Dornburg, A., Townsend, J. P., Brooks, W., Spriggs, E., Eytan, R. I., Moore, J. A., et al. (2017). New insights on the sister lineage of percomorph fishes with an anchored hybrid enrichment dataset. *Molecular Phylogenetics and Evolution*, *110*, 27–38.

Dornburg, A., Townsend, J. P., Friedman, M., & Near, T. J. (2014). Phylogenetic informativeness reconciles ray-finned fish molecular divergence times. *BMC Evolutionary Biology*, *14*, 169.

Drost, H.-G., Gabel, A., Grosse, I., & Quint, M. (2015). Evidence for active maintenance of phylotranscriptomic hourglass patterns in animal and plant embryogenesis. *Molecular Biology and Evolution*, *32*(5), 1221–1231.

Dujon, B. (2010). Yeast evolutionary genomics. *Nature Reviews. Genetics*, *11*(7), 512–524.

Dunn, C. W., Hejnol, A., Matus, D. Q., Pang, K., Browne, W. E., Smith, S. A., et al. (2008). Broad phylogenomic sampling improves resolution of the animal tree of life. *Nature*, *452*(7188), 745–749.

Ebersberger, I., Simoes, R. d. M., Kupczok, A., Gube, M., Kothe, E., Voigt, K., et al. (2012). A consistent phylogenetic backbone for the fungi. *Molecular Biology and Evolution, 29*(5), 1319–1334.

Edwards, D. L., & Knowles, L. L. (2014). Species detection and individual assignment in species delimitation: Can integrative data increase efficacy? *Proceedings. Biological Sciences, 281*(1777), 20132765.

Eytan, R. I., Evans, B. R., Dornburg, A., Lemmon, A. R., Lemmon, E. M., Wainwright, P. C., et al. (2015). Are 100 enough? Inferring acanthomorph teleost phylogeny using anchored hybrid enrichment. *BMC Evolutionary Biology, 15*, 113.

Feau, N., Decourcelle, T., Husson, C., Desprez-Loustau, M.-L., & Dutech, C. (2011). Finding single copy genes out of sequenced genomes for multilocus phylogenetics in non-model fungi. *PLoS One, 6*(4), e18803.

Federman, S., Dornburg, A., Daly, D. C., Downie, A., Perry, G. H., Yoder, A. D., et al. (2016). Implications of lemuriform extinctions for the malagasy flora. *Proceedings of the National Academy of Sciences of the United States of America, 113*(18), 5041–5046.

Federman, S., Dornburg, A., Downie, A., Richard, A. F., Daly, D. C., & Donoghue, M. J. (2015). The biogeographic origin of a radiation of trees in Madagascar: Implications for the assembly of a tropical forest biome. *BMC Evolutionary Biology, 15*, 216.

Felsenstein, J. (2004). *Inferring phylogenies*. Massachusetts, USA: Sinauer Associates Incorporated.

Fernández, R., Laumer, C. E., Vahtera, V., Libro, S., Kaluziak, S., Sharma, P. P., et al. (2014). Evaluating topological conflict in centipede phylogeny using transcriptomic data sets. *Molecular Biology and Evolution, 31*(6), 1500–1513.

Feurtey, A., Gladieux, P., Hood, M. E., Snirc, A., Cornille, A., Rosenthal, L., et al. (2016). Strong phylogeographic co-structure between the anther-smut fungus and its white campion host. *The New Phytologist, 212*(3), 668–679.

Floudas, D., Held, B. W., Riley, R., Nagy, L. G., Koehler, G., Ransdell, A. S., et al. (2015). Evolution of novel wood decay mechanisms in Agaricales revealed by the genome sequences of Fistulina hepatica and Cylindrobasidium torrendii. *Fungal Genetics and Biology: FG & B, 76*, 78–92.

Forrestel, E. J., Donoghue, M. J., & Smith, M. D. (2014). Convergent phylogenetic and functional responses to altered fire regimes in mesic savanna grasslands of North America and South Africa. *The New Phytologist, 203*(3), 1000–1011.

Forrestel, E. J., Donoghue, M. J., & Smith, M. D. (2015). Functional differences between dominant grasses drive divergent responses to large herbivore loss in mesic savanna grasslands of North America and South Africa. *The Journal of Ecology, 103*(3), 714–724.

Frisvad. (1998). *Chemical fungal taxonomy*. Florida, USA: CRC Press, Taylor & Francis.

Fukami, T., Beaumont, H. J. E., Zhang, X.-X., & Rainey, P. B. (2007). Immigration history controls diversification in experimental adaptive radiation. *Nature, 446*(7134), 436–439.

Gabaldón, T., & Marcet-Houben, M. (2014). 3 Phylogenomics for the study of fungal biology. In M. Nowrousian (Ed.), *Fungal genomics. The mycota (a comprehensive treatise on fungi as experimental systems for basic and applied research), Vol: 13*. Berlin, Heidelberg: Springer.

Gabaldón, T., Naranjo-Ortíz, M. A., & Marcet-Houben, M. (2016). Evolutionary genomics of yeast pathogens in the Saccharomycotina. *FEMS Yeast Research, 16*(6) (p. 10). https://doi.org/10.1093/femsyr/fow064.

Galagan, J. E., Henn, M. R., Ma, L.-J., Cuomo, C. A., & Birren, B. (2005). Genomics of the fungal kingdom: Insights into eukaryotic biology. *Genome Research, 15*(12), 1620–1631.

Garamszegi, L. Z. (2014). *Modern phylogenetic comparative methods and their application in evolutionary biology: Concepts and practice*. Berlin, Heidelberg: Springer-Verlag Berlin Heidelberg.

Gatesy, J., DeSalle, R., & Wahlberg, N. (2007). How many genes should a systematist sample? Conflicting insights from a phylogenomic matrix characterized by replicated incongruence. *Systematic Biology*, *56*(2), 355–363.

Ge, Z.-W., Yang, Z. L., Pfister, D. H., Carbone, M., Bau, T., & Smith, M. E. (2014). Multigene molecular phylogeny and biogeographic diversification of the earth tongue fungi in the genera *Cudonia* and *Spathularia* (Rhytismatales, Ascomycota). *PLoS One*, *9*(8), e103457.

Gerlinger, M., Horswell, S., Larkin, J., Rowan, A. J., Salm, M. P., Varela, I., et al. (2014). Genomic architecture and evolution of clear cell renal cell carcinomas defined by multiregion sequencing. *Nature Genetics*, *46*(3), 225–233.

Gherbawy, Y., & Voigt, K. (2010). *Molecular identification of fungi*. New York, NY: Springer Science & Business Media.

Gladieux, P., Ropars, J., Badouin, H., Branca, A., Aguileta, G., Vienne, D. M. d., et al. (2014). Fungal evolutionary genomics provides insight into the mechanisms of adaptive divergence in eukaryotes. *Molecular Ecology*, *23*(4), 753–773.

Goffeau, A. (2000). Four years of post-genomic life with 6000 yeast genes. *FEBS Letters*, *480*(1), 37–41.

Goremykin, V. V., Nikiforova, S. V., & Bininda-Emonds, O. R. P. (2010). Automated removal of noisy data in phylogenomic analyses. *Journal of Molecular Evolution*, *71*(5-6), 319–331.

Gouy, R., Baurain, D., & Philippe, H. (2015). Rooting the tree of life: The phylogenetic jury is still out. *Philosophical Transactions of the Royal Society of London. Series B, Biological Sciences*, *370*(1678), 20140329.

Graybeal, A. (1993). The phylogenetic utility of cytochrome B: Lessons from bufonid frogs. *Molecular Phylogenetics and Evolution*, *2*(3), 256–269.

Grenfell, B. T., Pybus, O. G., Gog, J. R., Wood, J. L. N., Daly, J. M., Mumford, J. A., et al. (2004). Unifying the epidemiological and evolutionary dynamics of pathogens. *Science*, *303*(5656), 327–332.

Grigoriev, Igor V. 2011. "JGI fungal genomics program." https://doi.org/10.2172/1012482.

Hansen, M. C., Potapov, P. V., Moore, R., Hancher, M., Turubanova, S. A., Tyukavina, A., et al. (2013). High-resolution global maps of 21st-century forest cover change. *Science*, *342*(6160), 850–853.

Hao, B., Qi, J. I., & Wang, B. (2003). Prokaryotic phylogeny based on complete genomes without sequence alignment. *Modern Physics Letters B*, *17*(03), 91–94.

Harish, A., & Kurland, C. G. (2017). Empirical genome evolution models root the tree of life. *Biochimie*, *138*, 137–155.

Harmon, L. J., Weir, J. T., Brock, C. D., Glor, R. E., & Challenger, W. (2008). GEIGER: Investigating evolutionary radiations. *Bioinformatics*, *24*(1), 129–131.

Hawksworth, D. L. (2001). The magnitude of fungal diversity: The 1.5 million species estimate revisited. *Mycological Research*, *105*(12), 1422–1432.

Heath, T. A., Huelsenbeck, J. P., & Stadler, T. (2014). The fossilized birth-death process for coherent calibration of divergence-time estimates. *Proceedings of the National Academy of Sciences of the United States of America*, *111*(29), E2957–66.

Heath, T. A., Zwickl, D. J., Kim, J., & Hillis, D. M. (2008). Taxon sampling affects inferences of macroevolutionary processes from phylogenetic trees. *Systematic Biology*, *57*(1), 160–166.

Heckman, D. S., Geiser, D. M., Eidell, B. R., Stauffer, R. L., Kardos, N. L., & Hedges, S. B. (2001). Molecular evidence for the early colonization of land by fungi and plants. *Science*, *293*(5532), 1129–1133.

Hejnol, A., Obst, M., Stamatakis, A., Ott, M., Rouse, G. W., Edgecombe, G. D., et al. (2009). Assessing the root of bilaterian animals with scalable phylogenomic methods. *Proceedings. Biological Sciences*, *276*(1677), 4261–4270.

Hellmann, J. J., & Pfrender, M. E. (2011). Future human intervention in ecosystems and the critical role for evolutionary biology. *Conservation Biology*, *25*(6), 1143–1147.

Helmus, M. R., Luke Mahler, D., & Losos, J. B. (2014). Island biogeography of the anthropocene. *Nature*, *513*(7519), 543–546.

Herendeen, P. S., Friis, E. M., Pedersen, K. R., & Crane, P. R. (2017). Palaeobotanical redux: Revisiting the age of the angiosperms. *Nature Plants*, *3*, 17015.

Hibbett, D. S. (2007). After the gold rush, or before the flood? Evolutionary morphology of mushroom-forming fungi (Agaricomycetes) in the early 21st century. *Mycological Research*, *111*(9), 1001–1018.

Hibbett, D. S., Binder, M., Bischoff, J. F., Blackwell, M., Cannon, P. F., Eriksson, O. E., et al. (2007). A higher-level phylogenetic classification of the fungi. *Mycological Research*, *111*(Pt. 5), 509–547.

Hibbett, D. S., & Donoghue, M. J. (2001). Analysis of character correlations among wood decay mechanisms, mating systems, and substrate ranges in homobasidiomycetes. *Systematic Biology*, *50*(2), 215–242.

Hibbett, D. S., Fukumasa-Nakai, Y., Tsuneda, A., & Donoghue, M. J. (1995). Phylogenetic diversity in shiitake inferred from nuclear ribosomal DNA sequences. *Mycologia*, *87*(5), 618.

Hibbett, D. S., Ohman, A., Glotzer, D., Nuhn, M., Kirk, P., & Henrik Nilsson, R. (2011). Progress in molecular and morphological taxon discovery in fungi and options for formal classification of environmental sequences. *Fungal Biology Reviews*, *25*(1), 38–47.

Hibbett, D. S., & Taylor, J. W. (2013). Fungal systematics: Is a new age of enlightenment at hand? *Nature Reviews. Microbiology*, *11*, 129–133. https://doi.org/10.1038/nrmicro2942.

Hirt, M. V., Vincent Hirt, M., Arratia, G., Chen, W.-J., Mayden, R. L., Tang, K. L., et al. (2017). Effects of gene choice, base composition and rate heterogeneity on inference and estimates of divergence times in Cypriniform fishes. *Biological Journal of the Linnean Society. Linnean Society of London*, *121*(2), 319–339.

Hooper, D. U., Carol Adair, E., Cardinale, B. J., Byrnes, J. E. K., Hungate, B. A., Matulich, K. L., et al. (2012). A global synthesis reveals biodiversity loss as a major driver of ecosystem change. *Nature*, *486*(7401), 105–108.

Hori, C., Gaskell, J., Igarashi, K., Samejima, M., Hibbett, D., Henrissat, B., et al. (2013). Genomewide analysis of polysaccharides degrading enzymes in 11 white- and brown-rot polyporales provides insight into mechanisms of wood decay. *Mycologia*, *105*(6), 1412–1427.

Howard, S. J. (2014). Multi-resistant aspergillosis due to cryptic species. *Mycopathologia*, *178*(5-6), 435–439.

Howlett, B. J., Lowe, R. G. T., Marcroft, S. J., & Wouw, A. P. v. d. (2015). Evolution of virulence in fungal plant pathogens: Exploiting fungal genomics to control plant disease. *Mycologia*, *107*(3), 441–451.

Hu, X., Xiao, G., Zheng, P., Shang, Y., Yao, S., Zhang, X., et al. (2014). Trajectory and genomic determinants of fungal-pathogen speciation and host adaptation. *Proceedings of the National Academy of Sciences of the United States of America*, *111*(47), 16796–16801.

Hug, L. A., Baker, B. J., Anantharaman, K., Brown, C. T., Probst, A. J., Castelle, C. J., et al. (2016). A new view of the tree of life. *Nature Microbiology*, *1*, 16048.

Hulsey, C. D., García de León, F. J., & Rodiles-Hernández, R. (2006). Micro- and macroevolutionary decoupling of cichlid jaws: A test of liem's key innovation hypothesis. *Evolution*, *60*(10), 2096.

Hwang, J., Qi, Z., Yang, Z. L., Wang, Z., & Townsend, J. P. (2015). Solving the ecological puzzle of mycorrhizal associations using data from annotated collections and environmental samples—An example of saddle fungi. *Environmental Microbiology Reports*, 7(4), 658–667.

Ikeda, D. H., Max, T. L., Allan, G. J., Lau, M. K., Shuster, S. M., & Whitham, T. G. (2017). Genetically informed ecological niche models improve climate change predictions. *Global Change Biology*, *23*(1), 164–176.

Inoue, Y., Vy, T. T. P., Yoshida, K., Asano, H., Mitsuoka, C., Asuke, S., et al. (2017). Evolution of the wheat blast fungus through functional losses in a host specificity determinant. *Science*, *357*(6346), 80–83.

Jackson, N. D., Carstens, B. C., Morales, A. E., & O'Meara, B. C. (2016). Species delimitation with gene flow. *Systematic Biology*, *66*(5), 799–812. https://doi.org/10.1093/sysbio/syw117.

James, T. Y., & Berbee, M. L. (2012). No jacket required—New fungal lineage defies dress code: Recently described zoosporic fungi lack a cell wall during trophic phase. *BioEssays: News and Reviews in Molecular, Cellular and Developmental Biology*, *34*(2), 94–102.

James, T. Y., Kauff, F., Schoch, C. L., Matheny, P. B., Hofstetter, V., Cox, C. J., et al. (2006). Reconstructing the early evolution of fungi using a six-gene phylogeny. *Nature*, *443*(7113), 818–822.

James, T. Y., Letcher, P. M., Longcore, J. E., Mozley-Standridge, S. E., Porter, D., Powell, M. J., et al. (2006). A molecular phylogeny of the flagellated fungi (Chytridiomycota) and description of a new phylum (Blastocladiomycota). *Mycologia*, *98*(6), 860–871.

James, T. Y., Pelin, A., Bonen, L., Ahrendt, S., Sain, D., Corradi, N., et al. (2013). Shared signatures of parasitism and Phylogenomics unite Cryptomycota and *Microsporidia*. *Current Biology: CB*, *23*(16), 1548–1553.

Janouškovec, J., Gavelis, G. S., Burki, F., Dinh, D., Bachvaroff, T. R., Gornik, S. G., et al. (2017). Major transitions in dinoflagellate evolution unveiled by phylotranscriptomics. *Proceedings of the National Academy of Sciences of the United States of America*, *114*(2), E171–80.

Jeffroy, O., Brinkmann, H., Delsuc, F., & Philippe, H. (2006). Phylogenomics: The beginning of incongruence? *Trends in Genetics*, *22*(4), 225–231.

Ješovnik, A., González, V. L., & Schultz, T. R. (2016). Phylogenomics and divergence dating of fungus-farming ants (Hymenoptera: Formicidae) of the genera *Sericomyrmex* and *Apterostigma*. *PLoS One*, *11*(7), e0151059.

Jiang, X., Xiang, M., & Liu, X. (2017). Nematode-trapping fungi. *Microbiology Spectrum*, *5*(1) (p. 12). https://doi.org/10.1128/microbiolspec.FUNK-0022-2016.

Jones, B. (2015). Phylogenomics: Building the insect tree-of-life. *Nature Reviews. Genetics*, *16*(1), 2.

Jones, M. D. M., Forn, I., Gadelha, C., Egan, M. J., Bass, D., Massana, R., et al. (2011). Discovery of novel intermediate forms redefines the fungal tree of life. *Nature*, *474*(7350), 200–203.

Kalinka, A. T., Varga, K. M., Gerrard, D. T., Preibisch, S., Corcoran, D. L., Jarrells, J., et al. (2010). Gene expression divergence recapitulates the developmental hourglass model. *Nature*, *468*(7325), 811–814.

Kennicutt, M. C., Steven L., C., 2nd, Cassano, J. J., Liggett, D., Massom, R., Peck, L. S., et al. (2014). Polar research: Six priorities for antarctic science. *Nature*, *512*(7512), 23–25.

Kepler, R., Ban, S., Nakagiri, A., Bischoff, J., Hywel-Jones, N., Owensby, C. A., et al. (2013). The phylogenetic placement of hypocrealean insect pathogens in the genus Polycephalomyces: An application of one fungus one name. *Fungal Biology*, *117*(9), 611–622.

Knope, M. L., Forde, S. E., & Fukami, T. (2011). Evolutionary history, immigration history, and the extent of diversification in community assembly. *Frontiers in Microbiology*, *2*, 273.

Kraichak, E., Crespo, A., Divakar, P. K., Leavitt, S. D., & Thorsten Lumbsch, H. (2017). A temporal banding approach for consistent taxonomic ranking above the species level. *Scientific Reports*, 7(1), 2297.

Krüger, M., Krüger, C., Walker, C., Stockinger, H., & Schüßler, A. (2011). Phylogenetic reference data for systematics and phylotaxonomy of Arbuscular Mycorrhizal fungi from phylum to species level. *The New Phytologist*, *193*(4), 970–984.

Kumar, S., Filipski, A. J., Battistuzzi, F. U., Pond, S. L. K., & Tamura, K. (2012). Statistics and truth in phylogenomics. *Molecular Biology and Evolution*, *29*(2), 457–472.

Kuo, C.-H., Wares, J. P., & Kissinger, J. C. (2008). The apicomplexan whole-genome phylogeny: An analysis of incongruence among gene trees. *Molecular Biology and Evolution*, *25*(12), 2689–2698.

Lanfear, R., Calcott, B., Kainer, D., Mayer, C., & Stamatakis, A. (2014). Selecting optimal partitioning schemes for phylogenomic datasets. *BMC Evolutionary Biology*, *14*, 82.

Lanier, K. A., & Williams, L. D. (2017). The origin of life: Models and data. *Journal of Molecular Evolution*, *84*(2-3), 85–92.

Last, P. R., White, W. T., Gledhill, D. C., Hobday, A. J., Brown, R., Edgar, G. J., et al. (2010). Long-term shifts in abundance and distribution of a temperate fish fauna: A response to climate change and fishing practices. *Global Ecology and Biogeography: A Journal of Macroecology*, *20*(1), 58–72.

Leavitt, S. D., Esslinger, T. L., Divakar, P. K., Crespo, A., & Thorsten Lumbsch, H. (2016). Hidden diversity before our eyes: Delimiting and describing cryptic lichen-forming fungal species in camouflage lichens (Parmeliaceae, Ascomycota). *Fungal Biology*, *120*(11), 1374–1391.

Leavitt, S. D., Esslinger, T. L., Spribille, T., Divakar, P. K., & Thorsten Lumbsch, H. (2013). Multilocus phylogeny of the lichen-forming fungal genus *Melanohalea* (Parmeliaceae, Ascomycota): Insights on diversity, distributions, and a comparison of species tree and concatenated topologies. *Molecular Phylogenetics and Evolution*, *66*(1), 138–152.

Levy, J. (1994). Sequencing the yeast genome: An international achievement. *Yeast*, *10*(13), 1689–1706.

Li, D.-W. (2016). *Biology of microfungi*. Switzerland: Springer International Publishing.

Lindahl, B. D., Henrik Nilsson, R., Tedersoo, L., Abarenkov, K., Carlsen, T., Kjøller, R., et al. (2013). Fungal community analysis by high-throughput sequencing of amplified markers—A user's guide. *The New Phytologist*, *199*(1), 288–299.

Littleford-Colquhoun, B. L., Clemente, C., Whiting, M. J., Ortiz-Barrientos, D., & Frère, C. H. (2017). Archipelagos of the anthropocene: Rapid and extensive differentiation of native terrestrial vertebrates in a single metropolis. *Molecular Ecology*, *26*(9), 2466–2481.

Liu, Y., Cox, C. J., Wang, W., & Goffinet, B. (2014). Mitochondrial phylogenomics of early land plants: Mitigating the effects of saturation, compositional heterogeneity, and codon-usage bias. *Systematic Biology*, *63*(6), 862–878.

Liu, Y. J., & Hall, B. D. (2004). Body plan evolution of ascomycetes, as inferred from an RNA polymerase II phylogeny. *Proceedings of the National Academy of Sciences of the United States of America*, *101*(13), 4507–4512.

Liu, Y., Medina, R., & Goffinet, B. (2014). 350 My of mitochondrial genome stasis in mosses, an early land plant lineage. *Molecular Biology and Evolution*, *31*(10), 2586–2591.

Liu, F., Wang, M., Damm, U., Crous, P. W., & Cai, L. (2016). Species boundaries in plant pathogenic fungi: A colletotrichum case study. *BMC Evolutionary Biology*, *16*, 81.

Liu, J., Wang, J., Gao, G., Bartlam, M. G., & Wang, Y. (2015). Distribution and diversity of fungi in freshwater sediments on a river catchment scale. *Frontiers in Microbiology*, *6*, 329.

Liu, P., Wen, J., & Yi, T.-S. (2017). Evolution of biogeographic disjunction between eastern Asia and north America in Chamaecyparis: Insights from ecological niche models. *Plant Diversity*, *39*, 111–116. https://doi.org/10.1016/j.pld.2017.04.001.

Liu, L., Wu, S., & Yu, L. (2015). Coalescent methods for estimating species trees from phylogenomic data. *Journal of Systematics and Evolution*, *53*(5), 380–390.

Loiselle, B. A., Jørgensen, P. M., Consiglio, T., Jiménez, I., Blake, J. G., Lohmann, L. G., et al. (2008). Predicting species distributions from herbarium collections: Does climate bias in collection sampling influence model outcomes? *Journal of Biogeography*, *35*(1), 105–116.

Looney, B. P., Ryberg, M., Hampe, F., Sánchez-García, M., & Matheny, P. B. (2016). Into and out of the tropics: Global diversification patterns in a hyperdiverse clade of ectomycorrhizal fungi. *Molecular Ecology*, *25*(2), 630–647.
López-Giráldez, F., & Townsend, J. P. (2011). PhyDesign: An online application for profiling phylogenetic informativeness. *BMC Evolutionary Biology*, *11*, 152.
Lopez-Llorca, L. V., Jansson, H.-B., Vicente, J. G. M., & Salinas, J. (2006). Nematophagous fungi as root endophytes. In *Soil biology* (pp. 191–206). Berlin, Heidelberg: Springer Berlin Heidelberg.
Lorch, J. M., Meteyer, C. U., Behr, M. J., Boyles, J. G., Cryan, P. M., Hicks, A. C., et al. (2011). Experimental infection of bats with geomyces destructans causes white-nose syndrome. *Nature*, *480*(7377), 376–378.
Losos, J. B. (2008). Phylogenetic niche conservatism, phylogenetic signal and the relationship between phylogenetic relatedness and ecological similarity among species. *Ecology Letters*, *11*(10), 995–1003.
Ludmir, E. B., & Enquist, L. W. (2009). Viral genomes are part of the phylogenetic tree of life. *Nature Reviews. Microbiology*, 7(8), 615.
Luo, Z.-L., Jayarama Bhat, D., Jeewon, R., Boonmee, S., Bao, D.-F., Zhao, Y.-C., et al. (2017). Molecular phylogeny and morphological characterization of asexual fungi (Tubeufiaceae) from freshwater habitats in Yunnan, China. *Cryptogamie. Mycologie*, *38*(1), 27–53.
Luo, J., Qiu, H., Cai, G., Wagner, N. E., Bhattacharya, D., & Zhang, N. (2015). Phylogenomic analysis uncovers the evolutionary history of nutrition and infection mode in rice blast fungus and other magnaporthales. *Scientific Reports*, *5*, 9448.
Lutzoni, F., Kauff, F., Cox, C. J., McLaughlin, D., Celio, G., Dentinger, B., et al. (2004). Assembling the fungal tree of life: Progress, classification, and evolution of subcellular traits. *American Journal of Botany*, *91*(10), 1446–1480.
Mahler, D. L., Ingram, T., Revell, L. J., & Losos, J. B. (2013). Exceptional convergence on the macroevolutionary landscape in island lizard radiations. *Science*, *341*(6143), 292–295.
Marcet-Houben, M., Marceddu, G., & Gabaldón, T. (2009). Phylogenomics of the oxidative phosphorylation in fungi reveals extensive gene duplication followed by functional divergence. *BMC Evolutionary Biology*, *9*, 295.
Mark, K., Saag, L., Leavitt, S. D., Will-Wolf, S., Nelsen, M. P., Tõrra, T., et al. (2016). Erratum to: Evaluation of traditionally circumscribed species in the lichen-forming genus Usnea, section usnea (Parmeliaceae, Ascomycota) using a six-locus dataset. *Organisms, Diversity & Evolution*, *17*(1), 321.
Martin, G. W. (1951). Morphology and taxonomy of fungi by Bessey E.A. *Mycologia*, *43*(1), 108.
Martin, R., Gazis, R., Skaltsas, D., Chaverri, P., & Hibbett, D. (2015). Unexpected diversity of basidiomycetous endophytes in sapwood and leaves of hevea. *Mycologia*, *107*(2), 284–297.
McCormack, J. E., Faircloth, B. C., Crawford, N. G., Gowaty, P. A., Brumfield, R. T., & Glenn, T. C. (2012). Ultraconserved elements are novel phylogenomic markers that resolve placental mammal phylogeny when combined with species-tree analysis. *Genome Research*, *22*(4), 746–754.
McLaughlin, D., & Spatafora, J. W. (2015). *Systematics and evolution*. Berlin, Heidelberg: Springer-Verlag Berlin Heidelberg.
Medina, E. M., Jones, G. W., & Fitzpatrick, D. A. (2011). Reconstructing the fungal tree of life using phylogenomics and a preliminary investigation of the distribution of yeast prion-like proteins in the fungal kingdom. *Journal of Molecular Evolution*, *73*(3-4), 116–133.
Misof, B., Liu, S., Meusemann, K., Peters, R. S., Donath, A., Mayer, C., et al. (2014). Phylogenomics resolves the timing and pattern of insect evolution. *Science*, *346*(6210), 763–767.

Mixão, V., & Gabaldón, T. (2017). Hybridization and emergence of virulence in opportunistic human yeast pathogens. *Yeast*. https://doi.org/10.1002/yea.3242.

Money, N. P. (2013). Against the naming of fungi. *Fungal Biology*, *117*(7-8), 463–465.

Money, N. P. (2016). Fungal cell biology and development. In *The fungi* (pp. 37–66). London, UK: Academic Press.

Moran, G. P., Coleman, D. C., & Sullivan, D. J. (2010). Comparative genomics and the evolution of pathogenicity in human pathogenic fungi. *Eukaryotic Cell*, *10*(1), 34–42.

Muers, M. (2011). Evolutionary genomics: Fission yeast compared and contrasted. *Nature Reviews. Genetics*, *12*(6), 381.

Nagy, L. G. (2017). Evolution: Complex multicellular life with 5,500 genes. *Current Biology: CB*, *27*(12), R609–12.

Nagy, L. G., Ohm, R. A., Kovács, G. M., Floudas, D., Riley, R., Gácser, A., et al. (2014). Latent homology and convergent regulatory evolution underlies the repeated emergence of yeasts. *Nature Communications*, *5*, 4471.

Nagy, L. G., Riley, R., Bergmann, P. J., Krizsán, K., Martin, F. M., Grigoriev, I. V., et al. (2017). Genetic bases of fungal white rot wood decay predicted by phylogenomic analysis of correlated gene-phenotype evolution. *Molecular Biology and Evolution*, *34*(1), 35–44.

Narechania, A., Baker, R., DeSalle, R., Mathema, B., Kolokotronis, S.-O., Kreiswirth, B., et al. (2016). Clusterflock: A flocking algorithm for isolating congruent phylogenomic datasets. *GigaScience*, *5*(1), 44.

Near, T. J., Dornburg, A., Kuhn, K. L., Eastman, J. T., Pennington, J. N., Patarnello, T., et al. (2012). Ancient climate change, antifreeze, and the evolutionary diversification of antarctic fishes. *Proceedings of the National Academy of Sciences of the United States of America*, *109*(9), 3434–3439.

Near, T. J., Dornburg, A., Tokita, M., Suzuki, D., Brandley, M. C., & Friedman, M. (2014). Boom and bust: Ancient and recent diversification in Bichirs (Polypteridae: Actinopterygii), a relictual lineage of ray-finned fishes. *Evolution*, *68*(4), 1014–1026.

Near, T. J., Eytan, R. I., Dornburg, A., Kuhn, K. L., Moore, J. A., Davis, M. P., et al. (2012). Resolution of ray-finned fish phylogeny and timing of diversification. *Proceedings of the National Academy of Sciences of the United States of America*, *109*(34), 13698–13703.

Nguyen, T. A., Cissé, O. H., Wong, J. Y., Zheng, P., Hewitt, D., Nowrousian, M., et al. (2017). Innovation and constraint leading to complex multicellularity in the ascomycota. *Nature Communications*, *8*, 14444.

Nguyen, N. H., Vellinga, E. C., Bruns, T. D., & Kennedy, P. G. (2016). Phylogenetic assessment of global Suillus ITS sequences supports morphologically defined species and reveals synonymous and undescribed taxa. *Mycologia*, *108*(6), 1216–1228.

Nilsson, R. H., Abarenkov, K., Larsson, K.-H., & Kõljalg, U. (2011). Molecular identification of fungi: Rationale, philosophical concerns, and the UNITE database. *The Open Applied Informatics Journal*, *5*(81), 81–86.

Oakley, T. H., Wolfe, J. M., Lindgren, A. R., & Zaharoff, A. K. (2013). Phylotranscriptomics to bring the understudied into the fold: Monophyletic Ostracoda, fossil placement, and pancrustacean phylogeny. *Molecular Biology and Evolution*, *30*(1), 215–233.

Oliverio, A. M., & Katz, L. A. (2014). The dynamic nature of genomes across the tree of life. *Genome Biology and Evolution*, *6*(3), 482–488.

Oono, R., Lutzoni, F., Elizabeth Arnold, A., Kaye, L., U'Ren, J. M., May, G., et al. (2014). Genetic variation in horizontally transmitted fungal endophytes of pine needles reveals population structure in cryptic species. *American Journal of Botany*, *101*(8), 1362–1374.

Pacheco-Arjona, J. R., & Ramirez-Prado, J. H. (2014). Large-scale phylogenetic classification of fungal chitin synthases and identification of a putative cell-wall metabolism gene cluster in *Aspergillus* genomes. *PLoS One*, *9*(8), e104920.

Paradis, E., Claude, J., & Strimmer, K. (2004). APE: Analyses of phylogenetics and evolution in R language. *Bioinformatics*, *20*(2), 289–290.

Parfrey, L. W., Lahr, D. J. G., & Katz, L. A. (2008). The dynamic nature of eukaryotic genomes. *Molecular Biology and Evolution*, *25*(4), 787–794.

Parnmen, S., Rangsiruji, A., Mongkolsuk, P., Boonpragob, K., Nutakki, A., & Thorsten Lumbsch, H. (2012). Using phylogenetic and coalescent methods to understand the species diversity in the *Cladia aggregata* complex (Ascomycota, Lecanorales). *PLoS One*, 7(12), e52245.

Pena, R., Lang, C., Lohaus, G., Boch, S., Schall, P., Schöning, I., et al. (2017). Phylogenetic and functional traits of ectomycorrhizal assemblages in top soil from different biogeographic regions and forest types. *Mycorrhiza*, *27*(3), 233–245.

Pennisi, E. (2008). Evolution: Building the tree of life, genome by genome. *Science*, *320*(5884), 1716–1717.

Perkins, S. L. (2000). Species concepts and malaria parasites: Detecting a cryptic species of *Plasmodium*. *Proceedings. Biological Sciences*, *267*(1459), 2345–2350.

Peterson, K. R., & Pfister, D. H. (2010). Phylogeny of *Cyttaria* inferred from nuclear and mitochondrial sequence and morphological data. *Mycologia*, *102*(6), 1398–1416.

Petit, R. J., & Excoffier, L. (2009). Gene flow and species delimitation. *Trends in Ecology & Evolution*, *24*(7), 386–393.

Philippe, H., Hervé, P., Henner, B., Lavrov, D. V., Littlewood, D. T. J., Michael, M., et al. (2011). Resolving difficult phylogenetic questions: Why more sequences are not enough. *PLoS Biology*, *9*(3), e1000602.

Phillips, M. J. (2009). Branch-length estimation bias misleads molecular dating for a vertebrate mitochondrial phylogeny. *Gene*, *441*(1-2), 132–140.

Pisani, D., Feuda, R., Peterson, K. J., & Smith, A. B. (2012). Resolving phylogenetic signal from noise when divergence is rapid: A new look at the old problem of echinoderm class relationships. *Molecular Phylogenetics and Evolution*, *62*(1), 27–34.

Ploch, S., Telle, S., Choi, Y.-J., Cunnington, J. H., Priest, M., Rost, C., et al. (2011). The molecular phylogeny of the white blister rust genus *Pustula* reveals a case of underestimated biodiversity with several undescribed species on ornamentals and crop plants. *Fungal Biology*, *115*(3), 214–219.

Poe, S., & Swofford, D. L. (1999). Taxon sampling revisited. *Nature*, *398*(6725), 299–300.

Pöggeler, S., & Wöstemeyer, J. (2011). *Evolution of fungi and fungal-like organisms*. New York, NY: Springer Science & Business Media.

Pollock, D. D., Zwickl, D. J., McGuire, J. A., & Hillis, D. M. (2002). Increased taxon sampling is advantageous for phylogenetic inference. *Systematic Biology*, *51*(4), 664–671.

Posada, D. (2016). Phylogenomics for systematic biology. *Systematic Biology*, *65*(3), 353–356.

Prum, R. O., Berv, J. S., Dornburg, A., Field, D. J., Townsend, J. P., Lemmon, E. M., et al. (2015). A Comprehensive phylogeny of birds (Aves) using targeted next-generation DNA sequencing. *Nature*, *526*(7574), 569–573.

Qiao, H., Townsend Peterson, A., Ji, L., & Junhua, H. (2017). Using data from related species to overcome spatial sampling bias and associated limitations in ecological niche modeling. *Methods in Ecology and Evolution* (p. 9). https://doi.org/10.1111/2041-210x.12832.

Qiu, Y.-L., Li, L., Wang, B., Chen, Z., Knoop, V., Groth-Malonek, M., et al. (2006). The deepest divergences in land plants inferred from phylogenomic evidence. *Proceedings of the National Academy of Sciences of the United States of America*, *103*(42), 15511–15516.

Quint, M., Drost, H.-G., Gabel, A., Ullrich, K. K., Bönn, M., & Grosse, I. (2012). A transcriptomic hourglass in plant embryogenesis. *Nature*, *490*(7418), 98–101.

Rabosky, D. L., & Adams, D. C. (2012). Rates of morphological evolution are correlated with species richness in salamanders. *Evolution*, *66*(6), 1807–1818.

Ragan, M. (2015). Faculty of 1000 evaluation for synthesis of phylogeny and taxonomy into a comprehensive tree of life. *F1000—Post-Publication Peer Review of the Biomedical Literature*. https://doi.org/10.3410/f.725798837.793511389.

Raghukumar, S. (2017). Origin and evolution of marine fungi. In *Fungi in coastal and oceanic marine ecosystems* (pp. 307–321). Cham: Springer.
Rannala, B. (1998). Taxon sampling and the accuracy of large phylogenies. *Systematic Biology, 47*(4), 702–710.
Rannala, B. (2016). Conceptual issues in Bayesian divergence time estimation. *Philosophical Transactions of the Royal Society of London. Series B, Biological Sciences, 371*(1699) (p. 6). https://doi.org/10.1098/rstb.2015.0134.
Reddy, S., & Dávalos, L. M. (2003). Geographical sampling bias and its implications for conservation priorities in Africa. *Journal of Biogeography, 30*(11), 1719–1727.
Reddy, S., Kimball, R. T., Pandey, A., Hosner, P. A., Braun, M. J., Hackett, S. J., et al. (2017). Why do phylogenomic data sets yield conflicting trees? Data type influences the avian tree of life more than taxon sampling. *Systematic Biology, 66*(5), 857–879. https://doi.org/10.1093/sysbio/syx041.
Reis, M. d., Inoue, J., Hasegawa, M., Asher, R. J., Donoghue, P. C. J., & Yang, Z. (2012). Phylogenomic datasets provide both precision and accuracy in estimating the timescale of placental mammal phylogeny. *Proceedings. Biological Sciences, 279*(1742), 3491–3500.
Ren, R., Sun, Y., Zhao, Y., Geiser, D., Ma, H., & Zhou, X. (2016). Phylogenetic resolution of deep eukaryotic and fungal relationships using highly conserved low-copy nuclear genes. *Genome Biology and Evolution, 8*(9), 2683–2701.
Revell, L. J. (2011). Phytools: An R package for phylogenetic comparative biology (and other things). *Methods in Ecology and Evolution, 3*(2), 217–223.
Revell, L. J., & Graham Reynolds, R. (2012). A new Bayesian method for fitting evolutionary models to comparative data with intraspecific variation. *Evolution, 66*(9), 2697–2707.
Ribeiro, G. V. T., Teixido, A. L., Barbosa, N. P. U., & Silveira, F. A. O. (2016). Assessing bias and knowledge gaps on seed ecology research: Implications for conservation agenda and policy. *Ecological Applications: A Publication of the Ecological Society of America, 26*(7), 2033–2043.
Richards, T. A., Jones, M. D. M., Leonard, G., & Bass, D. (2012). Marine fungi: Their ecology and molecular diversity. *Annual Review of Marine Science, 4*(1), 495–522.
Richards, T. A., Leonard, G., Mahé, F., Campo, J. D., Romac, S., Jones, M. D. M., et al. (2015). Molecular diversity and distribution of marine fungi across 130 European environmental samples. *Proceedings. Biological Sciences, 282*(1819) (p. 10). https://doi.org/10.1098/rspb.2015.2243.
Riley, R., Salamov, A. A., Brown, D. W., Nagy, L. G., Floudas, D., Held, B. W., et al. (2014). Extensive sampling of basidiomycete genomes demonstrates inadequacy of the white-rot/brown-rot paradigm for wood decay fungi. *Proceedings of the National Academy of Sciences of the United States of America, 111*(27), 9923–9928.
Rodríguez, F., Oliver, J. L., Marín, A., & Medina, J. R. (1990). The general stochastic model of nucleotide substitution. *Journal of Theoretical Biology, 142*(4), 485–501.
Roe, A. D., Rice, A. V., Bromilow, S. E., Cooke, J. E. K., & Sperling, F. A. H. (2010). Multilocus species identification and fungal DNA barcoding: Insights from blue stain fungal symbionts of the mountain pine beetle. *Molecular Ecology Resources, 10*(6), 946–959.
Rokas, A. (2005). More genes or more taxa? The relative contribution of gene number and taxon number to phylogenetic accuracy. *Molecular Biology and Evolution, 22*(5), 1337–1344.
Romiguier, J., Ranwez, V., Delsuc, F., Galtier, N., & Douzery, E. J. P. (2013). Less is more in mammalian phylogenomics: AT-rich genes minimize tree conflicts and unravel the root of placental mammals. *Molecular Biology and Evolution, 30*(9), 2134–2144.
Ropars, J., Rodríguez de la Vega, R. C., López-Villavicencio, M., Gouzy, J., Sallet, E., Dumas, É., et al. (2015). Adaptive horizontal gene transfers between multiple cheese-associated fungi. *Current Biology: CB, 25*(19), 2562–2569.

Rosling, A., Cox, F., Cruz-Martinez, K., Ihrmark, K., Grelet, G.-A., Lindahl, B. D., et al. (2011). Archaeorhizomycetes: Unearthing an ancient class of ubiquitous soil fungi. *Science*, *333*(6044), 876–879.
Saag, L., Mark, K., Saag, A., & Randlane, T. (2014). Species delimitation in the lichenized fungal genus *Vulpicida* (Parmeliaceae, Ascomycota) using gene concatenation and coalescent-based species tree approaches. *American Journal of Botany*, *101*(12), 2169–2182.
Saenz, G. S., Taylor, J. W., & Gargas, A. (1994). 18S rRNA gene sequences and supraordinal classification of the Erysiphales. *Mycologia*, *86*(2), 212.
Salichos, L., & Rokas, A. (2013). Inferring ancient divergences requires genes with strong phylogenetic signals. *Nature*, *497*(7449), 327–331.
Salichos, L., Stamatakis, A., & Rokas, A. (2014). Novel information theory-based measures for quantifying incongruence among phylogenetic trees. *Molecular Biology and Evolution*, *31*(5), 1261–1271.
Sánchez-García, M., & Matheny, P. B. (2016). Is the switch to an ectomycorrhizal state an evolutionary key innovation in mushroom-forming fungi? A case study in the Tricholomatineae (Agaricales). *Evolution*, *71*(1), 51–65.
Santini, F., Sorenson, L., & Alfaro, M. E. (2013). A new multi-locus timescale reveals the evolutionary basis of diversity patterns in triggerfishes and filefishes (Balistidae, Monacanthidae; Tetraodontiformes). *Molecular Phylogenetics and Evolution*, *69*(1), 165–176.
Sato, H., Yumoto, T., & Murakami, N. (2007). Cryptic species and host specificity in the Ectomycorrhizal genus Strobilomyces (Strobilomycetaceae). *American Journal of Botany*, *94*(10), 1630–1641.
Schoch, C. L., Seifert, K. A., Huhndorf, S., Robert, V., Spouge, J. L., André Levesque, C., et al. (2012). Nuclear ribosomal internal transcribed spacer (ITS) region as a universal DNA barcode marker for fungi. *Proceedings of the National Academy of Sciences of the United States of America*, *109*(16), 6241–6246.
Schoch, C. L., Sung, G.-H., López-Giráldez, F., Townsend, J. P., Miadlikowska, J., Hofstetter, V., et al. (2009). The ascomycota tree of life: A phylum-wide phylogeny clarifies the origin and evolution of fundamental reproductive and ecological traits. *Systematic Biology*, *58*(2), 224–239.
Seena, S., & Monroy, S. (2016). Preliminary insights into the evolutionary relationships of aquatic Hyphomycetes and Endophytic fungi. *Fungal Ecology*, *19*, 128–134.
Seto, K. C., Güneralp, B., & Hutyra, L. R. (2012). Global forecasts of urban expansion to 2030 and direct impacts on biodiversity and carbon pools. *Proceedings of the National Academy of Sciences of the United States of America*, *109*(40), 16083–16088.
Shang, Y., Xiao, G., Zheng, P., Cen, K., Zhan, S., & Wang, C. (2016). Divergent and convergent evolution of fungal pathogenicity. *Genome Biology and Evolution*, *8*(5), 1374–1387.
Sharma, P. P., Kaluziak, S. T., Pérez-Porro, A. R., González, V. L., Hormiga, G., Wheeler, W. C., et al. (2014). Phylogenomic interrogation of *Arachnida* reveals systemic conflicts in phylogenetic signal. *Molecular Biology and Evolution*, *31*(11), 2963–2984.
Shelest, E., & Voigt, K. (2014). 2 Genomics to study basal lineage fungal biology: Phylogenomics suggests a common origin. In M. Nowrousian (Ed.), *Fungal genomics*. (pp. 31–60). Berlin, Heidelberg: Springer Berlin Heidelberg.
Shen, X.-X., Hittinger, C. T., & Rokas, A. (2017). Contentious relationships in phylogenomic studies can be driven by a handful of genes. *Nature Ecology & Evolution*, *1*(5), 0126.
Sillo, F., Garbelotto, M., Friedman, M., & Gonthier, P. (2015). Comparative genomics of sibling fungal pathogenic taxa identifies adaptive evolution without divergence in pathogenicity genes or genomic structure. *Genome Biology and Evolution*, 7(12), 3190–3206.

Simion, P., Philippe, H., Baurain, D., Jager, M., Richter, D. J., Franco, A. D., et al. (2017). A large and consistent phylogenomic dataset supports sponges as the sister group to all other animals. *Current Biology: CB*, *27*(7), 958–967.

Simon, S., Narechania, A., Desalle, R., & Hadrys, H. (2012). Insect phylogenomics: Exploring the source of incongruence using new transcriptomic data. *Genome Biology and Evolution*, *4*(12), 1295–1309.

Simonson, A. B., Servin, J. A., Skophammer, R. G., Herbold, C. W., Rivera, M. C., & Lake, J. A. (2005). Decoding the genomic tree of life. *Proceedings of the National Academy of Sciences of the United States of America*, *102*(Suppl. 1), 6608–6613.

Singh, B. P., & Gupta, V. K. (2017). Molecular markers in mycology: Diagnostics and marker developments. In B. P. Singh & V. K. Gupta (Eds.), Switzerland: Springer International Publishing.

Skerratt, L. F., Berger, L., Speare, R., Cashins, S., McDonald, K. R., Phillott, A. D., et al. (2007). Spread of Chytridiomycosis has caused the rapid global decline and extinction of frogs. *EcoHealth*, *4*(2), 125–134.

Smith, K. L., Harmon, L. J., Shoo, L. P., & Melville, J. (2011). Evidence of constrained phenotypic evolution in a cryptic species complex of agamid lizards. *Evolution*, *65*(4), 976–992.

Soltis, P. S., Soltis, D. E., Savolainen, V., Crane, P. R., & Barraclough, T. G. (2002). Rate heterogeneity among lineages of Tracheophytes: Integration of molecular and fossil data and evidence for molecular living fossils. *Proceedings of the National Academy of Sciences of the United States of America*, *99*(7), 4430–4435.

Song, J., & Cui, B.-K. (2017). Phylogeny, divergence time and historical biogeography of *Laetiporus* (Basidiomycota, Polyporales). *BMC Evolutionary Biology*, *17*(1), 102.

Song, S., Liang, L., Edwards, S. V., & Shaoyuan, W. (2012). Resolving conflict in Eutherian mammal phylogeny using phylogenomics and the multispecies coalescent model. *Proceedings of the National Academy of Sciences of the United States of America*, *109*(37), 14942–14947.

Spatafora, J. W., & Bushley, K. E. (2015). Phylogenomics and evolution of secondary metabolism in plant-associated fungi. *Current Opinion in Plant Biology*, *26*, 37–44.

Spatafora, J. W., Chang, Y., Benny, G. L., Lazarus, K., Smith, M. E., Berbee, M. L., et al. (2016). A phylum-level phylogenetic classification of zygomycete fungi based on genome-scale data. *Mycologia*, *108*(5), 1028–1046.

Spatafora, J. W., Hughes, K. W., & Blackwell, M. (2006). *A phylogeny for kingdom fungi: Deep hypha issue.* USA: The Mycological Society of America.

Spriggs, E. L., Clement, W. L., Sweeney, P. W., Madriñán, S., Edwards, E. J., & Donoghue, M. J. (2015). Temperate radiations and dying embers of a tropical past: The diversification of Viburnum. *The New Phytologist*, *207*(2), 340–354.

Springer, M. S., Emerling, C. A., Meredith, R. W., Janečka, J. E., Eizirik, E., & Murphy, W. J. (2017). Waking the undead: Implications of a soft explosive model for the timing of placental mammal diversification. *Molecular Phylogenetics and Evolution*, *106*, 86–102.

Stajich, J. E., Berbee, M. L., Blackwell, M., Hibbett, D. S., James, T. Y., Spatafora, J. W., et al. (2009). The fungi. *Current Biology: CB*, *19*(18), R840–45.

Stajich, J. E., Wilke, S. K., Ahrén, D., Chun Hang, A., Birren, B. W., Borodovsky, M., et al. (2010). Insights into evolution of multicellular fungi from the assembled chromosomes of the mushroom *Coprinopsis cinerea* (*Coprinus cinereus*). *Proceedings of the National Academy of Sciences of the United States of America*, *107*(26), 11889–11894.

Stukenbrock, E. H., & Croll, D. (2014). The evolving fungal genome. *Fungal Biology Reviews*, *28*(1), 1–12.

Su, Z., & Townsend, J. P. (2015). Utility of characters evolving at diverse rates of evolution to resolve quartet trees with unequal branch lengths: Analytical predictions of long-branch effects. *BMC Evolutionary Biology*, *15*, 86.

Su, Z., Wang, Z., López-Giráldez, F., & Townsend, J. P. (2014). The impact of incorporating molecular evolutionary model into predictions of phylogenetic signal and noise. *Frontiers in Ecology and Evolution*, *2*, 11. https://doi.org/10.3389/fevo.2014.00011.

Sukumaran, J., & Knowles, L. L. (2017). Multispecies coalescent delimits structure, not species. *Proceedings of the National Academy of Sciences of the United States of America*, *114*(7), 1607–1612.

Sun, R., Zheng, H., Sottoriva, A., Graham, T. A., Harpak, A., Ma, Z., et al. (2017). Between-region genetic divergence reflects the mode and tempo of tumor evolution. *Nature Genetics*, *49*(7), 1015–1024.

Suvorov, A., Jensen, N. O., Sharkey, C. R., Stanley Fujimoto, M., Bodily, P., Wightman, H. M. C., et al. (2017). Opsins have evolved under the permanent heterozygote model: Insights from phylotranscriptomics of Odonata. *Molecular Ecology*, *26*(5), 1306–1322.

Suzán, G., García-Peña, G. E., Castro-Arellano, I., Rico, O., Rubio, A. V., Tolsá, M. J., et al. (2015). Metacommunity and phylogenetic structure determine wildlife and zoonotic infectious disease patterns in time and space. *Ecology and Evolution*, *5*(4), 865–873.

Swann, E. C., & Taylor, J. W. (1993). Higher taxa of basidiomycetes: An 18S rRNA gene perspective. *Mycologia*, *85*(6), 923.

Talbot, J. M., Bruns, T. D., Taylor, J. W., Smith, D. P., Branco, S., Glassman, S. I., et al. (2014). Endemism and functional convergence across the North American soil mycobiome. *Proceedings of the National Academy of Sciences of the United States of America*, *111*(17), 6341–6346.

Tanabe, Y., Watanabe, M. M., & Sugiyama, J. (2005). Evolutionary relationships among basal fungi (Chytridiomycota and Zygomycota): Insights from molecular phylogenetics. *The Journal of General and Applied Microbiology*, *51*(5), 267–276.

Tavaré, S. (1986). Some probabilistic and statistical problems in the analysis of DNA sequences. *Lectures on Mathematics in the Life Sciences*, *17*, 57–86.

Taylor, J. W. (2006). Evolution of human-pathogenic fungi: Phylogenies and species. In J. Heitman, S. Filler, J. Edwards, Jr., & A. Mitchell (Eds.), *Molecular principles of fungal pathogenesis* (pp. 113–132). Washington, DC: ASM Press.

Taylor, J. W., & Berbee, M. L. (2006). Dating divergences in the fungal tree of life: Review and new analyses. *Mycologia*, *98*(6), 838–849.

Taylor, J. W., Turner, E., Townsend, J. P., Dettman, J. R., & Jacobson, D. (2006). Eukaryotic microbes, species recognition and the geographic limits of species: Examples from the kingdom fungi. *Philosophical Transactions of the Royal Society of London. Series B, Biological Sciences*, *361*(1475), 1947–1963.

Tedersoo, L., Bahram, M., Põlme, S., Kõljalg, U., Yorou, N. S., Wijesundera, R., et al. (2014). Fungal biogeography. Global diversity and geography of soil fungi. *Science*, *346*(6213), 1256688.

Tedersoo, L., Bahram, M., Rasmus, P., Henrik Nilsson, R., & James, T. Y. (2017). Novel soil-inhabiting clades fill gaps in the fungal tree of life. *Microbiome*, *5*(1), 42.

Thuiller, W., Lavorel, S., & Araujo, M. B. (2005). Niche properties and geographical extent as predictors of species sensitivity to climate change. *Global Ecology and Biogeography: A Journal of Macroecology*, *14*(4), 347–357.

Torruella, G., Mendoza, A. d., Grau-Bové, X., Antó, M., Chaplin, M. A., Campo, J. d., et al. (2015). Phylogenomics reveals convergent evolution of lifestyles in close relatives of animals and fungi. *Current Biology: CB*, *25*(18), 2404–2410.

Townsend, J. P. (2007). Profiling phylogenetic informativeness. *Systematic Biology*, *56*(2), 222–231.

Townsend, J. P., & Leuenberger, C. (2011). Taxon sampling and the optimal rates of evolution for phylogenetic inference. *Systematic Biology*, *60*(3), 358–365.

Townsend, J. P., & Lopez-Giraldez, F. (2010). Optimal selection of gene and ingroup taxon sampling for resolving phylogenetic relationships. *Systematic Biology*, *59*(4), 446–457.

Townsend, J. P., Su, Z., & Tekle, Y. I. (2012). Phylogenetic signal and noise: Predicting the power of a data set to resolve phylogeny. *Systematic Biology*, *61*(5), 835–849.

Traeger, S., Altegoer, F., Freitag, M., Gabaldon, T., Kempken, F., Kumar, A., et al. (2013). The genome and development-dependent transcriptomes of *Pyronema confluens*: A window into fungal evolution. *PLoS Genetics*, *9*(9), e1003820.

Trail, F., Wang, Z., Stefanko, K., Cubba, C., & Townsend, J. P. (2017). The ancestral levels of transcription and the evolution of sexual phenotypes in filamentous fungi. *PLoS Genetics*, *13*(7), e1006867.

Unterseher, M., & Schnittler, M. (2010). Species richness analysis and ITS rDNA phylogeny revealed the majority of cultivable foliar endophytes from beech (*Fagus sylvatica*). *Fungal Ecology*, *3*(4), 366–378.

U'Ren, J. M., Dalling, J. W., Gallery, R. E., Maddison, D. R., Christine Davis, E., Gibson, C. M., et al. (2009). Diversity and evolutionary origins of fungi associated with seeds of a neotropical pioneer tree: A case study for analysing fungal environmental samples. *Mycological Research*, *113*(4), 432–449.

Varela, S., Anderson, R. P., García-Valdés, R., & Fernández-González, F. (2014). Environmental filters reduce the effects of sampling bias and improve predictions of ecological niche models. *Ecography*, 1084–1091.

Villarreal, A., Carlos, J., Juan Carlos Villarreal, A., Crandall-Stotler, B. J., Hart, M. L., Long, D. G., et al. (2015). Divergence times and the evolution of morphological complexity in an early land plant lineage (Marchantiopsida) with a slow molecular rate. *The New Phytologist*, *209*(4), 1734–1746.

Walker, A. (2014). Adding genomic 'foliage' to the tree of life. *Nature Reviews. Microbiology*, *12*(2), 78.

Walker, D. M., Castlebury, L. A., Rossman, A. Y., & White, J. F., Jr. (2012). New molecular markers for fungal phylogenetics: Two genes for species-level systematics in the Sordariomycetes (Ascomycota). *Molecular Phylogenetics and Evolution*, *64*(3), 500–512.

Wang, Z., Binder, M., Schoch, C. L., Johnston, P. R., Spatafora, J. W., & Hibbett, D. S. (2006). Evolution of helotialean fungi (Leotiomycetes, Pezizomycotina): A nuclear rDNA phylogeny. *Molecular Phylogenetics and Evolution*, *41*(2), 295–312.

Wang, Z., Henrik Nilsson, R., James, T. Y., Dai, Y., & Townsend, J. P. (2016). Future perspectives and challenges of fungal systematics in the age of big data. Biology of micro-fungi. In DW Li (Ed.), Switzerland: Springer International Publishing, pp. 25–46.

Wang, Z., Henrik Nilsson, R., Lopez-Giraldez, F., Zhuang, W.-Y., Dai, Y.-C., Johnston, P. R., et al. (2011). Tasting soil fungal diversity with earth tongues: Phylogenetic test of SATé alignments for environmental ITS data. *PLoS One*, *6*(4), e19039.

Wang, Z., Johnston, P. R., Yang, Z. L., & Townsend, J. P. (2009). Evolution of reproductive morphology in leaf endophytes. *PLoS One*, *4*(1), e4246.

Wang, Y., Tretter, E. D., Johnson, E. M., Kandel, P., Lichtwardt, R. W., Novak, S. J., et al. (2014). Using a five-gene phylogeny to test morphology-based hypotheses of Smittium and allies, endosymbiotic gut fungi (Harpellales) associated with arthropods. *Molecular Phylogenetics and Evolution*, *79*, 23–41.

Wang, H., Xu, Z., Gao, L., & Hao, B. (2009). A fungal phylogeny based on 82 complete genomes using the composition vector method. *BMC Evolutionary Biology*, *9*, 195.

Warnecke, L., Turner, J. M., Bollinger, T. K., Lorch, J. M., Misra, V., Cryan, P. M., et al. (2012). Inoculation of bats with European geomyces destructans supports the novel pathogen hypothesis for the origin of white-nose syndrome. *Proceedings of the National Academy of Sciences of the United States of America*, *109*(18), 6999–7003.

Warnock, R. C. M., Parham, J. F., Joyce, W. G., Lyson, T. R., & Donoghue, P. C. J. (2015). Calibration uncertainty in molecular dating analyses: There is no substitute for the prior evaluation of time priors. *Proceedings. Biological Sciences*, *282*(1798), 20141013.

Warren, D. L., Cardillo, M., Rosauer, D. F., & Bolnick, D. I. (2014). Mistaking geography for biology: Inferring processes from species distributions. *Trends in Ecology & Evolution*, *29*(10), 572–580.

Warren, D. L., Wright, A. N., Seifert, S. N., & Bradley Shaffer, H. (2013). Incorporating model complexity and spatial sampling bias into ecological niche models of climate change risks faced by 90 California vertebrate species of concern. *Diversity and Distributions*, *20*(3), 334–343.

Watanabe, M., Yonezawa, T., Lee, K.-I., Kumagai, S., Sugita-Konishi, Y., Goto, K., et al. (2011). Molecular phylogeny of the higher and lower taxonomy of the Fusarium genus and differences in the evolutionary histories of multiple genes. *BMC Evolutionary Biology*, *11*, 322.

Waterhouse, R. M., Tegenfeldt, F., Li, J., Zdobnov, E. M., & Kriventseva, E. V. (2013). OrthoDB: A hierarchical catalog of animal, fungal and bacterial orthologs. *Nucleic Acids Research*, *41*, D358–65.

Wei, X., McCune, B., Thorsten Lumbsch, H., Li, H., Leavitt, S., Yamamoto, Y., et al. (2016). Limitations of species delimitation based on phylogenetic analyses: A case study in the *Hypogymnia hypotrypa* group (Parmeliaceae, Ascomycota). *PLoS One*, *11*(11), e0163664.

Wellman, C. H., Osterloff, P. L., & Mohiuddin, U. (2003). Fragments of the earliest land plants. *Nature*, *425*(6955), 282–285.

Wen, J., Nie, Z.-L., & Ickert-Bond, S. M. (2016). Intercontinental disjunctions between eastern Asia and western North America in vascular plants highlight the biogeographic importance of the Bering land bridge from late Cretaceous to Neogene. *Journal of Systematics and Evolution*, *54*(5), 469–490.

Werner, G. D. A., & Kiers, E. T. (2012). Friends in fungi. *Science*, *337*(6101), 1452.

Wilke, T., Schultheiß, R., & Albrecht, C. (2009). As time goes by: A Simple fool's guide to molecular clock approaches in invertebrates. *American Malacological Bulletin*, *27*(1-2), 25–45.

Williams, C. M., Henry, H. A. L., & Sinclair, B. J. (2015). Cold truths: How winter drives responses of terrestrial organisms to climate change. *Biological Reviews of the Cambridge Philosophical Society*, *90*(1), 214–235.

Williams, T. A., Szöllősi, G. J., Spang, A., Foster, P. G., Heaps, S. E., Boussau, B., et al. (2017). Integrative modeling of gene and genome evolution roots the Archaeal tree of life. *Proceedings of the National Academy of Sciences of the United States of America*, *114*(23), E4602–11.

Willis, C. G., Ruhfel, B., Primack, R. B., Miller-Rushing, A. J., & Davis, C. C. (2008). Phylogenetic patterns of species loss in Thoreau's woods are driven by climate change. *Proceedings of the National Academy of Sciences of the United States of America*, *105*(44), 17029–17033.

Wilson, A. W., Hosaka, K., & Mueller, G. M. (2017). Evolution of ectomycorrhizas as a driver of diversification and biogeographic patterns in the model mycorrhizal mushroom genus *Laccaria*. *The New Phytologist*, *213*(4), 1862–1873.

Wingfield, M. J., Wilhelm de Beer, Z., Slippers, B., Wingfield, B. D., Groenewald, J. Z., Lombard, L., et al. (2011). One fungus, one name promotes progressive plant pathology. *Molecular Plant Pathology*, *13*(6), 604–613.

Wolfgang Wägele, J., & Bartolomaeus, T. (2014). *Deep metazoan phylogeny: The backbone of the tree of life: New insights from analyses of molecules, morphology, and theory of data analysis*. Berlin, Germany: Walter de Gruyter GmbH & Co KG.

Xia, X. (2013). DAMBE5: A comprehensive software package for data analysis in molecular biology and evolution. *Molecular Biology and Evolution, 30*(7), 1720–1728.

Yang, Z., & Rannala, B. (2010). Bayesian species delimitation using multilocus sequence data. *Proceedings of the National Academy of Sciences of the United States of America, 107*(20), 9264–9269.

Yang, Y., Yang, E., An, Z., & Liu, X. (2007). Evolution of nematode-trapping cells of predatory fungi of the Orbiliaceae based on evidence from rRNA-encoding DNA and multiprotein sequences. *Proceedings of the National Academy of Sciences of the United States of America, 104*(20), 8379–8384.

Young, J., & Peter, W. (2012). A molecular guide to the taxonomy of arbuscular mycorrhizal fungi. *The New Phytologist, 193*(4), 823–826.

Zeigler, D. (2014). Phylogeny—The tree of life. In *Evolution* (pp. 165–170). London, UK: Academic Press (Elsevier).

Zhang, N., & Wang, Z. (2015). 3 Pezizomycotina: Sordariomycetes and leotiomycetes. In D. J. McLaughlin & J. W. Spatafora (Eds.), *Systematics and evolution: Part B* (pp. 57–88). Berlin, Heidelberg: Springer Berlin Heidelberg.

Zhao, Z.-M., Zhao, B., Bai, Y., Iamarino, A., Gaffney, S. G., Schlessinger, J., et al. (2016). Early and multiple origins of metastatic lineages within primary tumors. *Proceedings of the National Academy of Sciences of the United States of America, 113*(8), 2140–2145.

Zwickl, D. J., & Hillis, D. M. (2002). Increased taxon sampling greatly reduces phylogenetic error. *Systematic Biology, 51*(4), 588–598.

CHAPTER TWO

Fungal Phylogeny in the Age of Genomics: Insights Into Phylogenetic Inference From Genome-Scale Datasets

László G. Nagy*,[1], Gergely Szöllősi†
*Hungarian Academy of Sciences, Szeged, Hungary
†Eötvös Loránd University, Budapest, Hungary
[1]Corresponding author: e-mail address: lnagy@fungenomelab.com

Contents

Abstract

The genomic era has been transformative for many fields, including our understanding of the phylogenetic relationships between organisms. The wide availability of whole-genome sequences practically eliminated data availability as a limiting factor for inferring phylogenetic trees, providing hundreds to thousands of loci for analyses, leading to molecular phylogenetics gradually being replaced by phylogenomics. The new era has also brought new challenges: systematic errors (resulting from, e.g., model violation) can be more pronounced in phylogenomic datasets and can lead to strongly supported incorrect relationships, creating significant incongruence among studies. Here, we review common practices, technical and biological challenges of phylogenomic

Advances in Genetics, Volume 100
ISSN 0065-2660
https://doi.org/10.1016/bs.adgen.2017.09.008

analyses, with examples illustrated from fungi. We compare major approaches of phylogenetic inference, and illustrate the advantages conferred and challenges presented in phylogenomic case studies across the fungal tree of life, including cases where genome-scale data could conclusively resolve contentious relationships, and others that remain challenging despite the flood of genomic data.

1. INTRODUCTION

Species alive today share a common history through their ancestry that is reflected in their phylogeny. Reconstructing this history by inferring phylogenetic relationships is of central interest in biology and a prerequisite of evolutionary study. Until the mid-2000s phylogenetic analyses relied on sequence data from a single or a few genes to infer phylogenetic relationships (most commonly between species). A common limitation of these studies was low resolution, due to which key relationships proved difficult or even impossible to resolve. The development of next-generation-sequencing technologies brought a revolution by making genome-scale sequence data widely accessible for several applications, including inferring species relationships. Today, phylogenomic datasets that are typically 10 × to 1000 × the size of previous phylogenetic datasets are routinely available, and have the potential to improve the resolution of phylogenetic trees through a dramatic reduction in stochastic error resulting from short alignments, that is, errors in phylogenetic inferences arising as consequences of finite alignment lengths. If the availability of sequence data was the only limitation, then this reduction in error would have led us to "the end of incongruence" (Gee, 2003). With the dawn of phylogenomics, however, it soon became apparent that analyzing large amounts of sequence data comes with its own challenges and pitfalls.

In this chapter, we review the most common approaches to the inference of genome-scale phylogenies, including concatenation and methods that model phylogenetic discord between gene trees and the species tree. We discuss potential sources of bias associated with them together with the challenges of phylogenomic inference in the context of fungal phylogenies. We present examples across the fungal tree of life in which phylogenomics has provided conclusive evidence and other examples in which further research remains necessary to be done to understand fungal relationships.

1.1 Are We Nearing the End of Incongruence?

In theory, genome-scale sequence data hold a promise for resolving all phylogenetic relationships with high statistical certainty. However, case studies highlighted early on that some relationships can remain difficult to resolve even when large amounts of data are available—while other results can be positively misleading by lending strong support to incorrect relationships (Delsuc, Brinkmann, & Philippe, 2005; Kumar, Filipski, Battistuzzi, Kosakovsky Pond, & Tamura, 2012; Philippe et al., 2011; Philippe, Delsuc, Brinkmann, & Lartillot, 2005). These contradictions can be explained by two important caveats of using genome-scale sequence data: (i) statistical inconsistency can lead to strong support for erroneous relationships and (ii) more fundamentally, bona fide phylogenetic differences exist between individual gene phylogenies as a result of evolutionary processes such as gene duplication, transfer, and loss (DTL) and incomplete lineage sorting (ILS). Consequently, distinguishing strong support for real relationships from that of incorrect ones can be daunting. First, systematic error becomes more pronounced for genome-scale datasets (Fig. 1), as a result

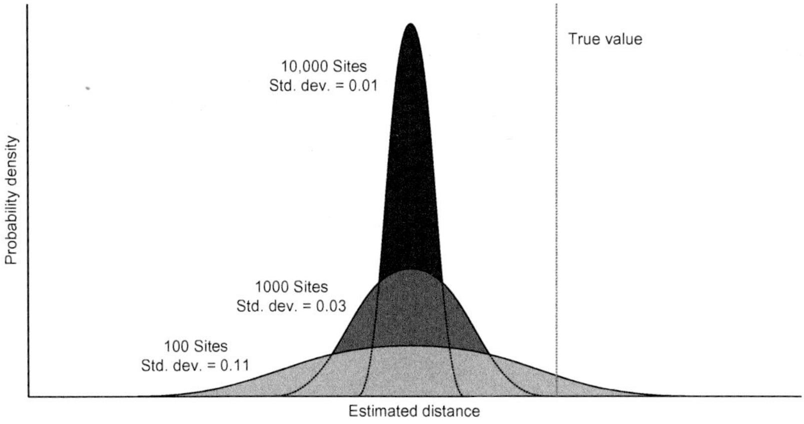

Fig. 1 A hypothetical example of how increasing the amount of data leads to increased statistical confidence in the wrong answer. Sequences ranging from 100 to 10,000 sites were simulated under the GTR model of evolution and then maximum likelihood distances estimated under the much simpler JC model. Because of model misspecification, the estimates are biased as shown by their distance from the distance used for the simulations. Note that, increasing the sequence length in the analysis does not reduce error, but increases our confidence in the incorrect estimate. *Adapted from Kumar, S., Filipski, A.J., Battistuzzi, F.U., Kosakovsky Pond, S.L., Tamura, K. (2012). Statistics and truth in phylogenomics.* Molecular Biology and Evolution 29, *457–472. https://doi.org/10.1093/molbev/msp123.*

of statistical inconsistency, i.e., the inference method converges toward an incorrect solution (e.g., wrong topology) with increasing support as more and more data are analyzed (Felsenstein, 1978; Roch & Steel, 2015). Because of this phenomenon, significant incongruence can arise among studies based on different datasets and inference methods, which has led phylogenomics seen as "the beginning of incongruence" (Roch & Steel, 2015). Second, the histories of genes and species are tightly linked, but seldom identical, because genes duplicate, are lost or horizontally transferred, and because alleles can coexist in populations for periods that may span several speciation events, which can also lead to bona fide phylogenetic incongruence (Szollosi, Tannier, Daubin, & Boussau, 2015). Therefore, processes of genome evolution and potential confounding factors of phylogenomic inference must be considered in order to reconstruct true relationships. If high-quality data are analyzed under appropriate models and methods, both concatenation and gene tree-based approaches can be highly accurate. However, this is often not the case, necessitating new approaches, especially ones that model gene DTL and ILS, to become available and offer improved precision and the promise of using lots of extra data (Patterson, Szollosi, Daubin, & Tannier, 2013). These new methods also harness new sources of information, e.g., rooting without an out-group and information on relative dates from gene transfer events. Perhaps even more importantly, these methods also provide reconstructions of individual gene histories, as a series of DTL and speciation events—and on the genome-scale provide phylogenetically informed ancestral gene content estimates that open the door to genome-enabled mycology.

2. CONCATENATION

The observation that single-gene analyses often result in poorly supported trees led to the idea of combining multiple genes to create a "supermatrix." This approach, called concatenation, results in a single-output tree that can be assumed to be the species tree. Although this assumption is reasonable as a generality, individual gene genealogies differ, which should be taken into account. In the mid-2000s, the flood of genome data turned multigene phylogenetics into phylogenomics, a development that improved the resolution of phylogenetic trees through a dramatic reduction in stochastic error (Delsuc et al., 2005; Philippe et al., 2005). By analyzing more and more data, statistical support is expected to eventually climb to maximal values and stochastic errors to average out across the dataset. While some saw this trend ending incongruence in phylogenetics (Gee, 2003), it soon became clear that analyzing large amounts of sequence data without an adequate model of sequence evolution can lead to phylogenetic artifacts. It is important to note that although the amount of data increases in genome-scale datasets, ratio of phylogenetic signal and "noise" remains unchanged or could even be worse than in traditional phylogenetics (Philippe et al., 2005). Systematic errors stemming from unmodeled aspects of the evolutionary process naturally become more apparent at the genomic scale (Kumar et al., 2012), that is, the inference becomes statistically inconsistent. This inconsistency can also lead to strongly supported incorrect trees. As a result, phylogenomics in a sense has been "the beginning of incongruence" (Jeffroy, Brinkmann, Delsuc, & Philippe, 2006) among studies based on different datasets and (especially) inference methods.

From a methodological perspective, concatenation-based approaches simply analyze phylogenomic supermatrices as scaled-up versions of traditional multigene alignments. With the flood of genomic data from large-scale genome sequencing projects, manual curation of each input gene alignment becomes impractical, leading to an additional error. These errors include issues arising as a result of the automation of the supermatrix construction process, e.g., contaminant sequences, missing data, sequencing errors, collectively termed "data errors" (Philippe et al., 2017). Although such errors are usually randomly distributed and are expected to average out across hundreds or thousands of genes, this scenario is not always the case. For example, accidental inclusion of contaminant sequences (e.g., pseudogenes or paralogs) varies from gene to gene which adds noise to the analysis but, unless pervasive across the entire gene set, will not bias the results. On the other hand, large amounts of missing data might lead to biased inferences, which made to missing data becoming an important consideration in phylogenomics. Simulation studies and empirical data show that missing data can be tolerated (Philippe et al., 2004; Wiens, 2003) in analyses up to surprisingly high levels, although very sparse supermatrices can be more sensitive to phylogenetic artifacts (Roure, Baurain, & Philippe, 2013) (e.g., long-branch attraction, LBA). Simulations have demonstrated that taxa with lots of missing data can harbor unstable positions in the tree topology because of the lack of data that would firmly place it relative to others rather than the effects of missing data per se (Wiens, 2003). This observation also means that adding taxa with highly incomplete data can often be beneficial to accuracy, for example, to avoid LBA by breaking up long internal branches (Wiens, 2005). Nonrandomly distributed missing data, however, might lead to phylogenetic biases. A special case of this arises when a reference species is used to "fish" for shared single-copy orthologs: in this case, the amount of missing data will increase as a function of phylogenetic distance from the reference species, often following a power law. This will result in a strongly biased matrix completeness, which could eventually lead to biased trees and should therefore be avoided.

Another important consideration for phylogenomics is taxon sampling—in particular, the balance between taxon sampling and alignment completeness. Although the relative importance of increasing taxon vs gene sampling itself is somewhat debated (Philippe et al., 2005), significant progress has been made recently (Geuten, Massingham, Darius, Smets, & Goldman, 2007; Townsend & Leuenberger, 2011; Townsend & Lopez-Giraldez, 2010). Breaking up long branches between distant groups and selecting out-groups so that they are genetically as close to the in-group as possible all contribute to mitigating LBA. Species with high rates of molecular evolution also create challenging situations that are rooted in LBA. It can be difficult to avoid such taxa if they are key to the analysis, as seen in the case of several secondarily reduced parasites (Haag et al., 2014; James et al., 2013; Mikhailov Kirill et al., 2016). However, long branches can also arise as a result of data errors. Alignment contamination by paralogs, horizontally transferred genes or pseudogenes, or genomic regions of low-sequencing quality for certain species can result in long branches and LBA. Several strategies have been developed to avoid contaminating genes, including screening for highly divergent sequences (see, e.g., dos Reis et al., 2012 or Nagy et al., 2016) or sophisticated mechanisms for selecting orthologous groups of genes. Although beyond the scope of this chapter, we note that the selection of single-copy orthologous genes is often complicated by deep paralogy across gene trees, and that approaches based on best reciprocal Blast hits developed originally for bacterial gene families cannot capture the complexity of eukaryotic multigene families. Approaches capable of distinguishing gene duplications from speciations (e.g., those based on gene trees) are thus critical for harvesting high-quality sets of orthologs from eukaryotic genomes.

Phylogenomic inference is a complex task even in the postgenomic era when the availability of genes/sites is virtually not a limiting factor for understanding species relationships. Despite of the complexity of the exercise, often in practice, the correctness of the results is often (in practice) simply justified by the sheer number of concatenated characters analyzed. This strategy is prone to phylogenetic biases and can be a hotbed of strongly supported incongruence. Therefore, a careful analysis of the sensitivity of results to assumptions about the model, taxon sampling, missing data, and other sources of bias is always essential. Concatenation is by far the most widespread and robustly applicable approach in current phylogenomic practice. Due to its long prehistory in traditional phylogenetics, many of its statistical properties, including good and bad, are well known. However, the understanding that concatenation-based analyses are prone to artifacts caused by systematic biases—in particular nonphylogenetic signals and model violation—led to increased interest in methods that bypass the joint analysis of concatenated data to infer species trees. Approaches that explicitly model differences in the evolutionary histories of individual genes are in theory immune to several of the factors that trouble concatenation, but are not as mature methodologically as concatenation-based approaches.

3. MODELING DIFFERENCES BETWEEN GENE TREES

3.1 The Problem Is That Gene Trees Are Not Species Trees

During the last 50 years, phylogeny has increasingly favored homologous molecular markers (amino acid and nucleotide sequences) over morphological characters. Phylogenomics has been extremely fruitful and has improved both the accuracy and resolution of phylogenetic reconstruction and our understanding of evolutionary processes at the molecular level. However, from a theoretical point of view we have known all along that we are barking up the wrong trees: we have used increasingly sophisticated models of sequence evolution to reconstruct trees that describe at best the history of fragments of genomic sequence, which here we will liberally call "genes," but never the history of species. Gene trees are not species trees (Maddison, 1997; Szollosi, Tannier, et al., 2015).

Each gene tree has a unique story, which is related to species history, but can be significantly different from it. In practice, the majority of phylogenomic analyses have used the "concatenation" approach described earlier. During such an analysis only a minority (1%–10%) of genes (Dagan & Martin, 2006)—those found in exactly one copy in each genome—are selected under the ad hoc assumption that all of them share a single-evolutionary history that can be equated with the history of species. This restriction to a minority of genes that are forced to share a common history not only neglects available information in a quantitative sense (by ignoring 90%–99% of genes) but also leaves unexploited a qualitatively different source of information encoded in true discrepancies among gene trees. Even more problematically, to orient the tree in time and identify the position of the root under the concatenation approach it is necessary to include sequences from out-group species (i.e., to find the position of the root). Using an out-group has the unintended effect of (i) further reducing the dataset as only genes present in both the out-group and the in-group of interest can be used and (ii) potentially distorting the phylogeny by spuriously grouping together divergent taxa due to the typically large divergence between the out- and in-groups and the consequent serious artifacts arising from by systematically mistaking convergent characters for shared ones. Dating the resulting rooted phylogeny relies on relaxed molecular clock approaches. However, when fossils are rare, or nearly completely absent, as is the case for fungi, molecular clock approaches suffer from an extreme lack of resolution. The resulting uncertainty leaves the timing of major events in the evolution of fungi essentially unknown.

3.2 The Solution Is to Model How Gene Trees Are Generated Along the Species Tree

If, however, the evolution of genomes is modeled as a series of DTL and the population level process of ILS, generating a plurality of gene histories, then gene and species phylogenies can be simultaneously reconstructed. Using dozens of complete genomes, it has been shown to be feasible to perform genome-scale joint inference of gene trees and the species tree while modeling DL (Boussau et al., 2013) and DTL (Szollosi, Boussau, Abby, Tannier, & Daubin, 2012). Similarly, a variety of methods exist that enable inference of species trees while explicitly modeling ILS (Liu, Wu, & Yu, 2015). Currently, however, no method is available that models DTL and ILS simultaneously.

From a practical point of view, it is important to note that in most cases considered in the two studies the unrooted species-tree topology recovered was identical to that obtained from a concatenation-based approaches. Given the high computational cost, and relative complexity of using the above gene tree-based methods, some of which are in early stages of software development, the full promise of joint inference remains to be realized in practice.

However, combining concatenation methods (used to infer the species phylogeny) with reconciliations in order to distinguish between competing phylogenetic hypotheses (including rooting without an out-group) are feasible for several dozen to a few hundred species. For example, Williams et al. recently defined the unrooted tree topology for the Archaeal tree of life using a concatenated approach and implemented a DTL-based method in the ALE package (Szollosi, Rosikiewicz, Boussau, Tannier, & Daubin, 2013) to root the Archaeal tree of life to root it without an out-group (Williams et al., 2017).

In addition, while joint inferences have to date yielded few surprises at the level of the species-tree topology, several papers demonstrated that species-tree aware reconstruction results in gene trees dramatically more accurate. For both mammals—where a DL model was used (Boussau et al., 2013) and for cyanobacteria—where a DTL model was used (Szollosi et al., 2012), gene trees were in general more similar to the species tree then traditional species-tree unaware gene phylogenies. For example, ancestral genome sizes for mammals based on species-tree aware reconstructions were a significantly more accurate than those based on the trees available in the authoritative database Ensembl (Boussau et al., 2013). Similarly, for cyanobacteria two out of three transfer events inferred by traditional species-tree unaware methods were found to be the results of reconstruction errors (Szollosi et al., 2013). In both erroneous cases, synteny reconstruction provided independent evidence of a significant and substantial gain in gene tree accuracy (Boussau et al., 2013; Patterson et al., 2013).

As a corollary, ancestral gene content reconstructions are expected to be dramatically more accurate when based on gene trees inferred using species-tree aware methods. This increase in accuracy extends to ancestral sequence reconstruction. For example, a study by Groussin, Hobbs, et al. (2015) showed using in vitro resurrection of the LeuB enzyme for the ancestor of the Firmicutes—a major and ancient bacterial phylum—that gene trees inferred using species-tree aware methods result in a biochemically more realistic and kinetically more stable ancestral protein.

It also follows that reconstructing the pattern and process of genome evolution over evolutionary time scales requires using information on gene phylogenies. For example, when gene tree topologies are considered, lineage-specific genome reduction in archaea (Csuros & Miklos, 2009) is not observed (Williams et al., 2017) and evidence for transfer to major archaeal clades from bacteria (Nelson-Sathi et al., 2015) do not hold (Groussin, Boussau, et al., 2015) when gene tree topologies are considered. For fungi—where species-tree aware reconstruction methods have only recently been applied—it was found that including gene tree topologies in the estimation of rates of gene transfer lead to estimates of transfer rates comparable to those in cyanobacteria (Szollosi, Davin, Tannier, Daubin, & Boussau, 2015).

3.3 Incomplete Lineage Sorting

The history of genes within a single genome can be different. Reflecting a series of speciation, duplication, loss, and horizontal transfer events the gene family has undergone (Fig. 2). Moreover, two gene trees can be different

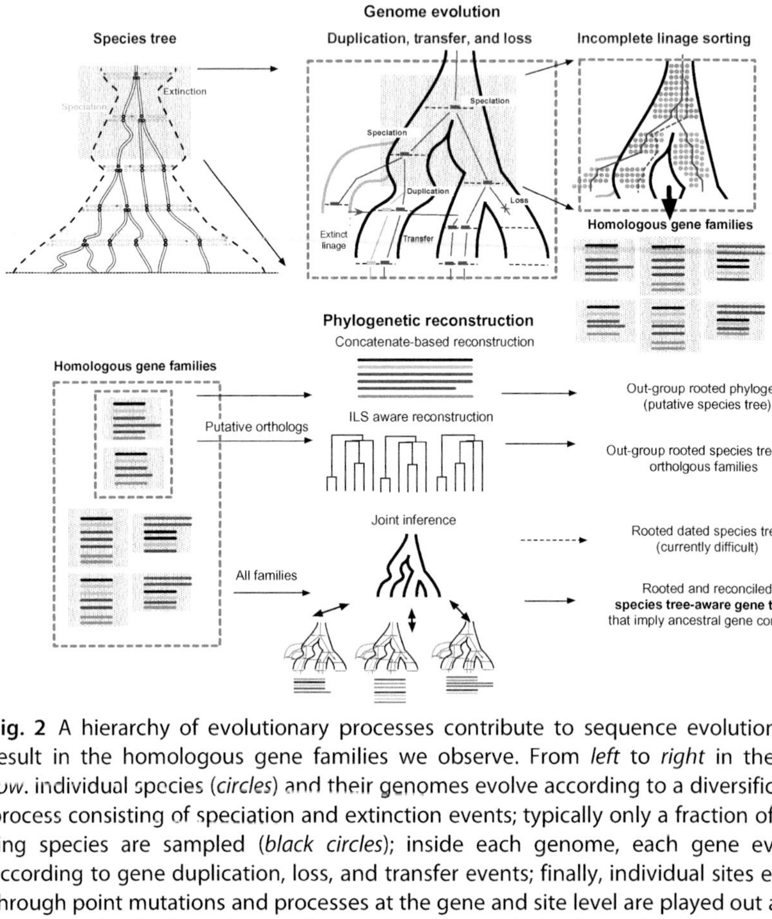

Fig. 2 A hierarchy of evolutionary processes contribute to sequence evolution and result in the homologous gene families we observe. From *left* to *right* in the *trop row*. individual species (*circles*) and their genomes evolve according to a diversification process consisting of speciation and extinction events; typically only a fraction of existing species are sampled (*black circles*); inside each genome, each gene evolves according to gene duplication, loss, and transfer events; finally, individual sites evolve through point mutations and processes at the gene and site level are played out at the population level, where changes fix or are lost, potentially leading to incomplete lineage sorting (ILS). Phylogenomic reconstruction takes as its input the homologous gene families by this hierarchy of processes. Concatenation-based approaches evaluate one sequence per species and in order to identify the inferred phylogeny with the species tree it is assumed that (i) the input contains only orthologous genes (i.e., phylogenetic differences produced by DTL events are assumed to be absent) and (ii) ILS can be safely ignored. ILS-based approaches also rely on putative orthologs, but the latter assumption is relaxed, and ILS is modeled explicitly. Finally, joint inference approaches aim to take as their input all gene families, and as a result must model the duplication, transfer, and loss of genes explicitly.

even if they share the same series of genome evolutionary events. This happens because genes evolve in populations, and different allelic forms of a gene can coexist for periods that may span several speciation events potentially leading to differences as a result of the process of ILS.

ILS has predominantly been considered in the context of putatively orthologous loci. It has been shown to bias supermatrix approaches, especially when internal branch lengths are short (Liu et al., 2015) and/or ancestral population sizes are large. In particular, unpartitioned ML analyses of concatenated data are statistically inconsistent in the presence of ILS (Roch & Steel, 2015; Warnow, 2015), i.e., it can yield strongly supported incorrect topologies that would not be falsified by additional gathering of similar data. ILS is modeled by the multispecies coalescent model (MSC). Several methods that can handle ILS under the MSC have been developed (Drummond, Suchard, Xie, & Rambaut, 2012; Liu, Yu, & Edwards, 2010; Mirarab & Warnow, 2015) that unlike unpartitioned concatenation approaches are statistically consistent under MSC. These methods either coestimate gene trees with species trees (e.g., BEAST) or summarize a priori estimated gene trees into a species tree. Note that traditional consensus or supertree approaches (e.g., matrix representation with parsimony) can also generate summary trees based on sets of input trees, but these are not modeling ILS explicitly.

Approaches based on precomputed gene trees assume that the gene trees are known with relative certainty; if individual gene trees are of poor quality (e.g., because of finite alignment lengths) then the accuracy of these methods drop. Statistical binning aims to evade this by combining the advantages of supermatrix and "summary-based" approaches (Bayzid, Mirarab, Boussau, & Warnow, 2015; Mirarab, Bayzid, Boussau, & Warnow, 2014): it first sorts individual genes into similarly behaving groups, concatenates them, and infers "supergene trees," which are then combined into a species tree under the MSC model. Statistical binning has been successfully applied to a number of challenging phylogenomic questions, although some debate remains as to how it compares to unbinned inferences (Liu & Edwards, 2015).

3.4 Gene Transfers as Molecular Fossils

Horizontal gene transfer events carry a record of the timing of species diversification because they have occurred between species that existed at the same time (Szollosi et al., 2012). As a consequence, a transfer event can be used to establish a relative age constraint between nodes in a phylogeny independently of any molecular clock hypothesis: the ancestor node of the donor lineage must predate the descendant node of the receiving lineage. Recent results (Davin et al., in press) show that there is abundant information in extant genomes on dating the tree of life waiting to be harvested from the reconstruction of genome evolution. This signal mostly contains information on the relative timing of diversification of groups that have exchanged genes through LGT, but opens up several avenues to relate this relative timing to the broader history of life on Earth. In particular, gene transfers between organisms with a poor fossil record (bacteria or fungi) and multicellular organisms that have left a more substantial trace in the fossil record will allow the propagation of absolute time calibrations along the tree of life.

4. FUNGAL RELATIONSHIPS: PHYLOGENOMICS

Fungi are an extremely diverse eukaryotic supergroup, with immense importance in most ecosystems and many industrial applications. Their role as plant and human pathogens, and industrial workhorses combined with their relatively compact genomes has led to the availability of a large number of genome sequences at the dawn of the genomic era, which made fungi to become one of the first model clades for the development of techniques for phylogenomic inference. Despite a large number of studies that have used fungal datasets to test phylogenomic questions, several unresolved regions remain in the fungal tree of life that await closer inspection. With the dawn of genome-enabled mycology (Hibbett, Stajich, & Spatafora, 2013), uncertainty the fungal tree is expected to diminish if the flood of genome data can be channeled through the appropriate analytical strategies. In this section, we discuss case studies where phylogenomics has clarified contentious relationships, as well as others that remain puzzling despite the use of genome-scale data (Fig. 3).

Early phylogenomic treatments of the fungi include analyses of yeast relationships and the effects of concatenation vs gene tree-based methods (Hess & Goldman, 2011; Jeffroy et al., 2006; Rokas, Williams, King, & Carroll, 2003). The sequencing of genomes of several filamentous fungi opened the way for phylogenomic analyses across all fungi and major clades therein (Aguileta et al., 2008; Dutilh et al., 2007; Fitzpatrick, Logue, Stajich, & Butler, 2006; Floudas et al., 2012; Kohler, Kuo, Nagy, Morin, et al., 2015; Leavitt et al., 2016; Medina, Jones, & Fitzpatrick, 2011;

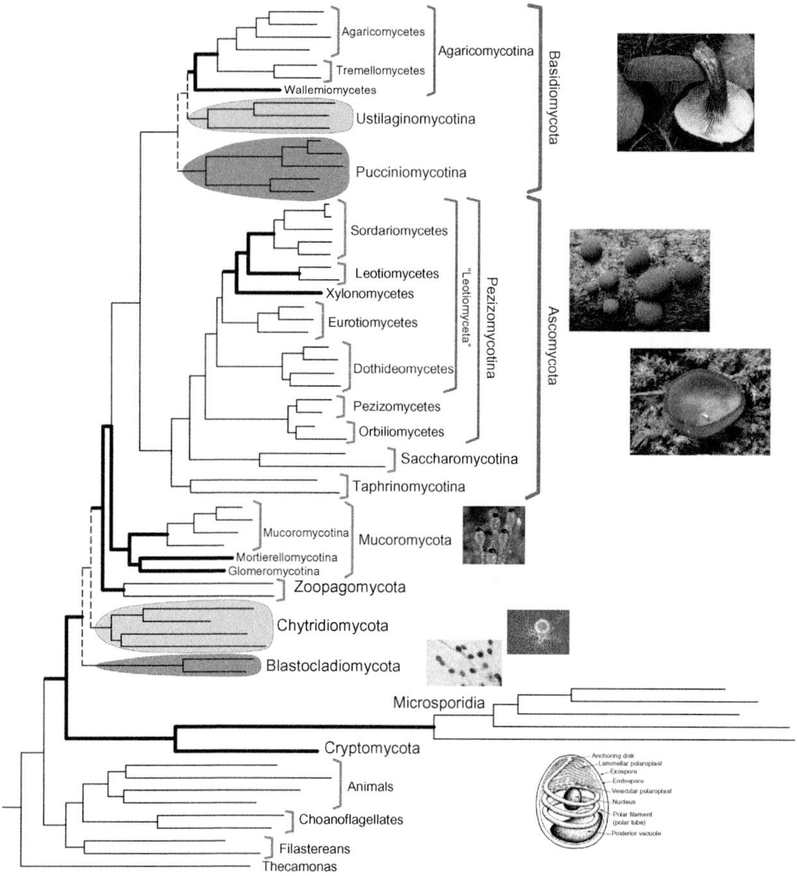

Fig. 3 Summary of the advances in understanding fungal relationships in the genomic era. Topology combined manually based on published tree topologies (Chang et al., 2015; Fitzpatrick et al., 2006; James et al., 2013; Padamsee et al., 2012; Spatafora et al., 2016), branch lengths estimated using a partitioned analysis of 361 concatenated single-copy genes under the WAG + G model in RAxML 7.2.3. Thickened branches identify relationships where phylogenomics resolved previously ambiguous nodes and *dashed lines* identify those that remain contentious despite the use of genome data. Note the uncertainty in basal Basidiomycota relationships and the branching order of the Blastocladiomycota and the Chytridiomycota (see text for details). The branch lengths of microsporidia have been reduced by 4 × for clarity.

Robbertse, Reeves, Schoch, & Spatafora, 2006). The accelerating pace of fungal genome sequencing by a number of large-scale sequencing projects soon paved the way for assembling larger and taxon-specific datasets that clarified some of the puzzling fungal relationships. For example, using a collection of 42 genomes and 153 universal orthologs, (Fitzpatrick et al. 2006) found conclusive evidence for the sister relationship between the Leotiomycetes and Sordariomycetes. In another study, Gazis et al. (2016) resolved the position of the Xylonomycetes, a small class of leaf endophytes, as a sister group to the Lecanoromycetes and Eurotiomycetes. In this case too, genome-scale data resolved the position of a class that has been difficult to place in a radiation comprising the "Leotiomyceta" using multilocus phylogenetics (Gazis, Miadlikowska, Lutzoni, Arnold, & Chaverri, 2012). Another small and lesser-known class with uncertain phylogenetic placement, the Wallemiomycetes, was inferred as the earliest diverging branch of the Agaricomycotina, using a dataset of 72 universally conserved genes (Padamsee et al., 2012). Beyond the numerous success stories of the application of genome-scale data, however, there are evolutionary questions, two of which discussed later, that seem resistant to simply increasing the amount of sequence data in phylogenetic analyses. Intricate evolutionary processes, known or unknown genetic mechanisms could underlie such cases, and the interference of hidden signals in the data with the models and methods used can lead to strongly supported but incorrect relationships and can open cases for statistically significant incongruence (Jeffroy et al., 2006). Obtaining robust topologies around recalcitrant nodes requires careful selection of data and in many cases the posthoc dissection of the phylogenetic (and non-phylogenetic) signals is inevitable to the appropriate evaluation of the models to use for the underlying evolutionary processes.

4.1 Early Diverging Fungi

Resolving ancient divergences poses significant challenges even for phylogenomic datasets (Jeffroy et al., 2006; Philippe et al., 2017, 2005). Studies of the early evolution of fungi have presented a number of phylogenetic puzzles, the resolution of which, has been a continuing quest in fungal biology with important implications for the reconstruction of the evolution of many organismal traits, such as the origins of plant cell wall decomposition that of terrestriality, chitinous cell wall and the loss of flagellated forms in the life cycle, to name a few. Classification of the early diverging fungal groups and our understanding of their evolution has been in a considerable flux since sequence-based phylogenetics became available in the early 1990s. Phylogenetic inference at these early nodes is challenging not only because of the temporal depth at which they diverged, but is also complicated by rapid species radiations, hard polytomies and a number of rare evolutionary events including extensive gene loss and extreme sequence divergence (Chang et al., 2015; James et al., 2013; Keeling & Fast, 2002; Spatafora & Robbertse, 2010). Despite these challenges, resolution of many of the relationships along the fungal backbone has benefited a lot from genome-scale phylogenetics.

Some relationships remain contentious. For example, the classic Zygomycota that appeared paraphyletic already in early multigene phylogenies could be resolved into two phyla, Zoopagomycota and Mucoromycota (Spatafora et al., 2016). Their branching order is now relatively stable (Fig. 3), with the Zoopagomycota branching off first, then the Mucoromycota. The latter includes the Mortierellomycotina, a group of ubiquitous soil fungi and the Mucoromycotina containing *Mucor, Rhizopus*, and some of the best-known opportunistic pathogens. It probably also includes the arbuscular mycorrhizal fungi (Glomeromycota), a group of plant symbionts with a previously uncertain position (James et al., 2006). The position of the Glomeromycota as the sister group to the Mortierellomycotina and Mucoromycotina has recently been suggested based on genome-scale phylogenies (Chang et al., 2015) and is in conflict with previous rDNA and multigene phylogenies that relatively consistently placed it as the sister group to the Dikarya (Hibbett et al., 2007; James et al., 2006; Spatafora & Robbertse, 2010).

The definition of the fungal kingdom and the placement of the Microsporidia as fungi or nonfungal eukaryotes have been debated (Haag et al., 2014; James & Berbee, 2012; James et al., 2013; Keeling & Fast, 2002). Multigene phylogenies have indicated affinities between the Microsporidia and *Rozella allomycis* (Capella-Gutiérrez, Marcet-Houben, & Gabaldón, 2012; James et al., 2013; Karpov et al., 2013), a parasite of filamentous fungi, although support for their grouping remained elusive. Both Microsporidia and *Rozella* are intracellular parasites and possess streamlined genomes with low-protein coding capacities and accelerated rates of molecular evolution. These are factors that have contributed to the difficulties resolving the phylogenetic position of the Microsporidia. Genomes of Microsporida comprise as few as 2000–4000 genes, and lack many of the core eukaryotic and single-copy housekeeping genes that are usually harvested for phylogenomic analyses. Therefore, genome-based datasets involving Microsporidia are often characterized by large proportions of missing data (Nagy et al., 2014), which, combined with their high rate of protein evolution can be a hotbed for LBA (James et al., 2013; Keeling & Fast, 2002). Using a whole genome sequence of *R. allomycis* and a 200-gene alignment, James and colleagues found that the removal of fast-evolving sites influenced the placement and support values for Microsporidia relative to *Rozella*. By gradually eliminating up to half of the fastest evolving sites, those that could contribute most to LBA, from the dataset they obtained increasing support values for the grouping of *Rozella* with Microsporidia. Both the elimination of fast-evolving sites and the inclusion of *Rozella*—to break up the long branch leading to Microsporidia—could have contributed to inferring a stable phylogeny at the very base of the fungal tree. These results were much in agreement with genomic and life-history traits, such as parasitism, the presence of chitin in the cell wall and the shared occurrence of nucleotide and nucleoside transporters, as well as class I chitinase genes. The Microsporidia thus—together with the Cryptomycota—found a place at the base of the fungi and probably combined with Aphelida make up the recently proposed Opisthosporidia supergroup (Karpov et al., 2014). This example demonstrates perfectly how increased sampling of genomes and sophisticated analytical methods can bring conclusive evidence to long-standing phylogenetic questions.

The branching order of the Blastocladiomycota, Chytridiomycota, and more derived fungi, on the other hand has remained difficult to resolve (Spatafora & Robbertse, 2010). These groups comprise some of the most distant fungal relatives of crown groups (Dikarya) and contain unicellular or primitively multicellular fungi. They have been subject to several phylogenomic studies, yet a conclusive answer to their branching order has so far failed to emerge. Some studies reported the Blastocladiomycota as the earlier branching clade, followed by chytrids forming a sister group to the rest of the fungi (Chang et al., 2015; Haag et al., 2014; Ren et al., 2016). Others reported the chytrids to diverge first followed by the Blastocladiomycota (James et al., 2013; Liu et al., 2009), yet others inferred a common clade uniting them (Ebersberger et al., 2012). Recently, Spatafora et al. performed phylogenomic analyses of the largest set of early diverging fungi to date (46 species) using 192 orthologous genes, and reconstructed the Blastocladiomycota as the sister group to other fungal clades, to the exclusion of the Chytridiomycota, albeit with weak support (56%). Although there is no consensus on which of these two branched off first from the fungal backbone, the published phylogenies are not necessarily conflict with each other, as support values remained low in almost all cases. Notably, these studies differed in the way data were collected, the phylogenetic inference methods used and the models applied, although apparently there is no correlation between the model or software used and the topology inferred. A recent, 32-gene dataset produced by concatenating conserved single-copy genes was found consistent with a hard polytomy (Chang et al., 2015), i.e., biologically relevant multifurcation that describes rapid or simultaneous species radiations. In such cases, too little time may pass between successive speciation events for phylogenetically informative substitutions to accumulate, although it should be emphasized that the lack of recoverable phylogenetic signal is also consistent with the phylogenetic signal having eroded over time or the models used not being able to capture it.

4.2 Basal Relationships in the Basidiomycota

The Basidiomycota is the second largest phylum of fungi, comprising some 35,000 species divided into three main groups, the Agaricomycotina (mushroom-forming fungi), Pucciniomycotina (rusts and allies), and the Ustilaginomycotina (smuts and allies). Although these groups have been recognized as monophyletic for almost a century, their relationships remained particularly recalcitrant (Aime et al., 2006; Hibbett et al., 2007; Kohler et al., 2015; Matheny, Gossmann, Zalar, Kumar, & Hibbett, 2006; Matheny et al., 2007; Nagy et al., 2016; Padamsee et al., 2012). Traditional phylogenetic analyses provided no resolution or weak support for each of the three possible conformations (Aime et al., 2006; Bauer et al., 2015; Hibbett et al., 2007; Matheny et al., 2007) with a slight edge toward a grouping of Ustilaginomycotina and Agaricomycotina. More recently, phylogenomic analyses yielded strongly supported but conflicting results, with support values and topologies varying across the phylogenetic inference methods and models used. Taxon-sampling density for smuts and rusts, however, has been low in all previous phylogenomic studies, which could contribute to artifacts such as LBA. As in traditional multilocus phylogenies, a grouping of smuts with mushroom-forming fungi seems marginally more probable (Ebersberger et al., 2012; Floudas et al., 2012; Kohler et al., 2015; Nagy et al., 2016; Padamsee et al., 2012), although other topologies have also shown up in publications (Kohler et al., 2015; Medina et al., 2011; Riley et al., 2014).

In an analysis focusing on the branching order of rusts, smuts, and mushroom-forming fungi, we found that increased sampling does not alleviate the ambiguity in support values. We assembled three datasets comprising 314, 824, and 901 protein families by varying the level of stringency for excluding ambiguously aligned regions from the single-gene alignments and concatenating only those alignments that contained >50 amino acid sites after trimming. Increasing the amount of sequence data in concatenated analyses lended more support to the Agaricomycotina plus Ustilaginomycotina hypothesis, whereas shorter but more conserved datasets resulted in the Agaricomycotina and Pucciniomycotina as sister groups (Nagy LG and Prasanna AN unpublished). Two factors, increasing concatenated sequence length and decreasing model complexity favored the grouping of Ustilaginomycotina with Agaricomycotina to the exclusion of Pucciniomycotina. Bootstrap support increased to 100% for the Ustilaginomycotina + Agaricomycotina grouping in the largest and most variable dataset as opposed to 37% in the smallest, most conserved dataset. Notably, both higher sequence divergence and simplistic models lead to a higher number of inappropriately modeled amino acid sites which, in turn can result in systematic errors that get more pronounced with increasing the amount of data. These observations are consistent with the notion of inflated bootstrap support (Felsenstein, 1978; Jeffroy et al., 2006; Phillips, Delsuc, & Penny, 2004) for a potentially incorrect grouping of Ustilaginomycotina with the Agaricomycotina. If only the most conserved dataset is considered, a grouping of Pucciniomycotina with Agaricomycotina is supported by low bootstrap frequencies (59%). Although support for this grouping depends on taxon sampling and the model used to some extent, it appears that strong support remains elusive. Could multifurcating evolution then provide a better explanation for early basidiomycete splits? The answer is an apparent no, as the hard polytomy hypothesis was rejected by Reversible-Jump MCMC analyses allowed to visit polytomous trees, even under polytomy-friendly priors (Nagy LG and Prasanna A. unpublished). Basal Basidiomycete relationships comprise a typical case of hard-to-resolve nodes that remain contentious even when large numbers of taxa and/or characters are used. Identifying the evolutionary processes that underlie such splits and developing models and methods that can take those into account probably represent the step we need to make. Although the failure to resolve basal relationships in the Basidiomycota to date is not satisfying from a biological point of view, it highlights the challenges associated with phylogenomic reconstruction deep in evolutionary time.

5. CONCLUSIONS

Phylogenomics is revolutionizing the field of evolutionary biology, making it possible to ask and resolve questions that were infeasible to tackle before. Genome-scale phylogenetic datasets yield a dramatic increase in our statistical confidence of inferred relationships, often yielding maximally supported species trees. Statistical confidence, however, can be high not only for the true species tree, in cases where systematically biased inferences yield strongly supported incorrect relationships. Such cases arise when there are aspects of the data (e.g., compositional heterogeneity, and heterotachy) that are not captured by the evolutionary model being used cause non-phylogenetic signals to show up in the inferred results. Different methods perform differently under various challenging circumstances, so it is important to test the robustness of the results to assumptions on the evolutionary process generating the data and understand potential interactions between the model and different subsets of the data. The strengths of concatenation and gene tree-based methods lie in different datasets with varying amounts of among gene incongruence (e.g., due to ILS) and support for individual gene trees. Novel methods that bypass the weaknesses of individual methods (e.g., statistical binning) and others that open up new sources of information for rooting phylogenetic trees and relative molecular dating are being developed and can be leveraged for refining our understanding of the tree of life.

During the relatively short time that phylogenomic analysis has been feasible, our understanding of fungal relationships has benefited enormously from the power supplied by genome-scale data. Nevertheless, some relationships, such as the branching order of Blastocladiomycota and Chytridiomycota, of basal relationships in the Basidiomycota, remain contentious even in the genomic era. In cases of weak support or strong support for mutually exclusive relationships, deciphering how the data, the model and potentially unknown background processes of evolution interact to generate the inferred tree is more important than ever. Whole-genome sequences might well be the cornucopia of phylogenetic information, which relaxes pressure on data collection, but at the same time makes smart dataset assembly and analytical strategies crucial for maximizing phylogenomic performance.

ACKNOWLEDGMENTS

L.G.N. was supported by the "Momentum Programme" of the Hungarian Academy of Sciences (No. LP2014/12) and by the ERC_HU scheme under contract No. 118722. G.J.Sz. received funding from the European Research Council (ERC) under the European Union's Horizon 2020 research and innovation programme under Grant agreement No. 714774.

REFERENCES

Aguileta, G., Marthey, S., Chiapello, H., Lebrun, M. H., Rodolphe, F., Fournier, E., et al. (2008). Assessing the performance of single-copy genes for recovering robust phylogenics. *Systematic Biology*, *57*, 613–627. https://doi.org/10.1080/10635150802306527.

Aime, M. C., Matheny, P. B., Henk, D. A., Frieders, E. M., Nilsson, R. H., Piepenbring, M., et al. (2006). An overview of the higher level classification of

Pucciniomycotina based on combined analyses of nuclear large and small subunit rDNA sequences. *Mycologia, 98*, 896–905.
Bauer, R., Garnica, S., Oberwinkler, F., Riess, K., Weiss, M., & Begerow, D. (2015). Entorrhizomycota: A new fungal phylum reveals new perspectives on the evolution of fungi. *PLoS One, 10*, e0128183. https://doi.org/10.1371/journal.pone.0128183.
Bayzid, M. S., Mirarab, S., Boussau, B., & Warnow, T. (2015). Weighted statistical binning: Enabling statistically consistent genome-scale phylogenetic analyses. *PLoS One, 10*, e0129183. https://doi.org/10.1371/journal.pone.0129183.
Boussau, B., Szollosi, G. J., Duret, L., Gouy, M., Tannier, E., & Daubin, V. (2013). Genome-scale coestimation of species and gene trees. *Genome Research, 23*, 323–330. https://doi.org/10.1101/gr.141978.112.
Capella-Gutiérrez, S., Marcet-Houben, M., & Gabaldón, T. (2012). Phylogenomics supports microsporidia as the earliest diverging clade of sequenced fungi. *BMC Biology, 10*, 47. https://doi.org/10.1186/1741-7007-10-47.
Chang, Y., Wang, S., Sekimoto, S., Aerts, A. L., Choi, C., Clum, A., et al. (2015). Phylogenomic analyses indicate that early fungi evolved digesting cell walls of algal ancestors of land plants. *Genome Biology and Evolution*, 7, 1590–1601. https://doi.org/10.1093/gbe/evv090.
Csuros, M., & Miklos, I. (2009). Streamlining and large ancestral genomes in Archaea inferred with a phylogenetic birth-and-death model. *Molecular Biology and Evolution, 26*, 2087–2095. https://doi.org/10.1093/molbev/msp123.
Dagan, T., & Martin, W. (2006). The tree of one percent. *Genome Biology*, 7, 118. https://doi.org/10.1186/gb-2006-7-10-118.
Davin AA, Tannier E, Williams TA, Boussau B, Daubin V, Szollosi GJ. Gene transfers, like fossils, can date the Tree of Life. BioArxiv, https://doi.org/10.1101/193813
Delsuc, F., Brinkmann, H., & Philippe, H. (2005). Phylogenomics and the reconstruction of the tree of life. *Nature Reviews. Genetics, 6*, 361–375.
dos Reis, M., Inoue, J., Hasegawa, M., Asher, R. J., Donoghue, P. C. J., & Yang, Z. (2012). Phylogenomic datasets provide both precision and accuracy in estimating the timescale of placental mammal phylogeny. *Proceedings of the Royal Society B: Biological Sciences, 279*, 3491–3500. https://doi.org/10.1098/rspb.2012.0683.
Drummond, A. J., Suchard, M. A., Xie, D., & Rambaut, A. (2012). Bayesian phylogenetics with BEAUti and the BEAST 1.7. *Molecular Biology and Evolution, 29*, 1969–1973. https://doi.org/10.1093/molbev/mss075.
Dutilh, B. E., van Noort, V., van der Heijden, R. T., Boekhout, T., Snel, B., & Huynen, M. A. (2007). Assessment of phylogenomic and orthology approaches for phylogenetic inference. *Bioinformatics, 23*, 815–824. https://doi.org/10.1093/bioinformatics/btm015.
Ebersberger, I., de Matos Simoes, R., Kupczok, A., Gube, M., Kothe, E., Voigt, K., et al. (2012). A consistent phylogenetic backbone for the fungi. *Molecular Biology and Evolution, 29*, 1319–1334. https://doi.org/10.1093/molbev/msr285.
Felsenstein, J. (1978). Cases in which parsimony or compatibility methods will be positively misleading. *Systematic Zoology*, 27, 401–410. https://doi.org/10.2307/2412923.
Fitzpatrick, D. A., Logue, M. E., Stajich, J. E., & Butler, G. (2006). A fungal phylogeny based on 42 complete genomes derived from supertree and combined gene analysis. *BMC Evolutionary Biology*, 6, 99. https://doi.org/10.1186/1471-2148-6-99.
Floudas, D., Binder, M., Riley, R., Barry, K., Blanchette, R. A., Henrissat, B., et al. (2012). The Paleozoic origin of enzymatic lignin decomposition reconstructed from 31 fungal genomes. *Science, 336*, 1715–1719. https://doi.org/10.1126/science.1221748.
Gazis, R., Kuo, A., Riley, R., LaButti, K., Lipzen, A., Lin, J., et al. (2016). The genome of Xylona heveae provides a window into fungal endophytism. *Fungal Biology, 120*, 26–42. https://doi.org/10.1016/j.funbio.2015.10.002.

Gazis, R., Miadlikowska, J., Lutzoni, F., Arnold, A. E., & Chaverri, P. (2012). Culture-based study of endophytes associated with rubber trees in Peru reveals a new class of Pezizomycotina: Xylonomycetes. *Molecular Phylogenetics and Evolution, 65*, 294–304. https://doi.org/10.1016/j.ympev.2012.06.019.

Gee, H. (2003). Evolution: Ending incongruence. *Nature, 425*, 782.

Geuten, K., Massingham, T., Darius, P., Smets, E., & Goldman, N. (2007). Experimental design criteria in phylogenetics: Where to add taxa. *Systematic Biology, 56*, 609–622. https://doi.org/10.1080/10635150701499563.

Groussin, M., Boussau, B., Szollosi, G., Eme, L., Gouy, M., Brochier-Armanet, C., et al. (2015). Gene acquisitions from bacteria at the origins of major archaeal clades are vastly overestimated. *Molecular Biology and Evolution, 33*, 305–310.

Groussin, M., Hobbs, J. K., Szollosi, G., Gribaldo, S., Arcus, V. L., & Gouy, M. (2015). Toward more accurate ancestral protein genotype–phenotype reconstructions with the use of species tree-aware gene trees. *Molecular Biology and Evolution, 32*, 13–22.

Haag, K. L., James, T. Y., Pombert, J. F., Larsson, R., Schaer, T. M., Refardt, D., et al. (2014). Evolution of a morphological novelty occurred before genome compaction in a lineage of extreme parasites. *Proceedings of the National Academy of Sciences of the United States of America, 111*, 15480–15485. https://doi.org/10.1073/pnas.1410442111.

Hess, J., & Goldman, N. (2011). Addressing inter-gene heterogeneity in maximum likelihood phylogenomic analysis: Yeasts revisited. *PLoS One, 6*, e22783. https://doi.org/10.1371/journal.pone.0022783.

Hibbett, D. S., Binder, M., Bischoff, J. F., Blackwell, M., Cannon, P. F., Eriksson, O. E., et al. (2007). A higher-level phylogenetic classification of the Fungi. *Mycological Research, 111*, 509–547. https://doi.org/10.1016/j.mycres.2007.03.004. S0953-7562(07)00061-5 [pii].

Hibbett, D. S., Stajich, J. E., & Spatafora, J. W. (2013). Toward genome-enabled mycology. *Mycologia, 105*, 1339–1349. https://doi.org/10.3852/13-196.

James, T. Y., & Berbee, M. L. (2012). No jacket required—New fungal lineage defies dress code: Recently described zoosporic fungi lack a cell wall during trophic phase. *BioEssays, 34*, 94–102. https://doi.org/10.1002/bies.201100110.

James, T. Y., Kauff, F., Schoch, C. L., Matheny, P. B., Hofstetter, V., Cox, C. J., et al. (2006). Reconstructing the early evolution of fungi using a six-gene phylogeny. *Nature, 443*, 818–822. https://doi.org/10.1038/nature05110. nature05110 [pii].

James, T. Y., Pelin, A., Bonen, L., Ahrendt, S., Sain, D., Corradi, N., et al. (2013). Shared signatures of parasitism and phylogenomics unite Cryptomycota and microsporidia. *Current Biology, 23*, 1548–1553. https://doi.org/10.1016/j.cub.2013.06.057.

Jeffroy, O., Brinkmann, H., Delsuc, F., & Philippe, H. (2006). Phylogenomics: The beginning of incongruence? *Trends in Genetics, 22*, 225–231. https://doi.org/10.1016/j.tig.2006.02.003.

Karpov, S., Mamkaeva, M., Aleoshin, V., Nassonova, E., Lilje, O., & Gleason, F. (2014). Morphology, phylogeny, and ecology of the aphelids (Aphelidea, Opisthokonta) and proposal for the new superphylum Opisthosporidia. *Frontiers in Microbiology, 5*, 112. https://doi.org/10.3389/fmicb.2014.00112.

Karpov, S. A., Mikhailov, K. V., Mirzaeva, G. S., Mirabdullaev, I. M., Mamkaeva, K. A., Titova, N. N., et al. (2013). Obligately phagotrophic aphelids turned out to branch with the earliest-diverging fungi. *Protist, 164*, 195–205. https://doi.org/10.1016/j.protis.2012.08.001.

Keeling, P. J., & Fast, N. M. (2002). Microsporidia: Biology and evolution of highly reduced intracellular parasites. *Annual Review of Microbiology, 56*, 93–116. https://doi.org/10.1146/annurev.micro.56.012302.160854.

Kohler, A., Kuo, A., Nagy, L. G., Morin, E., Barry, K. W., Buscot, F., et al. (2015). Convergent losses of decay mechanisms and rapid turnover of symbiosys genes in mycorrhizal mutualists. *Nature Genetics*, 47, 410–415. https://doi.org/10.1038/ng.3223.

Kumar, S., Filipski, A. J., Battistuzzi, F. U., Kosakovsky Pond, S. L., & Tamura, K. (2012). Statistics and truth in phylogenomics. *Molecular Biology and Evolution, 29*, 457–472. https://doi.org/10.1093/molbev/msr202.

Leavitt, S. D., Grewe, F., Widhelm, T., Muggia, L., Wray, B., & Lumbsch, H. T. (2016). Resolving evolutionary relationships in lichen-forming fungi using diverse phylogenomic datasets and analytical approaches. *Scientific Reports, 6*, 22262. https://doi.org/10.1038/Srep22262.

Liu, L., & Edwards, S. V. (2015). Comment on "Statistical binning enables an accurate coalescent-based estimation of the avian tree" *Science, 350*, 771. https://doi.org/10.1126/science.aaa7343.

Liu, Y., Steenkamp, E. T., Brinkmann, H., Forget, L., Philippe, H., & Lang, B. F. (2009). Phylogenomic analyses predict sistergroup relationship of nucleariids and Fungi and paraphyly of zygomycetes with significant support. *BMC Evolutionary Biology, 9*, 272. https://doi.org/10.1186/1471-2148-9-272.

Liu, L., Wu, S., & Yu, L. (2015). Coalescent methods for estimating species trees from phylogenomic data. *Journal of Systematics and Evolution, 53*, 380–390.

Liu, L. A., Yu, L. L., & Edwards, S. V. (2010). A maximum pseudo-likelihood approach for estimating species trees under the coalescent model. *BMC Evolutionary Biology, 10*, 302. https://doi.org/10.1186/1471-2148-10-302.

Maddison, W. (1997). Gene trees in species trees. *Systematic Biology, 46*, 523–536.

Matheny, P. B., Gossmann, J. A., Zalar, P., Kumar, T. K. A., & Hibbett, D. S. (2006). Resolving the phylogenetic position of the Wallemiomycetes: An enigmatic major lineage of Basidiomycota. *Canadian Journal of Botany, 84*, 1794–1805. https://doi.org/10.1139/b06-128.

Matheny, P. B., Wang, Z., Binder, M., Curtis, J. M., Lim, Y. W., Nilsson, R. H., et al. (2007). Contributions of rpb2 and tef1 to the phylogeny of mushrooms and allies (Basidiomycota, fungi). *Molecular Phylogenetics and Evolution, 43*, 430–451. https://doi.org/10.1016/j.ympev.2006.08.024.

Medina, E. M., Jones, G. W., & Fitzpatrick, D. A. (2011). Reconstructing the fungal tree of life using phylogenomics and a preliminary investigation of the distribution of yeast prion-like proteins in the fungal kingdom. *Journal of Molecular Evolution, 73*, 116–133. https://doi.org/10.1007/s00239-011-9461-4.

Mikhailov Kirill, V., Slyusarev Georgy, S., Nikitin Mikhail, A., Logacheva Maria, D., Penin Aleksey, A., Aleoshin Vladimir, V., et al. (2016). The genome of Intoshia linei affirms orthonectids as highly simplified spiralians. *Current Biology, 26*, 1768–1774. https://doi.org/10.1016/j.cub.2016.05.007.

Mirarab S, Bayzid MS, Boussau B, Warnow T 2014. Statistical binning enables an accurate coalescent-based estimation of the avian tree. Science 346: Artn 1250463 1337-+. https://doi.org/10.1126/Science.1250463.

Mirarab, S., & Warnow, T. (2015). ASTRAL-II: Coalescent-based species tree estimation with many hundreds of taxa and thousands of genes. *Bioinformatics, 31*, 44–52. https://doi.org/10.1093/bioinformatics/btv234.

Nagy, L. G., Ohm, R. A., Kovacs, G. M., Floudas, D., Riley, R., Gacser, A., et al. (2014). Latent homology and convergent regulatory evolution underlies the repeated emergence of yeasts. *Nature Communications, 5*, 4471. https://doi.org/10.1038/ncomms5471.

Nagy, L. G., Riley, R., Tritt, A., Adam, C., Daum, C., Floudas, D., et al. (2016). Comparative genomics of early-diverging mushroom-forming fungi provides insights into the origins of lignocellulose decay capabilities. *Molecular Biology and Evolution, 33*, 959–970. https://doi.org/10.1093/molbev/msv337.

Nelson-Sathi, S., Sousa, F. L., Roettger, M., Lozada-Chavez, N., Thiergart, T., Janssen, A., et al. (2015). Origins of major archaeal clades correspond to gene acquisitions from bacteria. *Nature, 517*, 77–80. https://doi.org/10.1038/nature13805.

Padamsee, M., Kumar, T. K., Riley, R., Binder, M., Boyd, A., Calvo, A. M., et al. (2012). The genome of the xerotolerant mold Wallemia sebi reveals adaptations to osmotic stress and suggests cryptic sexual reproduction. *Fungal Genetics and Biology*, *49*, 217–226. https://doi.org/10.1016/j.fgb.2012.01.007.

Patterson, M., Szollosi, G., Daubin, V., & Tannier, E. (2013). Lateral gene transfer, rearrangement, reconciliation. *BMC Bioinformatics*, *14*(Suppl. 15), S4. https://doi.org/10.1186/1471-2105-14-S15-S4.

Philippe, H., Brinkmann, H., Lavrov, D. V., Littlewood, D. T. J., Manuel, M., Wörheide, G., et al. (2011). Resolving difficult phylogenetic questions: Why more sequences are not enough. *PLoS Biology 9*, e1000602. https://doi.org/10.1371/journal.pbio.1000602.

Philippe, H., de Vienne, D. M., Ranwez, V., Roure, B., Baurain, D., & Delsuc, F. (2017). Pitfalls in supermatrix phylogenomics. *European Journal of Taxonomy*, (283), 1–25. https://doi.org/10.5852/ejt.2017.283.

Philippe, H., Delsuc, F., Brinkmann, H., & Lartillot, N. (2005). Phylogenomics. *36*, 541–562.

Philippe, H., Snell, E. A., Bapteste, E., Lopez, P., Holland, P. W., & Casane, D. (2004). Phylogenomics of eukaryotes: Impact of missing data on large alignments. *Molecular Biology and Evolution*, *21*, 1740–1752. https://doi.org/10.1093/molbev/msh182.

Phillips, M. J., Delsuc, F., & Penny, D. (2004). Genome-scale phylogeny and the detection of systematic biases. *Molecular Biology and Evolution*, *21*, 1455–1458. https://doi.org/10.1093/molbev/msh137.

Ren, R., Sun, Y., Zhao, Y., Geiser, D., Ma, H., & Zhou, X. (2016). Phylogenetic resolution of deep eukaryotic and fungal relationships using highly conserved low-copy nuclear genes. *Genome Biology and Evolution*, *8*, 2683–2701. https://doi.org/10.1093/gbe/evw196.

Riley, R., Salamov, A. A., Brown, D. W., Nagy, L. G., Floudas, D., Held, B. W., et al. (2014). Extensive sampling of basidiomycete genomes demonstrates inadequacy of the white-rot/brown-rot paradigm for wood decay fungi. *Proceedings of the National Academy of Sciences of the United States of America*, *111*, 9923–9928. https://doi.org/10.1073/pnas.1400592111.

Robbertse, B., Reeves, J. B., Schoch, C. L., & Spatafora, J. W. (2006). A phylogenomic analysis of the Ascomycota. *Fungal Genetics and Biology*, *43*, 715–725. https://doi.org/10.1016/j.fgb.2006.05.001.

Roch, S., & Steel, M. (2015). Likelihood-based tree reconstruction on a concatenation of alignments can be positively misleading. *Theoretical Population Biology*, *100*, 56–62.

Rokas, A., Williams, B. L., King, N., & Carroll, S. B. (2003). Genome-scale approaches to resolving incongruence in molecular phylogenies. *Nature*, *425*, 798–804. https://doi.org/10.1038/nature02053.

Roure, B., Baurain, D., & Philippe, H. (2013). Impact of missing data on phylogenies inferred from empirical phylogenomic data sets. *Molecular Biology and Evolution*, *30*, 197–214. https://doi.org/10.1093/molbev/mss208.

Spatafora, J. W., Chang, Y., Benny, G. L., Lazarus, K., Smith, M. E., Berbee, M. L., et al. (2016). A phylum-level phylogenetic classification of zygomycete fungi based on genome-scale data. *Mycologia*, *108*, 1028–1046. https://doi.org/10.3852/16-042.

Spatafora, J., & Robbertse, B. (2010). Phylogenetics and phylogenomics of the fungal tree of life. In K. Borkovich & D. Ebbole (Eds.), *Cellular and molecular biology of filamentous fungi* (pp. 36–49). Washington, DC: ASM Press. https://doi.org/10.1128/9781555816636.ch4.

Szollosi, G. J., Boussau, B., Abby, S. S., Tannier, E., & Daubin, V. (2012). Phylogenetic modeling of lateral gene transfer reconstructs the pattern and relative timing of speciations. *Proceedings of the National Academy of Sciences of the United States of America*, *109*, 17513–17518. https://doi.org/10.1073/pnas.1202997109.

Szollosi, G., Davin, A. A., Tannier, E., Daubin, V., & Boussau, B. (2015). Genome-scale phylogenetic analysis finds extensive gene transfer among fungi. *Philosophical Transactions of the Royal Society of London. Series B, Biological Sciences*, *370*, 20140335.

Szollosi, G. J., Rosikiewicz, W., Boussau, B., Tannier, E., & Daubin, V. (2013). Efficient exploration of the space of reconciled gene trees. *Systematic Biology*, *62*, 901–912. https://doi.org/10.1093/sysbio/syt054.

Szollosi, G. J., Tannier, E., Daubin, V., & Boussau, B. (2015). The inference of gene trees with species trees. *Systematic Biology*, *64*, e42–e62. https://doi.org/10.1093/sysbio/syu048.

Townsend, J. P., & Leuenberger, C. (2011). Taxon sampling and the optimal rates of evolution for phylogenetic inference. *Systematic Biology*, *60*, 358–365. https://doi.org/10.1093/sysbio/syq097.

Townsend, J. P., & Lopez-Giraldez, F. (2010). Optimal selection of gene and ingroup taxon sampling for resolving phylogenetic relationships. *Systematic Biology*, *59*, 446–457. https://doi.org/10.1093/sysbio/syq025.

Warnow, T. (2015). Concatenation analyses in the presence of incomplete lineage sorting. *PLoS Currents: Tree of Life*. https://doi.org/10.1371/currents.tol.8d41ac0f13d1abedf4c4a59f5d17b1f7.

Wiens, J. J. (2003). Missing data, incomplete taxa, and phylogenetic accuracy. *Systematic Biology*, *52*, 528–538.

Wiens, J. J. (2005). Can incomplete taxa rescue phylogenetic analyses from long-branch attraction? *Systematic Biology*, *54*, 731–742. https://doi.org/10.1080/10635150500234583.

Williams, T. A., Szollosi, G. J., Spang, A., Foster, P. G., Heaps, S. E., Boussau, B., et al. (2017). Integrative modeling of gene and genome evolution roots the archaeal tree of life. *Proceedings of the National Academy of Sciences of the United States of America*, *114*, E4602–E4611. https://doi.org/10.1073/pnas.1618463114.

CHAPTER THREE

Describing Genomic and Epigenomic Traits Underpinning Emerging Fungal Pathogens

Rhys A. Farrer[1], Matthew C. Fisher
Imperial College London, London, United Kingdom
[1]Corresponding author: e-mail address: rfarrer@broadinstitute.org

Contents

Abstract

An unprecedented number of pathogenic fungi are emerging and causing disease in animals and plants, putting the resilience of wild and managed ecosystems in jeopardy. While the past decades have seen an increase in the number of pathogenic fungi, they have also seen the birth of new big data technologies and analytical approaches to tackle these emerging pathogens. We review how the linked fields of genomics and epigenomics are transforming our ability to address the challenge of emerging fungal pathogens. We explore the methodologies and bioinformatic toolkits that currently exist to rapidly analyze the genomes of unknown fungi, then discuss how these data can be used to address key questions that shed light on their epidemiology. We show how genomic approaches are leading a revolution into our understanding of emerging fungal diseases and speculate on future approaches that will transform our ability to tackle this increasingly important class of emerging pathogens.

Advances in Genetics, Volume 100
ISSN 0065-2660
https://doi.org/10.1016/bs.adgen.2017.09.009

1. INTRODUCTION

The fungal kingdom, diverged from the animal and plant kingdoms around 1.5 million years ago (Wang, Kumar, & Hedges, 1999), is globally ubiquitous and taxonomically diverse with between 1.5 and 5 million species estimated to exist (Blackwell, 2011). Recent phylogenetic classifications (Hibbett et al., 2007; Spatafora et al., 2016) currently group fungi into eight separate phyla, with the zoosporic fungi (Cryptomycota, Chytridiomycota, and Blastocladiomycota) comprising the earliest lineages alongside the Microsporidia. The four remaining phyla include the Zoopagomycota, Mucoromycota, and the "Dikarya higher fungi," comprising the phylum Ascomycota and Basidiomycota. Spanning the breadth of the fungal kingdom are pathogenic fungi that infect animals, plants, and other fungi. Importantly, increasing numbers of fungi are emerging as aetiological agents of disease by either exhibiting newly acquired or increased pathogenicity, or invading new ecological niches (geographically or to new host species), or both (Cushion & Stringer, 2010; Fisher, Gow, & Gurr, 2016; Longo, Burrowes, & Zamudio, 2014).

Emerging fungal pathogens (EFPs) are infections that are rapidly increasing in their incidence, geographic or host range, and virulence (Morse, 1995). This class of pathogens are known to pose an increasing threat to the health of plants, humans, and other animals (Fisher et al., 2012; Fones, Fisher, & Gurr, 2017). Recently highlighted examples include the newly described chytrid fungus *Batrachochytrium salamandrivorans* causing rapid declines of fire salamanders across an expanding region of northern Europe (Martel et al., 2014; Stegen et al., 2017), the basidiomycete fungus *Puccinia graminis* f. sp. *tritici* (Ug99 race) now threatening wheat production and food security worldwide (Singh et al., 2011), the basidiomycete fungus *Cryptococcus gattii* expanding its range into nonendemic environments with a consequential increase of fatal disease in humans (Byrnes et al., 2010; Fraser et al., 2005), and the emergence of *Candida auris* in intensive care units worldwide (Chowdhary, Sharma, & Meis, 2017). The global threat of these and other related diseases is underpinned by fungi harboring complex, recombinogenic and dynamic genomes (Farrer, Henk, Garner, et al., 2013; Fisher et al., 2012). Genomic variability drives rapid macroevolutionary change that can overcome host defenses and allow colonization of new environments. Novel genetic diversity also leads to the genesis of new independently evolving pathogenic lineages. Consequently, there is a clear and urgent need to understand the mechanisms that drive the evolution of the phenotypic traits that underlie the virulence, pathogenicity, and geographic/host spread of EFPs.

EFPs of wildlife are generally detected following the observation of (initially "enigmatic") mass mortalities and species declines. For instance, population monitoring by ecologists led to the discovery of panzootic chytridiomycosis caused by novel species of *Batrachochytrium*, and bat white-nose syndrome caused by the novel species *Pseudogymnoascus destructans* (Blehert et al., 2009). In contrast, ongoing surveillance and genotyping of crop pathogens are used to detect and map the spread of phytopathogenic fungi and their lineages as they spread via trade and transportation, such as recently occurred with the spatial emergence of wheat blast *Magnaporthe oryzae* in Bangladesh (Islam et al., 2016). Crucially, in both animal and plant systems, rapid genome sequencing is essential to gain a greater understanding of the taxonomy, epidemiology, and evolutionary biology of EFPs and to inform possible mitigation efforts. A growing body of evidence is also meanwhile accumulating to show that epigenenomic processes (such as differential expression (Kuo et al., 2010), nucleosome positioning (Leach et al., 2016), and nucleic acid modifications (Jeon et al., 2015)), alongside genomic processes, influence both host and pathogen phenotypes. For example, in the aggressive phytopathogen *Botrytis cinerea*, small RNAs invade host cells and silence host immunity by hijacking the host RNA interference (RNAi) machinery leading to a virulent host/pathogen interaction (Weiberg et al., 2013). In this review, we discuss the experimental methodologies, and the discoveries they have enabled, that use genome variation within and between populations of EFPs, with a focus on future threats and the genomic resources that are needed to tackle them. Additionally, we discuss the methods and results emerging from experiments characterizing epigenomic variation within and between populations of EFPs, and show how this emerging field will contribute to a more nuanced understanding of the epidemiology of these infections. The toolkits and methodologies that we cover in this review are summarized in Fig. 1.

2. CHARACTERIZING GENOME VARIATION WITHIN AND BETWEEN POPULATIONS OF EFPS

Genome variation ultimately manifests, postsequencing, through the use of bioinformatics, where two or more individuals have subsections of their DNA aligned and compared, revealing single base changes (indicative

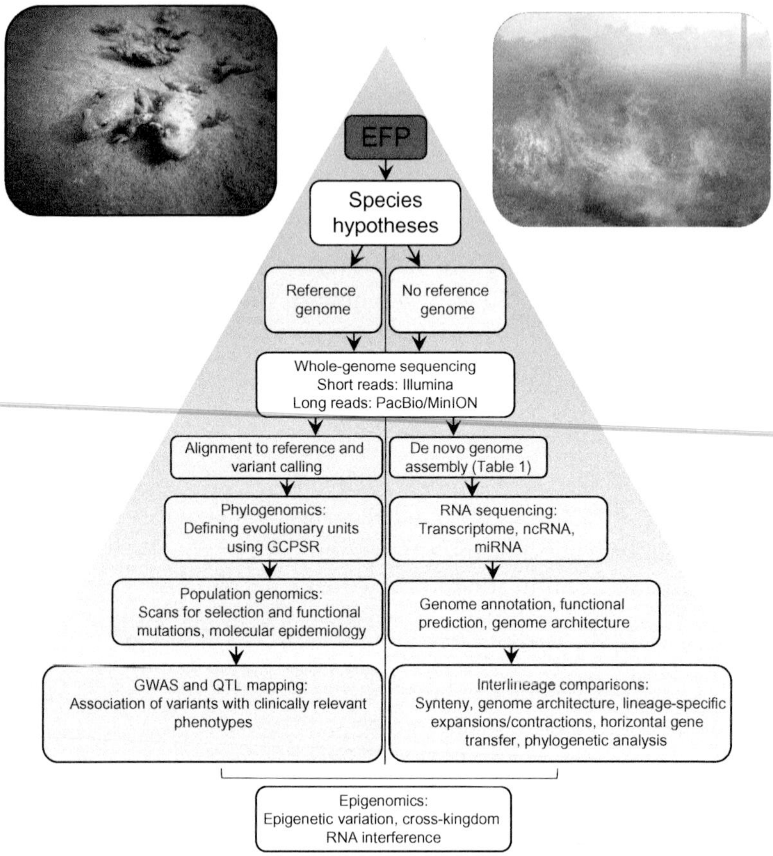

Fig. 1 A generalized workflow detailing the use of genomics to understand the genetic basis that underpins a novel EFP. *Images*: midwife toads with fatal chytridiomycosis caused by *Batrachochytrium dendrobatidis* (M.C. Fisher) and burning of a severely wheat blast (*Magnaporthe oryzae Triticum*) affected field in Meherpur district in Bangladesh, February 2016 (T. Islam, BSMR Agricultural University, Bangladesh).

of point mutations in one or more of the individuals), insertions and deletions (indels), and recombination (shuffling of sequences within and between genomes). Longer alignments and sequencing many times over (such as is often the case with next-generation and third-generation sequencing platforms) are required to identify additional features of genome variation. For example, changes in the depth of sequencing can suggest loss or gain of copy number variation (CNV) for single genes (gene duplication), regions (segmental aneuploidies), or entire chromosomal CNV (chromosomal aneuploidy; Fig. 2B) (Farrer, Henk, Garner, et al., 2013). Other

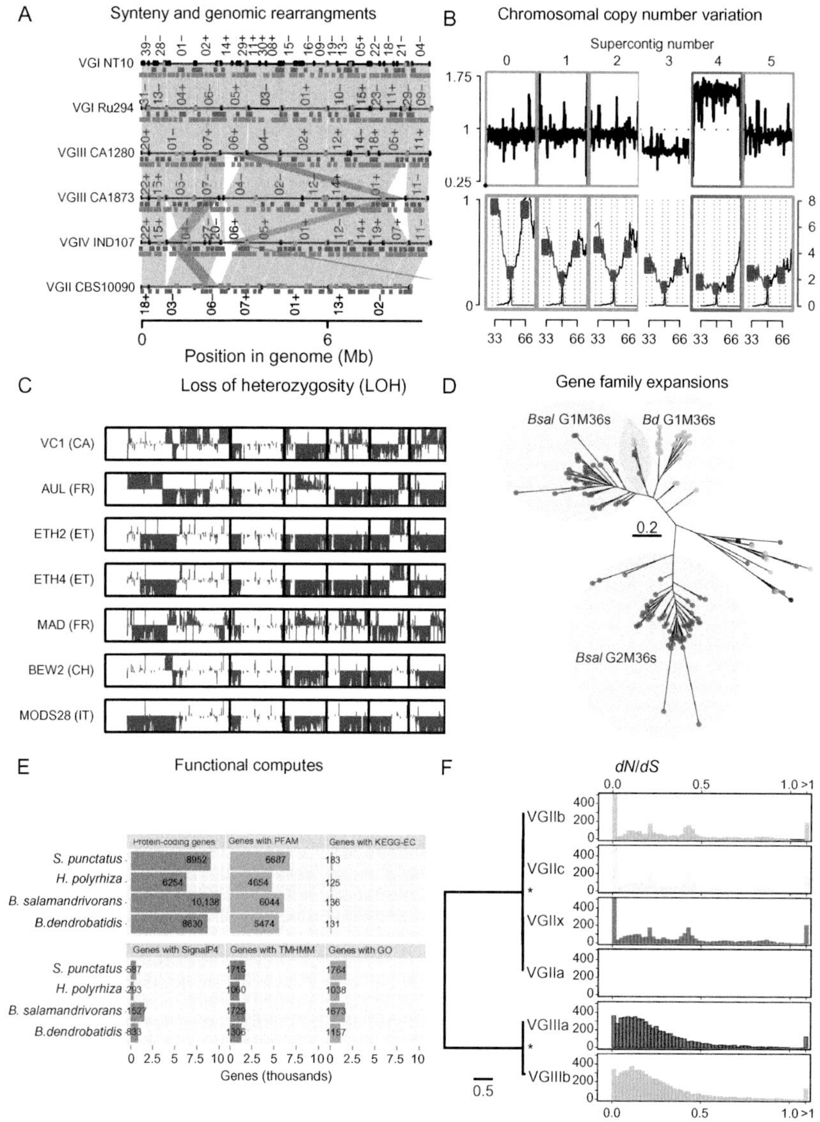

Fig. 2 Examples of genomic features that can be detected in EFPs. (A) Synteny and genomic rearrangements between and within two lineages of *C. gattii* (Farrer et al., 2015), (B) chromosomal copy number variation (CCNV) in *B. dendrobatidis* detected by average read depth from alignments (*top*) and allele frequencies (percent of bases agreeing with reference base vs tally in kilobases) (Farrer et al., 2011), (C) loss of heterozygosity in *B. dendrobatidis* detected using nonoverlapping sliding windows of SNPs minus heterozygous positions (*red*, predominately SNPs; *blue*, predominately heterozygous) (Farrer et al., 2011), (D) gene family expansion in *Batrachochytrium* spp. (Farrer et al., 2017), (E) gene annotation counts of gene types, and functional computes in *Batrachochytrium* spp. (Farrer et al., 2017), (F) measures of selection (i.e., d_N/d_S) across subclades of *C. gattii* using only fixed differences compared to VGIIa (Farrer et al., 2015).

examples include subsections of DNA that are reoriented to one another (inversions), subsections of DNA that occur in different locations in two individual genomes (translocations; Fig. 2A), genetic mosaics of two species in a single isolate (hybridizations) (Rhodes, Desjardins, et al., 2017), reduction of heterozygosity (gene conversion or loss of heterozygosity; Fig. 2C) (Farrer, Henk, Garner, et al., 2013), and changes to the ordering of genes (synteny). From these sources of genomic variation in fungal pathogens, epidemiological features of the outbreak can be inferred, virulence factors identified, and diagnostics and treatments devised.

Each source of genomic variation has unique and cumulative sources of uncertainty. These variants require careful detection and minimization, where possible. First, the quality of the sequence data can be highly variable between experiments, library-building protocols and different sequencing machines, containing low-throughput/depth sequencing, high levels of error in the base calls, or unexpected laboratory contamination (such as bacterial or host DNA). Uncontaminated high-quality samples may then be aligned to distantly related genomes resulting in decreasing accuracy of alignments and base calling. The quality of reference genomes themselves is variable, and they may contain inaccurate reference sequences (misassembled or containing sources of the prementioned errors), which can result in misleading comparisons. Second, variant calling from alignments against reference sequences may contain mistaken assumptions about ploidy, or inaccurately called bases. Alternatively, sequenced reads can be assembled into longer contiguous sections of chromosomes, which themselves can contain inaccurately assembled contigs or scaffolds. These are especially prone to occur over repetitive content or sequencing errors. From these assemblies, gene calling is often performed, which itself may include mistakes in intron/exon boundaries, and often absent or partial 5′ and/or 3′ untranslated regions (UTRs), for example. From comparisons of these gene predictions, analysis of patterns of natural selection could potentially identify unusually evolving genes that are artefacts caused by the aforementioned sources of errors. Fortunately, each of these errors has a range of hallmarks and remedial bioinformatics processes that can be used to ensure their accuracy or minimize those sources of error.

In the following sections, we will discuss the methods for identifying different sources of genomic variation with a focus on EFPs, and the manifestation of this variation within populations (population genetics approaches), and between populations (comparative genomics approaches). Importantly, we will be distinguishing between subgenomic approaches (PCR fingerprinting, microsatellites, restriction fragment length polymorphisms, etc.)

and whole-genomic approaches, focusing entirely on the latter. While subgenomic approaches are undoubtedly useful for characterizing EFPs (e.g., Hsueh et al., 2000; Mohammadi et al., 2015), approaches that are based on using whole genomes are increasingly being used for the detection and rapid characterization of novel pathogens (Hasman et al., 2014; Lecuit & Eloit, 2014). Indeed, beyond the identification of either a known or unknown fungus, the usage of full genome data provides far greater insights into the pathogens evolutionary history, population structure, and repertoire of virulence effectors.

2.1 Assemblies, Alignments, and Annotation

Many EFPs will be initially classified or identified based on their morphological traits, or host species, such as occurred with the amphibian-infecting chytrids *Batrachochytrium dendrobatidis* and *B. salamandrivorans* (Berger et al., 1998; Martel et al., 2013). Initially, subgenomic approaches using a taxonomic marker gene such as analysis of ribosomal DNA (rDNA) (Schoch et al., 2012) against global databases of known fungal sequences such as UNITE (Kõljalg et al., 2013) or the Ribosomal Database Project (RDP; Cole et al., 2014) are needed to define operational taxonomic units (OTUs). OTUs (also called "species hypotheses") are proxies for classically defined species and are used to ordinate the novel EFP taxonomically in the kingdom Fungi—bearing in mind however that the *Microsporidia* do not have the canonical rDNA structure (Dong, Shen, Xu, & Zhu, 2010). Subsequently, genome assembly (assembly de novo) is needed to provide a thorough examination of its genetic makeup and relatedness to known species.

Ideally, assembling a genome de novo will be preplanned, by implementing a long-read technology (such as third-generation sequencing platforms Oxford Nanopore's MinION, or Pacific Biosciences' single molecule real-time sequencing). Alternatively, multiple sequenced paired-end libraries of Illumina with short- and long (also known as "jump") insert sizes can be used by assembly tools optimized for such datasets (such as Allpaths; Butler et al., 2008). Other options for generating a high-quality assembly are the use of Fosmid libraries (fragmenting the genome then cloning into *E. coli*, and sequencing individual libraries separately) or constructing an optical map (a high-resolution restriction map of the genome to aid in assembling subsections of the genome).

Many assembly tools have been developed (Table 1), which may be optimized for different sequencing technologies (e.g., Allpaths for two libraries of paired-end Illumina (Butler et al., 2008), Canu for long reads such as

Table 1 Names, Versions and Descriptions of Popular Genomic Tools Used for Assembly de novo, Pairwise and Multiple Alignment, Gene Annotation and Variant Calling in EFPs

Purpose	Tool	Current Version	Input/Notes	Citations
Assembly	ALLPATHS-LG	v4.7	Two Illumina fragment (paired-end) libraries	Gnerre et al. (2011)
	Canu	v1.4	overlapping for noisy, long reads such as MinION or PacBio	Koren et al. (2017)
	DISCOVAR de novo	N/A	Single Illumina fragment (paired-end) library	Love, Weisenfeld, Jaffe, Besansky, and Neafsey (2016)
	Platanus	v1.2.4	De novo assembly of highly heterozygous genomes	Kajitani et al. (2014)
	SGA	N/A	Memory efficient tool for large genomes	Simpson and Durbin (2012)
	SOAPdenovo	v2	One or more single and/or paired-end libraries	Li et al. (2010)
	SPAdes	v3.5	Single-cell and standard (multicell) libraries, haploid or diploid	Bankevich et al. (2012)
	Trinity	v2.3.2	RNAseq data (optionally genome guided)	Haas et al. (2013)
Alignment	BLAST	Blast+	Fast searches against large databases	Altschul, Gish, Miller, Myers, and Lipman (1990)
	BLAT	N/A	Fast searches and connects homologous hits	Kent (2002)
	Bowtie	v2.3.0	Supports gapped, local, and paired-end alignment modes	Langmead and Salzberg (2012)

Table 1 Names, Versions and Descriptions of Popular Genomic Tools Used for Assembly de novo, Pairwise and Multiple Alignment, Gene Annotation and Variant Calling in EFPs—cont'd

Purpose	Tool	Current Version	Input/Notes	Citations
	BWA-mem	v0.7.15	Low-divergent sequences against a large reference genome	Li and Durbin (2010)
	HISAT2	v2.0.5	DNA or RNA to a population of genomes	Pertea, Kim, Pertea, Leek, and Salzberg (2016)
	MAFFT	v7	High speed multiple sequence alignment program	Katoh and Standley (2013)
	MAVID	v2.0.4	Multiple alignment program for large genomic sequences	Dewey (2007)
	MUMmer	v3.22	Aligns entire genomes	Kurtz et al. (2004)
	MUSCLE	v3.8.31	Multiple sequence alignment	Edgar (2004a)
	STAR	v2.5	Spliced transcripts (RNAseq) to a reference	Dobin et al. (2013)
	TBA MULTIZ	12109	Aligns highly rearranged or incompletely sequenced genomes	Blanchette et al. (2004)
Annotation	Augustus	v2.5.5	Ab initio gene-prediction program for eukaryotes	Stanke et al. (2006)
	EVM	v1.1.1	Combines diverse evidence types into single gene structures	Haas et al. (2008)
	FGENESH	v2.1	HMM-based ab initio gene-prediction program	Salamov and Solovyev (2000)
	GeneID	v1.4.4	Predicts genes in anonymous genomic sequences	Blanco, Parra, and Guigó (2007)

Continued

Table 1 Names, Versions and Descriptions of Popular Genomic Tools Used for Assembly de novo, Pairwise and Multiple Alignment, Gene Annotation and Variant Calling in EFPs—cont'd

Purpose	Tool	Current Version	Input/Notes	Citations
	GenemarkHmmEs	v2.3	Unsupervised training for identifying eukaryotic protein-coding genes	Lukashin and Borodovsky (1998)
	GlimmerHmm	v3.02b	gene finder based on interpolated Markov models (IMMs)	Majoros, Pertea, and Salzberg (2004)
	PASA	v2	spliced alignments of ESTs and RNAseq to model gene structures	Haas et al. (2008)
	RNAmmer	v1.2	Consistent and rapid annotation of ribosomal RNA genes	Lagesen et al. (2007)
	SNAP	2013	Ab initio gene finding program	Korf (2004)
	tRNAscan	v1.3.1	Transfer RNA detection	Lowe and Eddy (1997)
	Wise2 (GeneWise)	v2.4	Predicts gene structure using similar protein sequences.	Birney, Clamp, and Durbin (2004)
Variant calling	Biscap	v0.11	Variants called from Pileup format using binomial probabilities	Farrer, Henk, MacLean, Studholme, and Fisher (2013)
	FreeBayes	v0.9.10	Bayesian haplotype-based polymorphism discovery and genotyping	Garrison and Marth (2012)
	GATK	v3.7	Collection of tools with a focus on variant discovery	McKenna et al. (2010)
	Pilon	v1.5	Corrects draft assemblies and calls sequence variants	Walker et al. (2014)

PacBio or MinION (Koren et al., 2017)), sequencing libraries (e.g., Spades for DNA, Bankevich et al., 2012; Trinity for RNA, Haas et al., 2013); high levels of heterozygosity (e.g., Platinus (Kajitani et al., 2014), estimated ploidies or repeat content, scalability, or computational speeds given differing computational resources. Reviews of methodologies and tools include Assemblethon 2 (Bradnam et al., 2013) and Genome Assembly Gold-standard Evaluations (Salzberg et al., 2012). However, most tools make use of one of two underlying algorithms: Overlap of reads to construct contiguous stretches of sequences, or *k*-mers (subread sequences of length *k*) organized into deBruijn graphs. In both cases, the longest path through the graph is considered correct, and bubbles (loops caused by repeats) cut or removed. Usually the initial reads are organized into contigs, which are separately orientated and connected to one another into scaffolds (connected by Ns, representing ambiguous bases of the estimated length between the two contigs). The finished assembly should be assessed using a variety of metrics, as the result may be suboptimal or inaccurate—thereby negatively impacting any downstream analysis.

A genome assembly will usually aim to represent the nucleotide sequence of a single isolate, separated into individual chromosomes. However, it is always (unless from single-cell sequencing), a consensus from a colony or even population of cells, meaning that the assembly represents a range of individual genotypes. This is especially relevant when fungal cells are heterokaryotic, containing multiple nuclei such as is the case with filamentous ascomycetes and arbuscular mycorrhizal fungi (Pawlowska & Taylor, 2004). Sometimes, such a consensus may even be intentional such as with pan-genomes of *Saccharomyces cerevisiae* (Song et al., 2015). In nonhaploid EFPs, including many *Candida* isolates that represent over 50% of human mycoses (Nucci & Marr, 2005), the assembly will consist of a consensus (and arbitrary connection) of haplotypes. Therefore, even a nonerroneous assembly could easily be wrongly interpreted for various downstream analysis including genetic variation, linkage disequilibrium, and recombination.

A simple metric to assess the quality of genome assemblies is the total assembly size—which itself can be informative in identifying contaminants and miss-assemblies (Studholme, 2016). For example, a genome that is far larger or smaller than expected for the genus can indicate multiple sequenced species (i.e., contamination with other organisms), high error rates in the sequencing, highly repetitive sequence, low sequencing depth, or unsuitable parameters. To control quality, an important step is to BLAST (Altschul et al., 1990) the scaffolds against the online or local nonredundant database in order to identify whether contamination by another species is contributing to an erroneously assembled genome. Such a search, in addition to non-uniform GC content and other assembly metrics, can also be computed by such tools as the Genome Assembly Evaluation Metrics and Reporting (GAEMR) package (http://software.broadinstitute.org/software/gaemr/) or REAPR (Hunt et al., 2013). Given sufficient evidence of contamination, it is often beneficial to reassemble the reads after excluding any reads aligning (and therefore originating from) the source of the contamination—which can lead to improved contiguity and accuracy by excluding erroneous chimeric genomic regions. An assembly should then be assessed for its contiguity; a common measure of assembly contiguity is its N50 (meaning 50% of the assembly is in contigs of this length or larger). Similarly, N90 and N25 are sometimes also reported for assemblies. The N50 can be normalized for comparisons between multiple assemblies by using the estimated genome size instead of total assembly size (denoted NG instead of N (Bradnam et al., 2013)). These metrics can however both be misleading given, for example, a single very long scaffold above the N(G)50 length, which will then ignore the remaining assembly which may occur as highly fragmented scaffolds. Another proposed metric is the proportion of the assembly that has a length of at least the average eukaryotic gene (2.5 kb) (Bradnam et al., 2013) and will therefore be approximately the minimum length necessary for annotation—which may be the primary use for the assembly.

In the past year, genome assemblies from EFPs have included the chytrid fungus *B. salamandrivorans*, which is devastating fire salamanders in Europe (Farrer et al., 2017; Martel et al., 2014). Here, the genome was assembled using Illumina paired-end reads and SPAdes (Bankevich et al., 2012) into a draft assembly, revealing a substantial increase in genome length and expansion of metalloprotease M36 involved in skin destruction compared with its closest relative *B. dendrobatidis* (Farrer et al., 2017) (Fig. 2D). The ascomyceteous fungus *Sarocladium oryzae* is emerging as major threat for rice production (Bigirimana, Hua, Nyamangyoku, & Höfte, 2015) and was assembled by Illumina paired-end reads and SPAdes assembler (Bankevich et al., 2012), revealing a range of expanded gene families including the pathway for steroidal antibiotic helvolic acid thought to be a pathogenicity determinant (Hittalmani, Mahesh, Mahadevaiah, & Prasannakumar, 2016). The ascomyceteous fungus *C. auris* is emerging as a multidrug-resistant human pathogen in intensive care settings across the world and has been Illumina sequenced and assembled using Velvet (Zerbino & Birney, 2008) and scaffolded using SSPACE (Boetzer & Pirovano, 2014) and more recently using Oxford Nanopore Technology and Illumina (Rhodes et al., 2017). These assemblies are now being used to determine the genetic mechanisms that underpin the multidrug-resistant nature of this species to fluconazole, voriconazole, amphotericin B, and caspofungin (Sharma, Kumar, Meis, Pandey, & Chowdhary, 2015).

Short- or long-read alignments against a presequenced reference are more commonly used for NGS datasets than assemblies, providing that a suitable reference sequence is already available. There are several reasons for opting for alignment over multiple assemblies. First, it is almost always quicker in terms of computation time. Second, alignments negate the necessity to identify orthologous regions of the genome needed to make comparisons. Indeed, orthology finding from assemblies is hindered by the necessity for relaxed sequence similarity thresholds in global sequence alignment algorithms. Furthermore, the steps from alignment to variant calling, gene cataloging, and selection analysis are well established. However, it needs to be born in mind that suboptimal tools, parameters, or quality checks can lead to misleading results.

Sequence alignment tools arrange two (i.e., pairwise) or more (i.e., multiple alignment) nucleic acids or protein sequences, with the intent of identifying regions of similarity that may indicate functional, or evolutionary relationships. Pairwise alignment algorithms are often based on either the Smith–Waterman algorithm (local/subsequences) or the Needleman–Wunsch algorithm (global/complete sequences). Both create a substitution matrix based on a scoring scheme (e.g., +3 for match, −2 for mismatch, −2 for indel) for each nucleotide or amino acid, and then trace back from the highest score. Tools such as EMBOSS' Needle and Water implement these algorithms directly, while others use them for extending seeds (prior screen for short matches), e.g., BWA-mem (Li & Durbin, 2010).

Many heuristic alignment algorithms and tools have been developed to improve on the speed of the Smith–Waterman and Needleman–Wunsch algorithms, such as the Basic Local Alignment Search Tool (BLAST) (Altschul et al., 1990) and the BLAST-like alignment tools (BLAT) (Kent, 2002), which removes low-complexity regions and makes *k*-letter "words" from the query sequence for searching the database of sequences using a scoring matrix. Mash is another recent and promising ultra-fast genomic distance estimation tool (Ondov et al., 2016). BLAST and BLAT are most commonly used for querying against a very large database, while NGS aligners such as BWA-mem (Li & Durbin, 2010) or Bowtie2 (Langmead & Salzberg, 2012, p. 2) are optimized for memory and time-efficient alignment of a huge number of reads to a genome (Table 1). Global alignment tools for whole genomes include MAVID (Dewey, 2007), MUMMER (Kurtz et al., 2004), and TBA MULTIZ (Blanchette et al., 2004), which generally identify seeds that are joined (or removed) to form anchor regions for the final alignment. Others such as STAR (Dobin et al., 2013) are optimized for alignment of spliced transcripts (e.g., RNAseq data) to a genome.

To determine the overall accuracies of an input read dataset, alignment, and SNP calling method, one method is a comparison of false discovery rates (cFDR) (Farrer, Henk, MacLean, et al., 2013). In this procedure, a specified number of random single base positions in the reference sequence are randomly changes to one of the other three possible nucleotides. Sequence data is then aligned to this modified reference sequence, variant-calling performed, and a comparison made to those known changes to ascertain the overall accuracy. Multiple alignments or variant-calling tools or parameters can be used iteratively to identify the most suitable bioinformatics pipeline. Another metric for assessing the alignment quality is to assess the coverage across the genome by visualization. Some SNP calls such as GATK (McKenna et al., 2010) include the ability to perform local indel realignment (realign reads around indels). Importantly, multiple BAMs relating to related isolates (such as parent and progeny, or those closely related) should be realigned together to avoid newly introduced discrepancies, i.e., regions where one isolate is realigned at a region and another is not. However, this process was recently removed from the best practices of HaplotypeCaller but retained for UnifiedGenotyper. Other tools such as MUMSA (Lassmann & Sonnhammer, 2005) compare multiple alignments using an average overlap score and a multiple overlap score to assess the accuracy of alignments. Both alignments and assemblies can be improved via preprocessing of reads such as by removing low quality reads or 3′ ends.

There are a diversity of potential variant calling tools available, which are mostly used postalignment. Many SNP callers consider homozygous and heterozygous (biallelic) sequences, but often not trialleles, for example, which can be present at low numbers in genomes that exhibit aneuploidy, or polyploidy such as that which marks *B. dendrobatidis* (Farrer, Henk, Garner, et al., 2013). GATK's HaplotypeCaller and UnifiedGenotyper (McKenna et al., 2010) currently require a ploidy to be given as a parameter to inform its genotype likelihood and variant quality calculations. This prior setting is therefore poorly suited for the investigation of aneuploidy in EFPs. Recently the cancer genome variant calling tool MuTect2 (Cibulskis et al.,

2013) has been incorporated into the GATK and allows for a varying allelic fraction for each variant—which could provide a work-around for polyploidy or aneuploidy. Separately, FreeBayes (Garrison & Marth, 2012) calls variants based on a Bayesian statistical framework and is also capable of modeling multiallelic loci in sets of individuals with nonuniform copy number. A computationally inexpensive variant caller is BISCAP (Farrer, Henk, MacLean, et al., 2013), which uses binomial probabilities for an expected error rate following alignment. The tool Pilon (Walker et al., 2014) calls variants of multiple sizes, including very large insertions and deletions, while also able to use them for correcting draft assemblies. Indeed, many other SNP callers have been developed, which may be tailored to data types or expected levels of variation. The number of possible tools and their rate of development make benchmarking an issue that needs to be frequently readdressed to ensure their accuracy and therefore all the downstream analysis based on it.

For any newly sequenced genome including those of EFPs, one of the key features to query will be its gene content, which require gene prediction and annotation protocols. Many tools have been developed (Haas, Zeng, Pearson, Cuomo, & Wortman, 2011) and may be more or less suited to different genomes, genome fragmentation, repeat content, or gene characteristics such as intron lengths. Genomic research institutes such as the Broad Institute of MIT and Harvard, or the U.S. Department of Energy Joint Genome Institute (DOE JGI) automate their pipelines. However, in practice, only partial automation is obtainable due to genomic or gene idiosyncrasies, which require different tools, different parameters, or at a minimum, a manual check of certain gene-prediction outputs. Further, various methods are required for the prediction of protein-coding genes compared to RNA genes, to determine whether the genes are located on the nuclear or mitochondrial genome, and it is usually necessary to separately identify repetitive regions in the genome. However, from a well-annotated genome sequence, numerous aspects of an EFPs biology can be determined such as biological pathways, life cycle, mechanisms of pathogenicity, and its relationship to other species through ortholog detection.

The first step usually taken for gene annotation is repeat identification and masking (replacing sequence with the ambiguity IUPAC code "N"). Repetitive sequences within genomes constitute a range of functional and nonfunctional (in the evolutionary conserved sense) regions of the genomes. For example, if a genome assembly is finished to the level of whole linear chromosomes, the ends will contain tandem (consecutive) repeat sequences found within telomeres, ranging from 5-mer to 27-mer repeated several thousand times, which both protect the end of the chromosome from deterioration, chromosomal fusion, or recombination, and as a mechanism for senescence and triggering apoptosis. Other tandem repeats are found in centromeres, which are involved in kinetochore formation during mitosis. Centromeres in fungi are diverse sequences ranging from a few kilobases in *Candida albicans* (Sanyal, Baum, & Carbon, 2004) up to 75 kb in *Schizosaccharomyces pombe* (Fishel, Amstutz, Baum, Carbon, & Clarke, 1988), and due to their diverse sequences, are best detected by the binding of the specialized nucleosomes that contain the centromere-specific histone H3, CenH3. Interestingly, *Neurospora* centromeres are composed of degenerate (following repeat-induced point (RIP) mutations) transposons, mostly retrotransposons, and simple sequence repeats (Smith, Galazka, Phatale, Connolly, & Freitag, 2012). Tandem repeats such as those found in telomeres and centromeres are grouped into microsatellites (also known as short tandem repeats or simple sequence repeats), which comprise 2–5mers, and minisatellite repeats comprising 10–50mers. Micro- and minisatellites are useful as genomic markers and are also studied for their role in disease. Tools such as the tandem repeat finder (Benson, 1999) and microsatellite identification tool (Thiel, Michalek, Varshney, & Graner, 2003) can be used to identify tandem repeats, while the Tandem Repeat Database is a public repository of those already identified (Gelfand, Rodriguez, & Benson, 2007).

Repetitive regions of a genome also include mobile DNA elements such as retrotransposons, DNA transposons, and miniature inverted-repeat transposable elements. Transposable element content varies in the fungal kingdom from between 3% (e.g., *Aspergillus nidulans*, *Aspergillus fumigatus*, and *Aspergillus oryzae*) and 10% (e.g., *Neurospora*, *Magnaporthe*) (Galagan, Henn, Ma, Cuomo, & Birren, 2005), but also as much as 76.4% for species such as the ascomycetous *Blumeria graminis f.* sp. *hordei* (Barley powdery mildew) (Amselem, Lebrun, & Quesneville, 2015; Spanu et al., 2010). Retrotransposons usually have long terminal repeats (LTRs) encoding a reverse transcriptase necessary to convert their transcribed RNA back to DNA which is inserted back into a new position of the genome. Others belong to the long interspersed nuclear elements encode a reverse transcriptase and an RNA polymerase II promoter, but lack LTRs. Retrotransposons lacking reverse transcriptase genes and relying on other mobile elements for transposition are called short interspersed elements. DNA transposons, in comparison, do not involve an RNA intermediate and usually encodes transposase enzymes in order to bind and incorporate itself into a new position in the genome. These genes can be erroneously incorporated into gene models during prediction and provide nonuniform numbers of predicted genes compared with closely related isolates or species.

Following repeat masking by such software as RepeatMasker (http://www.repeatmasker.org/), protein-coding genes are predicted by both ab initio (based on sequence provided only) and homology based (based on similarity to known sequences). Ab initio methods rely on probabilistic models, such as generalized hidden Markov models (GHMMs) or neural networks (NN) to combine information from sequences that indicate the presence of a nearby gene (promoters and other regulatory signals) or protein-coding sequences. Most have individual models to assess, for example, splice donor sites (5′ end of the intron), splice acceptor sites (3′ end of the intron), intron and exon length distributions, open reading frame length, and transcriptional start and stop sites. Programs such as Augustus (Stanke et al., 2006), FGENESH (Salamov & Solovyev, 2000), GeneID (Blanco et al., 2007), GeneMark (Besemer & Borodovsky, 1999), GlimmerHMM (Majoros et al., 2004), and SNAP (Korf, 2004) are used for ab initio gene calling by first generating a training set (taking the highest scoring predictions from their GHMM) and then running across the genome sequence. Others such as GeneMark.hmm-ES (Lukashin & Borodovsky, 1998) is self-training. While any one of these methods could provide a modest initial assessment of gene content, it is worth running a number of tools in order to get a greater range (and therefore sensitivity) of predictions.

Homology-based (empirical) gene finding methods search for sequences that have sequence similarity to previously found genes in other organisms. These methods provide evidence for gene locations, which are both standalone, and compliment ab initio gene finding methods. This requires translating regions of the genomes (ideally in all six possible reading frames), using, for example, a translated BLAST (tblastn) against a database such as the nonredundant sequences from GenBank (Benson et al., 2012) and/or Uniprot (Wu et al., 2006). While this step is very computationally expensive, it provides likely protein hits which can then be assessed more rigorously. One such package for providing spliced gene models from these hits is Wise2 (Birney et al., 2004).

Transcript sequences from the same organism (such as RNAseq, expressed sequence tags/subsequence of cDNA) provide very strong evidence for gene structures in the genome sequence. A common first step is to assemble de novo the RNAseq reads into longer transcripts, via programs such as Trinity (Haas et al., 2013). The accuracy of Trinity can be improved with strand-specific RNAseq libraries, the genome-guided parameter (both where available), and *k*-mer depth should be increased to at least 2 for improved specificity. Next, tools such as the Program to Assemble Spliced Alignment (PASA) (Haas et al., 2008) map these assembled transcripts (or unprocessed RNAseq reads) to the genome using GMAP (Wu & Watanabe, 2005) or BLAT (Kent, 2002), filtering alignments, grouping alternatively spliced isoforms and output candidate gene structures based on the longest open reading frame (FASTA file and GFF3). In addition, PASA can update prepredicted gene structures.

Finally, tools such as EvidenceModeller (EVM) (Haas et al., 2008) or Maker (Cantarel et al., 2008) assess the evidence provided for gene calls from a range of gene predictions (ab initio, homology-based, transcript data), and output a single set of consensus gene structures. Maker also contains a complete pipeline for identifying repeats, aligning ESTs and proteins to the genome, and ab initio gene prediction, before assessing their evidence. The final gene set following evidence assessment should be given a final check for various issues that may remain (coding length nonmodular 3, genes >50 amino acids, genes with in-frame stops, contain Ns indicative of spanning contig gaps, covering predicted repeats, etc.) prior to finalizing these predicted protein-coding genes with annotation, and gene identifiers such as unique locus tags.

The correct genetic code should be used throughout the entire process of predicting protein-encoding genes. The standard code (which should be the default for most tools) is suitable for most fungal nuclear genomes, although some species such as various *Candida* species in the CTG Clade have CUG codons that encode the amino acid serine instead of leucine (Santos, Keith, & Tuite, 1993), and therefore require the alternative Yeast nuclear code (Osawa, Jukes, Watanabe, & Muto, 1992). Mitochondrial genomes all use separate nonstandard codes, a difference that needs to be accounted for when translating genes in silico as part of the operation of these gene-prediction methods.

Genes that encode tRNA and rRNA are normally found in large numbers throughout a well-assembled genome. rDNA encoding rRNA are usually found entirely occupying large sections of one or more chromosomes, comprising both structural rRNA for small (18S) and large (5S, 5.8S, and 28S) components of ribosomes separated by internal transcribed spacer (ITS) units. These regions of the genome tend to be among the most poorly resolved due to their repetitive nature—culminating in noncomplete regions that underestimates their number, but result in a region of unusually high depth of coverage following read alignment. Indeed, the rRNA completeness of a genome can be a proxy of genome assembly quality. Separately, ITS1 and ITS2 spacer regions tend to be useful for diagnostic PCR, fungal abundance (qPCR), and even rudimentary phylogenetics due to their ubiquity and genetic diversity in most fungal genomes (however, excluding the microsporidia) (Schoch et al., 2012).

Like protein sequences, RNA families have some level of conserved sequence, but a more highly conserved secondary structure, which is more integral to its function than that imposed by its primary sequence. Unlike protein sequences, ncRNA lack all features apparent from codons and gene structures (e.g., start, termination, codon bias, acceptor and donor splice sites, etc.) that are used for gene prediction, making the structure not only more relevant for its function (or predicted pseudogenization) but also for its prediction. RNA secondary structure arises from base-pairing interactions resulting in stems and loops, e.g., the cloverleaf structure of tRNA comprising several stem-loops, or the pseudoknot also comprising several stem-loops in the RNA component of telomerases (which incidentally is essential for telomerase activity (Chen & Greider, 2005)).

RNA secondary structure prediction based on the lowest free energy structure is a nondeterministic polynomial-time hard (NP-hard) problem (Lyngsø & Pedersen, 2000), and therefore tools based on heuristic algorithms are required for de novo RNA secondary structure (such as HotKnots (Ren, Rastegari, Condon, & Hoos, 2005) and ProbKnot (Bellaousov & Mathews, 2010)). However, prediction in a new genome is usually based on previously identified and characterized ncRNA families, which are often stored in covariance models (CMs) (analogous to hidden Markov models (HMMs)) describing both secondary structure and primary sequence consensus (Eddy & Durbin, 1994). For example, the INFERNAL (Inference of RNA Alignment) software (Nawrocki & Eddy, 2013, p. 1) searches a custom or collection of ncRNA CMs such as the RNA family database (RFam) (Griffiths-Jones, Bateman, Marshall, Khanna, & Eddy, 2003) comprising tRNAs, small nuclear RNAs (snRNA), and small nucleolar RNAs (snoRNAs). snRNAs are involved in splicing and RNA processing, while snoRNAs either methylate (C/D box snoRNA) or pseudouridylate (H/ACA box snoRNAs) other RNAs (rRNA, tRNA and snRNA). Separately, the CM-based tRNAscan-SE (Lowe & Eddy, 1997) can identify tRNA genes with extremely high sensitivity and specificity, and the HMM-based

RNAmmer predicts rRNA genes in the nuclear genome (Lagesen et al., 2007).

It is important to assess the quality of the final gene calls for multiple potential erroneous calls, such as genes that have a length that is not modulus 3 (i.e., sequences not entirely comprised of codons), genes with STOP codons within the sequence, or those ending without a STOP codon are likely errors. Other issues can include very distant exons (e.g., >15 kb) from the remainder of the gene will be likely inaccurate. Gene calls that are supported by only one of multiple gene-prediction methods may also be more dubious than those supported by multiple methods and tools. A simple postannotation metric is the total number of genes predicted. Too many or too few predicted genes for a given genus or species can be indicative of a failed step in the annotation pipeline, or suggest a problem with the genome assembly, e.g., species contamination. In addition to gene count, the completeness of gene sets can be assessed by the coverage of conserved gene sets such as CEGMA (Parra, Bradnam, & Korf, 2007) and BUSCO (Simão, Waterhouse, Ioannidis, Kriventseva, & Zdobnov, 2015), which will give a good indicator of the quality of both the annotation and the assembly protocols. The measure of gene completeness should complement other metrics of genome assembly, and be performed before functional predictions and other downstream analyses are performed.

2.2 Functional Predictions and Gene Family Expansion

Functional genomics describes the relationship between an organism's genome and its phenotype, and is widely used to determine novel pathogenicity-related traits in EFPs. There are numerous experimental ways to study these traits including gene knockouts, gene silencing, transposon or chemical mutagenesis, and QTL mapping. Computational methods for identifying pathogenicity-related gene functions includes Genome Wide Association Studies (GWAS), which commonly compares two large groups of individuals that differ by a pathogenicity-related trait, and to then search for a significant association (low *P*-value from a chi-squared test of the odds ratio). GWAS have been used to successfully identify a wide range of candidate genes and alleles implicated in disease or pathogenicity-related phenotypes, including in a broad range of fungal applications (Plissonneau et al., 2017). For example, a putative Avirulence gene (virulence factors that are detected by the host, and thereby prevent or reduce disease) was recently detected using a GWAS of *Zymoseptoria tritici*, the ascomycetous fungi responsible for septoria lead blotch in wheat (Hartmann, Sánchez-Vallet, McDonald, & Croll, 2017).

GWAS have several limitations including the necessity for very large sample sizes, which is commonly not available for EFPs, and the need to account for the large numbers of multiple comparisons that inevitably lead to false associations. Furthermore, many populations of fungal pathogens contain a large clonal component to their life cycle—with the consequence that variants are physically linked on the chromosome (high linkage disequilibrium). Clonality therefore impinges on the ability to identify individual variants that are associated with a trait. Finally, specific functions of a protein-coding gene (e.g., those encoding chloride channels) are relatively easy to predict, compared with predicting phenotypes and pathologies linked to mutations or protein misfolding (e.g., those causing cystic fibrosis in humans). This section will focus on ab initio and in silico methods of functional genomics that rely only on a single or very few isolates—such as might be available from the outbreak of an EFP.

Following (or as part of) gene prediction, functional annotation can be assigned to each protein-coding gene, and thereby provide a prediction of its function in the organism. Perhaps the most common method to do this is to assign Protein Family (PFAM) domains (Finn et al., 2014), which as of the current v31.0 (10/2016) has defined 16,712 protein families, and to a lesser extent, assigning TIGRFAM domains (Haft, Selengut, & White, 2003), which as of the current v15.0 (10/2014) has defined 4488 protein families—both of which are generated using HMMs. Each protein family is composed of one or more functional regions termed domains—which are found in multiple proteins and protein families. Both PFAM and TIGRFAM databases provide profile HMMs for each protein families, which are built from multiple sequence alignments and are searched either online via web servers (Finn, Clements, & Eddy, 2011) or local copies using the HMMER3 software (Eddy, 2011). Separately, the Kyoto Encyclopedia of Genes and Genomes (KEGG) database (Kanehisa & Goto, 2000) can be searched using the predicted gene sequences using a BLASTx (and a suitably stringent *e*-value, i.e., $e < 1 \times 10^{-10}$) to identify various functional information on gene functions, their role in biological pathways and cellular processes. Any successful matches with sufficiently stringent *e* values can provide compelling evidence toward the function of that gene. However, not all families or domains are contemporaneously informative regarding the function. For example, many domains of unknown function (DUF) are present in the database, which have been identified as a conserved domain across multiple species, but no known function has yet been identified.

Gene Ontologies (GO) provide a parallel and complimentary gene prediction along with PFAMs/TIGRFAM/etc. assignment. GO terms represent a controlled vocabulary and defined set of relationships between them, as part of the Open Biomedical Ontologies project by the National Center for Biomedical Ontology (NCBO). GO terms cover three domains: cellular components, molecular functions, and biological processes, for which a given gene is often assigned multiple terms, ranging from the very specific (low hierarchical/child terms) such as molecular function GO:0004375 (glycine dehydrogenase (decarboxylating) activity), to the very generic (high hierarchical/parent terms) such as molecular function GO:0003824 (catalytic activity). There are a wide range of tools for working with sets of GO terms, including Blast2GO (Conesa et al., 2005), which uses a stringent BLAST ($e < 1 \times 10^{-10}$) to identify genes with assigned GO terms, which can then be reassigned. Once assigned, GO terms can be very useful for predicted function, grouping genes into functionally relevant categories and ultimately performing enrichment statistical tests between groups of genes.

Some functions such as the secretion signal/peptide found at the N-terminus of newly synthesized proteins destined for the secretion pathway are best predicted by its biochemical properties rather than its poorly conserved primary sequence alone i.e., via sequence similarity or homology. SignalP is a popular tool that predicts the presence of type I signal peptidase cleavage sites from preprotein sequences in bacteria, archaea, fungi, plants, and animals (Petersen, Brunak, von Heijne, & Nielsen, 2011; Tuteja, 2005). Conversely, type II and type IV signal peptidases are restricted to prokaryotes and require prediction by other methods. In SignalP 3.0 and 4.0, type I signal peptidase cleavage sites are detected by neural networks, which are trained on real and negative data from SwissProt (Bairoch & Apweiler, 2000). The two neural networks used in SignalP recognize cleavage sites and determine if a given amino acid belongs to the signal peptide, respectively. SignalP is also informed by filtering propeptide cleavage sites, window length across the protein, and a discrimination score (D-score). The authors of SignalP also assessed the isoelectric point (pH(I); the pH at which the protein carries no net electrical charge) difference between the signal peptide and mature protein, which they found to be distinct in prokaryotes, but not eukaryotes—possibly owing to the much shorter length in eukaryotes. SignalP 4.0 updates the method by distinguishing between signal peptides and N-terminal transmembrane helices, which can be incorrectly identified. Limitations with SignalP include imperfect sensitivity and specificity (albeit the best method from their own comparisons to other tools and methods).

Some proteins predicted to have a signal peptide may nevertheless be retained intracellularly, i.e., in the endoplasmic reticulum or Golgi. For example, if the protein contains a C-terminal ER retention signal (KDEL or KKXX sequence), or via protein–protein interactions in the Golgi, the protein will not become extracellular (Banfield, 2011; Stornaiuolo et al., 2003). Furthermore, there are additional nonclassical secretion mechanisms in eukaryotes, such as via specific membrane transporters (Nickel & Seedorf, 2008). Tools such as SecretomeP (Bendtsen, Jensen, Blom, Von Heijne, & Brunak, 2004) predict secretory proteins that lack an N-terminal signal peptide. However, this method is tailored primarily to mammalian proteins, and when recently applied to four chytrid genomes, most proteins were identified as being nonclassically secreted (6523/10,128 *B. salamandrivorans* genes; 4478/8644 *B. dendrobatidis* genes; 4581/8952 *Spizellomyces punctatus* genes; 2991/6254 *Homolaphylictis polyrhiza* genes). This finding suggests that the mammalian-trained pipeline is, at least in this case, overpredicting nonclassical secretion motifs and needs to be retrained specifically to the fungal secretome (Farrer et al., 2017).

Secreted proteins are often of paramount importance to pathogens in acquiring nutrients, and interactions with the environment and host. An illuminating example of this are virulence effectors, which are secreted either into the environment or directly into the host, where they selectively bind to a host protein to regulate or modify its intended function (Hogenhout, Van der Hoorn, Terauchi, & Kamoun, 2009). Effectors are produced by a wide range of organisms including many fungal and bacterial pathogens, but also some animals (parasitic nematodes), as well as protists (*Plasmodium* sp. and Oomycetes). For example, effectors may encode proteins that target host defense mechanisms to enable the microbe to gain access to the host cell or avoid detection (either innate or acquired immunity, for example). One example is the gene AVR3a (belonging to a group that have an RXLR or RXLQ motif, and collectively known as RXLRs), which is found in the *Phytophthora* genus. AVR3a (specifically AVR3aKI that contains amino acids C19, K80, and I103) causes suppression of a hypersensitive response (apoptosis) in potatoes that lack the necessary resistance gene R3a (Bos et al., 2006), thereby facilitating its initial biotrophic stage of growth. Changes in the genetic backgrounds upon which virulence effectors are found can directly drive EFPs, as has been clearly shown by the change of virulence due to the horizontal gene transfer (HGT) of *ToxA* from *Phaeosphaeria nodorum* into *Pyrenophora tritici-repentis* in the 1940s, causing aggressive tan spot disease in wheat (Friesen et al., 2006).

A large number or fraction of secreted genes in the genome can be an indicator that a fungus is pathogenic when compared with their related saprobic relatives; this is clearly the case for the species of *Batrachochytrium* (Rosenblum, Stajich, Maddox, & Eisen, 2008). For example, amplifications of secreted protein repertoires are clearly seen in the genomes of the EFPs *B. salamandrivorans* ($n=1527$) and *B. dendrobatidis* ($n=833$) compared to the related free-living saprobic *S. punctatus* ($n=587$) and *H. polyrhiza* ($n=293$) (Farrer et al., 2017). Here, it was shown experimentally that of the chytrid genes that were significantly upregulated in vivo ($n=550$), a large proportion was unique to *B. salamandrivorans* ($n=327$; 60%), unique to *B. dendrobatidis* ($n=43$; 8%) or unique to the genus *Batrachochytrium* ($n=44$; 8%). Furthermore, around half of the *B. salamandrivorans* and *B. dendrobatidis* upregulated genes were secreted (55% and 47%, respectively). The fact that these secreted proteins are both largely not present in the saprobic chytrids based on ortholog identification, and that they show increased transcription during host colonization, suggests that the transcriptional response is focused on a unique host-interaction strategy in each species.

Separately, several genes from a class called Crinkler and Necrosis (CRN)-like genes can either trigger cell death (such as PsCRN63) or inhibit cell death (such as PsCRN115) when expressed inside plant cells (Liu et al., 2011) by pathogens belonging to the *Phytophthora* and *Lagenidium* genera of Oomycetes (Quiroz Velasquez et al., 2014; Schornack et al., 2010). Crinklers are often located in gene-sparse, repeat rich, regions of the genome in well-studied eukaryotic plant pathogens (Haas et al., 2009). A recent study examined gene-sparse regions of the amphibian-infecting chytrid pathogen *B. dendrobatidis* (Farrer et al., 2017). Notably, it was found regions of low gene density include homologs of CRNs. Chytrid CRNs were identified via BLASTp to those in *Phytophthora infestans* T30-4 (Haas et al., 2009), and CD-hit (Li & Godzik, 2006) under a number of sequence similarity identities, as well as trimming the more divergent C-terminal to 35aa, 40aa, 45aa, and 50aa, followed by, or proceeded by, a MUSCLE alignment (Edgar, 2004b) with or without removing excess gaps using trimAl gappyout (Capella-Gutiérrez, Silla-Martínez, & Gabaldón, 2009). Motif searching was performed using GLAM2 (Frith,

Saunders, Kobe, & Bailey, 2008). Searching all of the sequences together after trimming to 50aa did not yield a convincing single domain. Instead, it was found that manually separating genes with two overrepresented sequences obtained the highest confidence alignments spanning the most number of CRNs (Farrer et al., 2017). This process illustrates the trial-and-error approaches that can be required for investigating and classifying novel protein families in EFPs, particularly those with low sequence similarity or small proteins.

CRN-like genes in *B. dendrobatidis* are characterized by having long intergenic regions that are consistent with a gene-poor repeat-rich environment (averaging 1.4 kb)—a trait shared with *P. infestans* T30-4 (Haas et al., 2009). Farrer et al. (2017) showed that the CRN-like family is more widely distributed among the Chytridiomycota than previously realized. Specifically, this study identified 162 CRN-like genes in *B. dendrobatidis*, 10 in *B. salamandrivorans*, 11 in *H. polyrhiza*, and 6 in *S. punctatus*, many of which ($n=55$) belong to a single subfamily (known as DXX). Besides some sequence similarity, there are multiple differences between CRNs found in the chytrid genomes compared with Oomycete genomes. For example, only two chytrid CRNs had predicted secretion signals (via SignalP4 (Petersen et al., 2011))—one in each of the free-living saprobe chytrids *H. polyrhiza* and *S. punctatus*, which contrasts with CRNs in *Phytophthora* species that are mostly intracellular effectors that target the host nucleus during infection (Stam et al., 2013). Another discrepancy is that CRN-like genes appeared to be downregulated during advanced infection of a susceptible salamander species (*Tylototriton wenxianensis*) compared with in vitro conditions, while many Oomycete CRNs are upregulated *in planta* (Chen, Xing, Li, Tong, & Xu, 2013). In both *B. dendrobatidis* and *B. salamandrivorans*, some CRN-like genes were more highly expressed in the zoospore life stage compared to the sporangia life stage (Farrer et al., 2017). However, incubation of *B. dendrobatidis* zoospores with *T. wenxianensis* tissue for 2 h showed an increased expression of CRN genes, whereas *B. salamandrivorans* zoospores were associated with decreased expression, indicating that CRN genes are possibly of greater interest in the early infection stage of *B. dendrobatidis*, but not *B. salamandrivorans*; the notable expansion of CRN-like genes in *B. dendrobatidis* may suggest that they are of importance; however, their function currently remains unknown.

To ascertain if the secreted proteins in four species of chytrids included any large families (in addition to metalloproteases, for example), clustering was used to predict secreted genes using the Markov Cluster Algorithm tool (Enright, Van Dongen, & Ouzounis, 2002) with recommended settings (Farrer et al., 2017). Associated PFAM domains were found in all or nearly all members of some tribes, including the second largest, which contained protease M36 domains, or the sixth largest, which contained the peptidase S41 domain. The largest tribe had 105 proteins, and belonged entirely to *B. salamandrivorans*, as did the fourth largest tribe. Many of the members of these secreted tribes were significantly differentially expressed between in vivo and in vitro conditions, including in "Tribe 1" (48% of genes). Furthermore, these tribes are located almost exclusively in nonsyntenic, unique regions of the *B. salamandrivorans* genome. However, this study was unable to identify any sequence similarity to previously described proteins or recognizable functional domains (BLAST, GO terms, PFAM, TIGRFAM, etc.) showing that these putative virulence factors require further work to understand their possible function. This study illustrates an important point: the constellation of virulence factors that lie within the biodiverse fungal kingdom has only been touched on, and future (yet undescribed) EFPs will likely harbor a wealth of undescribed virulence factors that are not represented in today's databases, and need (sometimes urgent) investigation.

A further class of proteins that are often of interest in EFPs are transmembrane (TM) proteins. An HMM-based method for predicting TMs is TMHMM, which purports to correctly predict 97%–98% of the transmembrane helices (Krogh, Larsson, von Heijne, & Sonnhammer, 2001), and with 99% specificity and sensitivity. However, the authors note that the accuracy drops when signal peptides are present. TM proteins function as gateways for substances to move between the environment and intracellular (such as voltage-gated and ligand-gated ion channels), and as such are integral to the functioning of the cell, and at the host–pathogen interface. Toll-like receptors and receptor kinases are examples of conserved TM pattern recognition receptors of the innate immune system of animals and plants, respectively. In *Aspergillus*, the seven-transmembrane domain protein PalH is a putative pH sensor required for virulence on mice (Grice, Bertuzzi, & Bignell, 2013). Another TM protein, TmpL, is necessary for regulation of intracellular ROS levels and tolerance to external ROS, and is required for infection of plants by the necrotroph *Alternaria brassicicola* and for infection of mammals by the human pathogen *A. fumigatus* (Kim et al., 2009).

The repertoire of proteases in EFPs is of interest, owing to their importance in physiology, development, survival, and growth. Furthermore, extracellular serine, aspartic, and metalloproteases are considered virulence factors in many pathogenic species (Yike, 2011). Proteases can be identified, either by generic databases (e.g., nonredundant BLAST database) or by specialized protease databases (e.g., Merops (Rawlings, Barrett, & Finn, 2016), which as of release 11, contains 447,156 protein sequences of all the peptidases and peptidase inhibitors), both of which can be searched using BLAST.

The metalloproteases of the M36 fungalysin family are important pathogenicity determinants in a number of dermatophytes, which cause cutaneous infections and grow exclusively in the outermost layer of skin, nails, and hair of human and animals. Here, skin-infecting organisms, such as *Trichophyton* spp. that cause Tinea corporis/ringworm, Tinea pedis/athletes foot, secrete M36 proteases that are important for causing disease (Zhang et al., 2014). Again, the chytrid pathogens provide a further good example of M36 proteases and pathogenicity. Metalloproteases are dramatically expanded in *B. salamandrivorans* (Farrer et al., 2017), concordant with the aggressive necrotic pathology that this pathogen causes. Both *B. salamandrivorans* ($n=110$) and *B. dendrobatidis* ($n=35$) have expanded M36 families compared to lower counts in the free-living saprobic chytrids *S. punctatus* and *H. polyrhiza* ($n=2$ and $n=3$, respectively). Phylogenetic analysis revealed a subclass of closely related M36 metalloproteases that are shared across both pathogens that we termed the Batra Group 1 M36s (G1M36) (Fig. 2D).

Species-specific gene family expansion in chytrid pathogens is illustrated by the presence of a novel secreted clade of M36 genes ($n=57$) unique to *B. salamandrivorans*, which were termed the *Bsal* Group 2 M36s (G2M36) (Farrer et al., 2017). These G2M36s are entirely encoded by nonsyntenic regions of the *B. salamandrivorans* genome, supporting a recent species-specific expansion. Although most G1M36s and G2M36s are strongly upregulated in salamander skin, eight *Bsal* G1M36s (19%) appear more highly expressed in vitro, suggesting more complex regulatory circuits underlie this subclass of protease in *B. salamandrivorans*. Furthermore, G1M36s showed greater expression in *B. dendrobatidis* zoospores compared to sporangia, pointing to a crucial role of these proteases during early host colonization in *B. dendrobatidis*, for example, during insertion of their germ tube into the epidermal cells (Van Rooij et al., 2012). In contrast, the low protease activity in *B. salamandrivorans* zoospores, but high activity in the maturing sporangia, suggests a role during later stages of pathogenesis, for example, in breaching the sporangial wall of developing sporangia and subsequent spread to neighboring host cells (Martel et al., 2013).

Carbohydrate-binding modules (CBM), including CBM18, are expanded in *B. dendrobatidis* (Farrer et al., 2017) and been implicated in host–pathogen interactions (Abramyan & Stajich, 2012). To study individual protein families defined by PFAM domains, HMMs can be downloaded from the PFAM database (Finn et al., 2014) that are then used to search through a set of proteins using the HMMER3 (Eddy, 2011) application hmmsearch (with an *e* value cutoff of 0.01 or lower). CBM18s are markedly expanded in both *B. dendrobatidis* and *B. salamandrivorans* compared to the free-living chytrids (Farrer et al., 2017). CBM18 containing proteins are predicted to bind chitin and most copies of these proteins contain secretion signals that will target them to the cell surface or extracellular space. Species-specific differences are notable in the pronounced truncation of the lectin-like CBM18s of *B. salamandrivorans*, suggesting a fundamental difference in capacity to bind some chitin-like molecules. In comparison, CBM18 genes in *B. dendrobatidis* are threefold longer and harbor on average eight CBM18 domains compared with only 2.6 for *B. salamandrivorans*. However, their expression was not significantly altered upon exposure of sporangia to chitinases, suggesting their role in protecting the fungi from host chitinase activity by fencing off the fungal chitin unlikely. Rather, it was hypothesized that the CBM18s play a role in fungal adhesion to the host skin or in dampening the chitin-recognition host response.

CBM18 genes fall into three large groups among chytrids (Abramyan & Stajich, 2012). CBM18s containing carbohydrate esterase 4 (CE4) superfamily mainly includes chitin deacetylases clustered together, and called deacetylase-like. Another group of CBM18s contains a common central domain of tyrosinase, and called tyrosinase-like. A third group consisted of genes with no secondary domains is described as lectin-like. The six *B. dendrobatidis* LL CBM18 had a total of 48 CBM18 modules (averaging 8 per gene), while the six *B. salamandrivorans* lectin-like CBM18s had only 16 CBM18 modules (averaging 2.6 per gene) (Farrer et al., 2017). *B. salamandrivorans* lectin-like CBM18s are also considerably truncated compared with those of *B. dendrobatidis* (mean *B. salamandrivorans* protein length = 606, mean *B. dendrobatidis* protein length = 206). Most *B. dendrobatidis* CBM18s (17/21; 81%) are upregulated in vivo, although mostly nonsignificantly (2 DE, 1TL). In contrast, 7/15 (47%) *B. salamandrivorans* CBM18s are upregulated in vivo. However, five of these genes are significantly upregulated including two tyrosinase-like, two deacetylase-like and one lectin-like. The importance and function of these genes remain to be fully demonstrated, however, appear to be involved in recognizing and binding host ligands as part of the infection process (Liu & Stajich, 2015).

Identifying gene families in EFPs is a necessary precursor to quantify increases or decreases relative to close relatives, and that may indicate why changes in pathogenicity-related traits have occured. For example, gene family expansions in pathogens compared with closely related nonpathogens can provide candidate virulence determinants. A common way for comparing genes and identifying gene family expansions is to first identify single copy orthologs between two or more species, especially when one of those genomes is well characterized. Recently, substantial investment has been made into developing online fungal genomic resources, such as FungiDB (Stajich et al., 2012) which can be used to assist in the categorization of orthologs. However, the protein-coding genes and gene family expansions make up only one aspect of the EFP's genome, where other aspects such as chromosome number may also be important.

2.3 Chromosomal CNV

Pathogenic fungi often manifest highly plastic genome architecture in the form of variable numbers of individual chromosomes, known as chromosomal copy number variation (CCNV) or aneuploidy. CCNV has been identified across the fungal kingdom in both EFP and nonpathogens alike. For example, among ascomycetous fungi, CCNV has been identified in the generalist plant pathogen *Botrytis cinerea* (Büttner et al., 1994), the human pathogen *Histoplasma capsulatum* (Carr & Shearer, 1998), bakers/brewer's yeast (and an occasional opportunist) *S. cerevisiae* (Sheltzer et al., 2011), and the human pathogen *C. albicans* (Abbey, Hickman, Gresham, & Berman, 2011). The occurrence of stress due to either the host response or exposure to antifungal drugs has been linked to a rapid rate of CCNV in *Candida* spp. (Forche, Magee, Selmecki, Berman, & May, 2009) and, within the Basidiomycota, the human pathogens *Cryptococcus neoformans* and *C. gattii* are both found exhibiting CCNV (Hu et al., 2011; Lengeler, Cox, & Heitman, 2001; Sionov, Lee, Chang, & Kwon-Chung, 2010). Even among the Chytridiomycota, *B. dendrobatidis* shows widespread heterogeneity in ploidy among genomes and among chromosomes within a single genome (Farrer, Henk, Garner, et al., 2013). The mechanism(s) generating chromosomal CCNV in fungi are not yet well understood, but are thought to occur because of nondisjunction following meiotic or mitotic segregation (Reedy, Floyd, & Heitman, 2009), followed by selection operating to stabilize the chromosomal aneuploidies (Hu et al., 2011).

Dynamic numbers of chromosomes could offer routes to potentially advantageous phenotypic changes via several mechanisms such as overexpression of virulence factors (Hu et al., 2011) or drug efflux pumps (Kwon-Chung & Chang, 2012). CCNV contributes to the maintenance of diversity through homologous recombination (Forche et al., 2008), and increased rates of mutation and larger effective population sizes (Arnold, Bomblies, & Wakeley, 2012). CCNV may also provide the advantage of purging deleterious mutations through nondisjunction during chromosomal segregation (Schoustra, Debets, Slakhorst, & Hoekstra, 2007). Thus, CCNV likely represents an important, yet uncharacterized, source of de novo variation and adaptive potential in many fungi and other nonmodel eukaryote microbial pathogens. By mapping read depth and SNPs across *B. dendrobatidis* genomes, it was discovered that widespread genomic variation occurs in ploidy among genomes and among chromosomes within a single genome (Farrer, Henk, Garner, et al., 2013). Individuals from all three lineages harbored CCNV along with predominantly or even entirely diploid, triploid, and tetraploid genomes. Another study also identified widespread CCNV across diverse lineages of *B. dendrobatidis* recovered largely from infected amphibians in the Americas, including a single haploid chromosome in a global panzootic lineage (GPL) isolate (Rosenblum et al., 2013). This variation may itself, reflect only part of the full diversity in *B. dendrobatidis*, as +2/+3 shifts in ploidy, whole genomes in tetraploid, or chromosomes in pentaploid or greater, may occur and await discovery.

Chromosomal genotype in *B. dendrobatidis* was shown to be highly plastic as significant changes in CCNV occurred in as few as 40 generations in culture (Fig. 2B) (Farrer, Henk, Garner, et al., 2013). It is not known whether other chytrid species such as *B. salamandrivorans* also undergo CCNV, or if this is a unique feature of *B. dendrobatidis*, or even unique of just chytrid pathogens—and hence may be intrinsic to their parasitic mode of life. Currently, CCNV is known to occur in a variety of protist microbial pathogens, including fungi; however, it is currently not known whether this genomic feature is specific to a parasitic lifestyle or is a more general feature of eukaryote microbes; identifying the ubiquity of CCNV or otherwise across nonpathogenic species will therefore be of great interest. Further, the manner by which plasticity of CCNV in *B. dendrobatidis* affects patterns of global transcription and hence the phenotype of each isolate also remains to be studied. However, it is clear from research on yeast, *Candida* and *Cryptococcus*, that CCNV significantly contributes to generating altered transcriptomic profiles, phenotypic diversity, and rates of adaptive evolution even in the face of quantifiable costs; understanding the relationship between CCNV and the phenotype of *B. dendrobatidis* will therefore likely be key to understanding its patterns of evolution at both micro- and macroscales.

CCNV has also been identified in three isolates belonging to *C. gattii* VGII and VGIII using read coverage (Farrer et al., 2015). Specifically, an additional (disomic) copy of scaffold 13 in VGII veterinary isolate B8828 was identified, and a disomy of scaffold II in VGIII clinical isolate CA1280 (syntenic to the first half of WM276 chromosome cgba). Variation in chromosome copy number has previously been shown to influence the virulence of *Cryptococcus* (Hu et al., 2011) and can further provide resistance to azole drugs by increasing the copy number of the azole drug target (ERG11) or transporter (AFR1) commonly amplified in drug-resistant *Cryptococcus* (Kwon-Chung & Chang, 2012) and across the time-scale of a single infection (Rhodes, Beale, et al., 2017). However, neither gene appears in these aneuploidies, suggesting they are not associated with known drug resistance mechanisms, although may have other effects on those isolates. Separately, a 60 kb intrachromosomal duplication was found in the middle of scaffold 1 of VGII clinical isolate LA55 (also syntenic to WM276 chromosome cgba), which interestingly did not appear in the closely related isolate CBS10090, suggesting it was of a recent origin. This 60 kb region covers 24 protein-coding genes that are not known to influence drug resistance in *Cryptococcus*.

In addition to CCNV, chromosomes of EFPs can fuse, split, and undergo inversions and translocations—which can have a dramatic effect on their phenotype. One method to study this in silico is to identify orthologs, and then to map their synteny. Recently, chromosome structure was compared in detail among the lineages of *C. gattii* (Farrer et al., 2015). Chromosome structure was found to be highly conserved between the four lineages, and very highly conserved within VGII. Almost all syntenic variation was identified among the three closely related lineages, VGI, VGIII, and VGIV (Fig. 2A). In total, 15 large (greater than 100 kb) chromosomal rearrangements were identified, such that on average, only 2.6% of each of the 16 genomes was rearranged with respect to the others. These 15 rearrangements included 10 translocations (7 interchromosomal and 3 intrachromosomal) and 5 scaffold fusions, most of which (13 of the 15) associated with clusters of predicted *Cryptococcus*-specific TcN transposons found at centromeres (Janbon et al., 2014), suggesting these are primarily whole chromosome arm rearrangements. Four of the rearrangements were supported by multiple isolates, including one chromosomal fusion unique to VGII, two translocations unique to VGIII (700 and 140 kb, respectively), and one 450 kb translocation unique to VGIV. These changes may impact the ability for interlineage genetic exchange, as some crossover events will generate missing chromosomal regions or other aneuploidies and nonviable progeny.

2.4 Natural Selection

The widespread emergence of EFPs is testament to their ability to successfully adapt to infect diverse hosts and ecological niches, suggesting that their genomes are able to respond rapidly to natural selection. Characterizing variants in the genome by the type of selection acting upon them requires population genetics approaches. Some possible scenarios for variants in a population include those that are becoming fixed or rapidly evolving due to positive or diversifying selection, being purged due to purifying selection, being maintained in a population due to stabilizing selection, or accumulating mutations due to relaxed selection. In addition to selection pressures, knowledge of rates of recombination, ploidy, life cycle, population structure, and effective population size are all necessary to accurately assess the processes regulating and influencing allele frequencies in a population. Furthermore, a knowledge of how multiple loci or genes contribute to a given phenotype (epistasis) or are masked by others (pleiotropy), as well as random chance, e.g., genetic drift, gene flow, and HGT between populations all contribute to their genetic makeup.

One approach to study selection from genomic data is to look at patterns of synonymous mutations (those that maintain the amino acid sequence of the protein) and nonsynonymous mutations (those that change the amino acid sequence of the protein). An informative approach is to calculate the number of synonymous mutations per synonymous site (positions in the codon that can undergo synonymous mutations) (d_S) and the number of nonsynonymous mutations per nonsynonymous site (d_N) (Fig. 2F). However, the d_N/d_S ratio was originally developed for distantly diverged sequences, i.e., species, where the differences represent substitutions that have fixed along independent lineages, and is therefore unsuitable for identifying selection within a population (Kryazhimskiy & Plotkin, 2008).

The identification of variants in an alignment can be the result of multiple substitutions (increasingly with age since most recent common ancestor (MRCA)), and therefore substitution models (Markov model) are usually used when calculating d_N and d_S values (also denoted K_a and K_s, respectively). Different substation models may also differentially weight transitions (T_s) with respect to transversions (T_v) as T_s are more common at the third position in the codon, as well as GC and base/codon bias inherent to some genomes. Higher T_s/T_v ratios are also caused by spontaneous or cytidine deaminase-mediated deamination of methylated cytosines (Cooper, Mort, Stenson, Ball, & Chuzhanova, 2010), with differences even between animal mitochondrial genomes compared with their nuclear genomes (Belle, Piganeau, Gardner, & Eyre-Walker, 2005). Finally, suitable substitution models can be used by phylogenetic applications such as PAML (Yang, 2007), which estimates d_N, d_S, and $d_N/d_S = \omega$ by maximum likelihood.

When comparing two sequences (i.e., reference and consensus), any selection detected using d_N/d_S will not reveal where on the phylogenetic tree that selection has occurred, or even which of the two sequences or isolates are under selection. A more comprehensive test is to distinguish between selection on the reference sequence vs selection on the consensus sequence by using the branch-site model (BSM) of selection in the Codeml program of PAML (Yang, 2007) to calculate ω across genes and branches/lineages. Multiple corrections are then used to improve specificity for positive selection (such as Benjamini–Hochberg (Benjamini & Hochberg, 1995), Bonferroni correction (Dunn, 1959) or Storey-Tibshirani (Storey & Tibshirani, 2003)).

Comparing ω values for different gene categories, or individual genes, is indicative of the net selective pressures acting upon these loci. For example, in *Paracoccidioides*, the set of genes evolving under positive selection includes the surface antigen gene GP43, the superoxide dismutase gene SOD3, the alternative oxidase gene AOX, and the thioredoxin gene (Muñoz et al., 2016). Each are virulence-associated genes of fundamental importance in *Paracoccidioides* and other dimorphic fungi. In *Phytophthora* clade 1c, a high proportion of genes annotated as effector genes show signatures of positive selection (300 out of 796) (Raffaele et al., 2010). In *B. dendrobatidis*, CRN-like genes in both *Bd*CAPE and *Bd*CH had the greatest median, upper quartile, and upper tail values of ω compared with other gene categories tested (Farrer, Henk, Garner, et al., 2013). These tests are therefore very useful for identifying selection pressures acting on different genes between populations.

When attempting to understand recent selective processes, alternative methods need to be applied such as the direction of selection (DoS) measure for genes with few substitutions (Stoletzki & Eyre-Walker, 2011). DoS is based on the McDonald–Kreitman test, where the count of fixed synonymous (D_s) and fixed nonsynonymous (D_n) is used in conjunction with the numbers of polymorphisms (in this test defined as sites with any variation within species) and denoted P_n for nonsilent and P_s for silent polymorphisms. Next, using an 2×2 contingency table (McDonald & Kreitman, 1991), deviation from the neutrality index ($NI = D_sP_n/D_nP_s$ or $(P_n/P_s)/(D_n/D_s)$) can be detected and will indicate positive selection where $D_n/P_n > D_s/P_s$. However, being a ratio of two ratios, the neutrality index is undefined when either D_n or P_s is 0 and tends to be biased and to have a large variance, especially when numbers of observations are small (Stoletzki & Eyre-Walker, 2011). The DoS measure does not have these issues, and so is suitable when the data is sparse. More recently, powerful approaches have been developed that utilize generalized mixed models to estimate selection coefficients for new mutations at a locus and including the synonymous and nonsynonymous mutation rates alongside species divergence times (Eilertson, Booth, & Bustamante, 2012). Such approaches have further been extended to take into account intragenic heterogeneity in the intensity of natural selection (Zhao et al., 2017).

Using the BSM in Codeml, genes with very small *Q*-values are evidence for positive selection. For example, in *C. gattii*, multiple subclades had low *Q* values for the cell wall integrity protein SCW1 and the iron regulator 1, while other subclades such as VGI excluding a more divergent isolate had a low *Q* value for heat shock protein (HSP) 70 (Farrer et al., 2015)—all of which may play roles at the host–pathogen interface. Additionally, two genes (CDR ABC transporter, and ABC-2 type transporter) were independently identified in four subclades of *C. gattii*. Additionally, the PFAM domain "ABC transporter" belonging to a third gene was independently enriched in three of these subclades. Each of these transporters belongs to a single paralog cluster of six genes, which includes the ABC transporter-encoding gene AFR1. This class of gene includes multidrug transporters with azole and fluconozole transporter activity in *C. neoformans* (Sanguinetti et al., 2006), *C. albicans* (Gauthier et al., 2003) and *Penicillium digitatum* (Nakaune, Hamamoto, Imada, Akutsu, & Hibi, 2002). However, the closest *C. gattii* ortholog to AFR1 was not one of the three under selection. While it is likely that selection pressures driving genetic variation in the *C. gattii* population are occurring predominantly in the environment

(*Cryptococcus* is nontransmissable between hosts), they might also result in key pathophysiological differences in humans.

Within-species data on allele frequency spectra are used to detect impacts on natural selection that occur within more recent timeframes. These methods include Tajima's *D*, which is commonly used to describe genome-wide allele frequency distributions. Tajima's *D* is a widely used metric that distinguishes between genomic regions that are evolving neutrally (i.e., are under mutation/drift equilibrium) to those that are evolving nonneutrally through the action of selection or demographic processes selection (Tajima, 1989). The biological interpretation of Tajima's *D* however is not straightforward as divergence from neutral expectations ($D=0$) can be due to different processes that include demographic events alongside the intensity of natural selection. For example, on one hand a negative value of $D<0$ (equating to an excess of rare alleles) can owe to sweeps on a selected polymorphism or population expansion following a genetic bottleneck. On the other hand, a positive value of $D>0$ (equating to a scarcity of rare alleles) can owe to either balancing selection or a demographic contraction. In both cases, correct interpretation of *D* requires further population genetic analysis. A range of other methods for intrapopulation selection have been developed or used to infer selection, including Fu and Li's *D* and *F*, Fay and Wu's *H* test, long range haplotype test, iHS, LD decay, and F_{ST} (Biswas & Akey, 2006). Different methods may have benefits over others depending on sample size, sequence similarity or distance, population structure, population size, or recombination rates (along with other population-specific traits). Determining the best tools and methods usually requires some benchmarking on the data, testing the effect parameters has on results, and often comparing the results between tests to look for consistency. Ultimately, identifying genes or gene categories that are rapidly changing or are unusually conserved can offer new insights into the biology and pathology of EFPs.

A striking example of the response of fungi to directional selection leading to a novel emerging trait is seen when antifungal drugs are used to treat infectious fungi. Fungicides are an essential component in our armamentarium against fungal disease with sterol demethylation inhibitor (DMI) compounds, such as triazoles and imidazoles, representing the largest class of fungicides that are used in agriculture. These compounds are widely deployed for crop protection with, for instance, over 250,000 kg being used to protect UK crops each year (European Centre for Disease Prevention and Control, 2013) and the global usage being in the thousands of tonnes. In parallel, azoles are used as frontline drugs to protect humans and other animals against pathogenic fungi. However, the dual-use of DMIs in both clinical and agricultural settings may come at a considerable human cost as in recent years' multiazole resistance in fungi that infect humans has been observed as a widely emerging phenomenon across Europe and beyond. This emergence of resistance has led to the hypothesis that the deployment of azoles in agriculture has led to selection for antifungal resistance not only in target crop pathogens (Cools & Fraaije, 2008; European Centre for Disease Prevention and Control, 2013) but also those fungal species that cooccur in their environment, and that can opportunistically infect humans, specifically the saprophytic genus *Aspergillus* (European Centre for Disease Prevention and Control, 2013). Ergosterol is an essential component of the fungal cell membrane and is the target of triazoles that inhibit its biosynthesis, thereby interfering with the integrity of the fungal cell membrane (Diaz-Guerra, Mellado, Cuenca-Estrella, & Rodriguez-Tudela, 2003). In *Aspergillus*, azole resistance can be an intrinsic phenotype, as it is known to occur in cryptic *Aspergillus* species related to *A. fumigatus*, specifically *A. lentulus* and *A. pseudofischeri* (Van Der Linden, Warris, & Verweij, 2011), whereas wild-type *A. fumigatus* and *A. flavus* are sensitive to these drugs. In *A. fumigatus*, azole resistance is known to be an acquired trait that occurs after azole exposure during medical treatment, or after fungicide exposure in the field where *A. fumigatus* widely occurs in the soil. While a spectrum of resistance mechanisms to azoles has been characterized in *A. fumigatus* (Fraczek et al., 2013; Meneau, Coste, & Sanglard, 2016), azole resistance is frequently the result of mutations in the *cyp51A* gene. Many azole-resistant isolates have nonsynonymous point mutations at codons in this gene, for example, at positions G54, M220, and G138 (Chowdhary, Sharma, Hagen, & Meis, 2014), which are primarily found in patients who have been treated for long periods with azoles (Verweij, Chowdhary, Melchers, & Meis, 2016). However, in addition to mutations that are commonly associated with the de novo acquisition of resistance in the patient, an increasingly large constellation of *cyp51A* mutations are found to occur in "wild" *A. fumigatus*. These mutations are largely characterized by having a tandem repeat (TR) duplication in the promotor region of *cyp51A* linked to structurally important nonsynonymous SNPs (Meis, Chowdhary, Rhodes, Fisher, & Verweij, 2016).

It is now evident that triazole resistance in *Aspergillus* has a global distribution and constitutes a worldwide EFP with important consequences to human health. In some regions, up to 7% of patients are culture-positive for *Aspergillus* now harbor environmentally associated azole-resistance and azoles are increasingly failing in their role as frontline choices of therapy. Population genomic analysis has been used to show that the most frequently occurring environmental-resistance allele, known as TR_{34}/L98H, occurs on a subset of the observed genetic diversity of *A. fumigatus* with strong linkage disequilibrium being observed, and associations to clonal population sweeps in regions of high azole usage such as India. The balance of evidence suggests that TR_{34}/L98H is a relatively recent and novel evolutionary innovation, and that it is perturbing the natural population genetic structure of *A. fumigatus* in nature as selective sweeps imposed by this allele occur. Fitness costs that are associated with azole-resistance alleles appear to be negligible, and diversification in nature is known to occur as mating occurs leading to the genesis of new combinations of azole-resistance alleles (Abdolrasouli et al., 2015). Thus, strong direction selection through the global usage of azoles appears to have irrevocably perturbed the worldwide population genetic structure of *Aspergillus,* alongside many other plant pathogenic fungi, leading to worldwide breakdown in our ability to use this important class of drugs to secure our health and food security (European Centre for Disease Prevention and Control, 2013).

2.5 Genomic Approaches to Detecting Reproductive Modes, Demographic and Epidemiological Processes in EFPs

2.5.1 Know Your Enemy

Key to the genomic analysis of an EFP is to "know your enemy." Within this context, is the (often novel) EFP a single genotype, a lineage, a species, or a set of species? These distinctions are important as they determine the evolutionary trajectory of the organism by determining the type and rate of evolutionary changes that will occur through time, and how these need to be analyzed within an epidemiological context. Wiley (1978) used an evolutionary concept to define species as "··· a single lineage of ancestor descendent populations which maintains its identity from other such lineages and which has its own evolutionary tendencies and historical fate." The evolutionary species concept has been used as the framework that species of fungi can be identified using operational species concepts that use the genealogies inferred from DNA sequences. Of most benefit to analyses of fungal diversity, the system of genealogical concordance phylogenetic species recognition (GCPSR) has been widely used to define evolutionary significant units by identifying the transition from genealogical concordance to conflict (also known as reticulate genealogies) as a means of determining the limits of species (Dettman, Jacobson, & Taylor, 2003; Taylor et al., 2000). An important use of whole-genome data therefore is to determine the extent that evolutionarily significant units occur within the EFP, be these on a global scale or within a localized outbreak setting.

There are two fundamental means by which fungi and other organisms transmit genes vertically to the next generation, either via clonal reproduction or via mating and recombination (Taylor, Jacobson, & Fisher, 1999). Under a purely clonal reproductive mode, each progeny has as single parent with its genome being an exact mitotic copy of the parental one, and all parts of the parental and progeny genomes share the same evolutionary history. At the other extreme are genetically novel progeny formed by the mating and meiotic recombination of genetically different parental nuclei, events that cause different regions of the progeny genome to have different evolutionary histories. However, many fungi do not fit neatly into these two categories. For instance, on one hand recombination need not be meiotic or sexual because mitotic recombination via parasexuality can mix parental genomes. On the other hand, clonality need not be solely mitotic and asexual, because self-fertilizing or homothallic fungi make meiospores with identical parental and progeny genomes. In addition to the observation that reproductive mode (clonal or recombining) may be uncoupled from reproductive morphology (meiosporic or mitosporic), there is the complication that the same fungus may display different reproductive modes in different localities at different times. These are important distinctions from the point of view of EFPs, as many fungi are flexible in their ability to undergo genetic recombination, hybridization, or HGT (Taylor et al., 1999). This flexibility in life histories allows not only the clonal emergence of pathogenic lineages from their sexual parental species, but can also allow the formation of novel genetic diversity by generating mosaic genomes that may lead to the genesis of new pathogens (Stukenbrock & McDonald, 2008).

Reproductive barriers in fungi are known to evolve more rapidly between sympatric lineages that are in the nascent stages of divergence than between geographically separated allopatric lineages, in a process known as reinforcement (Turner, Jacobson, & Taylor, 2011). As a consequence, the anthropogenic (human-associated) mixing of previously allopatric fungal lineages that still retain the potential for genetic exchange across large genetic distances has the potential to drive rapid macroevolutionary change. Although many outcrossed individuals, or genuine species hybrids, are inviable owing to genome incompatibilities, large phenotypic leaps can be achieved by the resulting "hopeful monsters," potentially leading to host jumps and increased virulence. Therefore, a nuanced understanding of gene flow within and among fungal lineages is important as recombination is known to novel new interspecific hybrids with novel pathogenic phenotypes as lineages come into contact (Giraud, Gladieux, & Gavrilets, 2010; Inderbitzin, Davis, Bostock, & Subbarao, 2011).

The sequencing of the brewer's yeast *S. cerevisiae* represented a genetic landmark as it was the first fully sequenced eukaryotic genome. From this initial assembled sequence, over a thousand resequenced genomes have now been generated for *S. cerevisiae* and its close relatives leading to an unparalleled genomic description of the evolution of this model globalized fungal species across different spatial and temporal scales (Dujon & Louis, 2017; Liti et al., 2009). Descriptions of global patterns of *S. cerevisiae* genome-wide diversity are now identifying ancestral populations found in South East Asia (Wang, Liu, Liti, Wang, & Bai, 2012) alongside lineages which have undergone global spread through comigrating with humans (Liti et al., 2009). While many genotypes of *S. cerevisiae* are "clean lineages," others show widespread outcrossing that has resulted in gene flow generating mosaic genomes that are characterized by genetic introgressions from other lineages of *S. cerevisiae*, and also via hybridization with other related species of closely wild yeasts such as *Saccharomyces paradoxus*. Therefore, a GCPSR analysis, although not yet formally done, would likely show that the genomes that comprise the *Saccharomyces* clade are evolving in a reticulate manner rather than in a strictly genealogically concordant manner (Dujon & Louis, 2017). Reticulate evolution is likely to be the case for many fungal lineages that we currently recognize as species, and represents a fundamental challenge for the modern fungal taxonomist as well as fungal epidemiologist.

2.5.2 Occurrence of EFPs Caused by Clonal Through to Reticulate Evolution

The correct interpretation of the genetic epidemiology of a fungal outbreak critically depends on understanding how the outbreak isolates are related to the species-wide diversity across the realized global range of the pathogen. A key question is to determine whether the EFP represents the long-distance dispersal of a species resulting in host shifts and the loss of population diversity—clonal evolution, or is a genetic recombinant with novel phenotypic traits—reticulate evolution. Often in the context of an emergence of a novel fungal pathogen, these data can take months, years, or even decades to accrue (but see Islam et al., 2016). However, phylogenomic analysis is likely to provide crucial understanding of the evolutionary and epidemiological drivers leading to a mycotic outbreak. For example, genetic evidence of the clonal evolution of an EFP following a phylogeographic "leap" from its parental, sexual, population has been forcefully illustrated by the emergence of the aetiological agent of bat white-nose syndrome, *P. destructans* (Blehert et al., 2009). This mycosis emerged in 2006–07 from a single index outbreak site, spreading and devastating multiple species of bats across North America. However, while bats across Europe are infected by this fungus, they appear thus-far unscathed suggesting that European bats have a longer history of coevolution with *P. destructans* compared to their North America conspecifics. Support for this hypothesis initially came from multilocus evidence showing that European isolates of *P. destructans* are highly polymorphic at all loci examined (Leopardi, Blake, & Puechmaille, 2015) and are heterothallic with both mating types coexisting within single bat hibernacula (Palmer et al., 2014). In comparison, recent comparative genome analyses of North American outbreak isolates of *P. destructans* show that they are not only genetically highly homogenous but also comprise a single mating type (Palmer et al., 2014; Trivedi et al., 2017) and show no evidence of recombination. These data strongly support the hypothesis that a single genome of *P. destructans* contaminated North America from a thus-far unidentified location in Europe, followed by clonal amplification and continent-wide spatial expansion of this single genotype.

While the emergence of *P. destructans* presents a dramatic example of a contemporary clonal spatial escape, many other species of EFP show strong similarities to the basic process described earlier. Human-mediated intercontinental trade has been linked clearly to the spread of animal-pathogenic fungi through the transportation of infected vector species. *B. dendrobatidis* has been introduced repeatedly to naive populations worldwide as a consequence of the trade in the infected, yet disease-tolerant species such as North American bullfrogs (*Lithobates catesbeiana*) (Garner et al., 2006) and African clawed frogs (*Xenopus laevis*) (Walker et al., 2008). Recent genome sequencing of a global collection of over 250 genomes of *B. dendrobatidis* has been used to prove that a single genotype, *Bd*GPL, globally emerged in the early 20th century to cause the patterns of amphibian decline seen to date. Analogous to *P. destructans*, population genomic comparisons of sequenced *B. dendrobatidis* isolates show clear patterns of emergence from a defined geographic location, in this case East Asia (O'Hanlon and Fisher, unpublished observation), where isolates of *B. dendrobatidis* show levels of nucleotide diversity that are many fold higher than are seen across other global regions. However, in contrast to *P. destructans*, the emergence of amphibian chytridiomycosis across over half a century has allowed substantial diversification of the outbreak lineage *Bd*GPL to occur, including the homogenization of large tracts of the polyploid genome through losses of heterozygosity caused by mitotic recombination (Farrer et al., 2011; James et al., 2009). Furthermore, consecutive waves of expansion by *Bd* out of its East Asian home range has allowed globalized lineages to recontact and form recombinant genotypes many decades later (O'Hanlon and Fisher, unpublished observation).

As the rate of interlineage recombination between fungi will be proportional to their contact rates, a prediction is that the globalization of pathogenic fungi will increase the frequency that recombinant genotypes are generated. Confirming this hypothesis, outcrossing to generate novel mosaic genomes among lineages is now increasingly observed for sequenced isolates of *B. dendrobatidis* in regions where lineages are found to occur in sympatry. The process of recombination through secondary contact is potentially important in an epidemiological context as theory and experimentation have shown that virulent lineages can have a competitive advantage that results in increased transmission (de Roode et al., 2005; Karvonen, Rellstab, Louhi, & Jokela, 2012). This implies that the generation of novel genotypes with varied virulence phenotypes may force the epidemiological characteristics of a disease system as well as allowing the generation of novel interlineage recombinant mosaic genomes with novel phenotypes. A case in point here is the formation of a novel pathogen of triticale, *B. graminis triticale*, which evolved through the hybridization of two *formae specialis* from wheat and rye hosts (Menardo et al., 2016) clearly demonstrating that new evolutionarily significant units, and thus EFPs, can be generated through outcrossing.

The use of population genomics is increasingly widely used to map phylogeographic escapes that have led to outbreaks of EFPs. Owing to its ability to cause severe disease in humans, the basidiomycete yeasts *C. neoformans* and *C. gattii* have been subjected to detailed genomic scrutiny. Both species show the existence of strong genetic subdivision into lineages with high statistical support (Farrer et al., 2016; Rhodes, Desjardins, et al., 2017). Whether these represent evolutionary species or not is currently a subject of wide debate as genome-wide tests (Hagen et al., 2015; Menardo et al., 2016) of genealogical concordance have not been performed to date (Hagen et al., 2015; Kwon-Chung et al., 2017). Certainly, the realized potential for interlineage recombination is apparent as hybrid ancestry is readily detected based on the detection of large blocks of shared ancestry among all three lineages of *C. neoformans* var. *grubii* (lineages VNI, VNII, and VNB) (Rhodes, Desjardins, et al., 2017) and interlineage hybrids between *C. neoformans* and *C. gattii* have been described (Bovers et al., 2006; Engelthaler et al., 2014). However, superimposed upon this background of mosaic lineages, population genomic analysis of both species show very clear evidence of clonal expansions that are associated with clinical disease. *C. neoformans* lineage VNI appears to have expanded globally (likely anciently) due to widening avian host distributions (Litvintseva et al., 2011; Rhodes, Desjardins, et al., 2017), and the emergence of *C. gattii* lineage VGIIa in the Pacific Northwest has recently caused a widely studied outbreak of aggressive clinical disease (Engelthaler et al., 2014; Fraser et al., 2005; Hagen et al., 2013). For fungi that cause disease in plants, clonal expansion causing epidemic outbreaks following long-distance dispersal of infectious propagules has relentlessly attacked agriculturally important crops and damages our ability to safely feed humanity on an annual basis (Fisher et al., 2016, 2012; Fones et al., 2017). Examples here are many (Fisher et al., 2016, 2012; Fones et al., 2017; McDonald & Stukenbrock, 2016) and include the recent emergence of wheat blast caused by a clonal outbreak of *M. oryzae* vectored from South America to Bangladesh with associated catastrophic losses (Islam et al., 2016). This study is notable in that the team was able to sequence and assemble an open-access genome-wide dataset of SNPs derived from a broad global set of isolates within a matter of months in order to identify the likely geographic source of the Bangladesh outbreak, thus illustrating how the future of rapid population genomic analysis of EFPs may unfold.

Beyond describing the spatiotemporal phylodynamic aspects that underpin EFPs, population genomics is leading to an increasingly nuanced understand of how fungi acquire novel pathogenicity traits through the process of HGT. HGT is a special case of hybridization, where a defined genetic locus is transferred between large genetic distances that range from interspecies through to inter-kingdom transfers. An arresting example of a locus-specific HGT leading to the evolution of an EFP was determined through sequencing the genome of the wheat pathogen *P. nodorum* where a gene encoding a host-specific protein toxin (*ToxA*) was identified by homology to a known toxin from another wheat pathogen *P. tritici-repentis*. It is now known that *ToxA* jumped from *P. nodorum* into *P. tritici-repentis* through close genetic linkage to a retrotransposon, sometime in the 1940s resulting in the rapid emergence of aggressive tan spot disease of wheat caused by *P. tritici-repentis* (Friesen et al., 2006). More recent advances in other species have further detailed the acquisition of novel virulence-associated loci via HGT in *Fusarium pseudograminearum* where horizontal transfers from bacterial and other fungal species were discovered that were clearly associated with virulence in this EFP (Gardiner et al., 2012).

2.5.3 Mutation Rates, Molecular Clocks, and EFPs

A key question that needs to be asked early on when analyzing an outbreak of an EFP is to determine when the genotype (or phenotype) of interest evolved. This question is currently being addressed for a wide variety of EFPs including *Batrachochytrium* and *Cryptococcus* sp. For the latter, comparative genomics has been used to compare orthologous-coding regions in order to determine the proportion of nucleotide sites that have undergone substitutions. Such analyses were recently used to show that ~17% of sites were polymorphic when representative genomes of *C. gattii* and *C. neoformans* were compared against one another. If fungal mutation rates lie between 0.9×10^{-9} and 16.7×10^{-9} substitutions per nucleotide per year as has been calculated across a range of filamentous fungi (Kasuga, White, & Taylor, 2002; Sharpton, Neafsey, Galagan, & Taylor, 2008), then the divergence time between these species would lie between 5.2×10^{6} and 96.7×10^{6} years ago, which is concordant with the breakup of the Pangean supercontinent causing allopatric speciation of *C. neoformans* and *C. gattii* through a model of vicariance (Casadevall, Freij, Hann-Soden, & Taylor, 2017). However, a cautionary note needs to be interjected here: Accurate estimates of substitution rates are crucial in order to investigate the evolutionary history of virtually any species. It becoming increasingly apparent that "the molecular clock" is not a one-size-fits-all and in fact can vary by two orders of magnitude even within a single lineage. A case in point here are recent investigations into the population genomics of microevolution in serially collected isolates of *C. neoformans* from HIV/AIDs patients with cryptococcal meningitis in South Africa. While comparisons revealed a clonal relationship for most pairs of isolates recovered before and after relapse of the original infection, one pair of isolates manifested a substitution rate that was greatly inflated above that of the others. Further investigation showed the occurrence of nonsense mutations in DNA mismatch repair pathways leading to the evolution of a hypermutator phenotype (Rhodes, Beale, et al., 2017).

The occurrence of hypermutators in fungal populations is now being described more widely, not only in *Cryptococcus* (Boyce et al., 2017; Rhodes, Beale, et al., 2017) but also species of *Candida* (Healey et al.,

2016). This means that there is a real need to carefully scrutinize the range of substitution rates within and between species, and to not assume a "one-size-fits-all" approach as this is almost certainly incorrect. A further complexity is that nuclear genomes that have undergone recombination are mosaics of gene genealogies with varied evolutionary histories, which can have the effect of creating a false signal of mutation. Therefore, in order to accurately estimate substitution rates, efforts need to be made to control for the effects of recombination, either by directly partitioning the data around recombining sites as was done to date the origin of the *Batrachochytrium* hypervirulent lineage *Bd*GPL to the 20th century (Farrer et al., 2011) or by choosing a nonrecombining section of the genome, such as the mitochondrial DNA.

Once appropriate genomic regions have been identified, then the most direct approach is to use root-to-tip estimations of substitution rates for collections, where the MRCAs are known from either a fossil record or time-dated biological events such as date of isolation. Critically, for root-to-tip estimations of rates to work, studies need to be able to access time-stamped genomic data that is measurably evolving through time (Rieux & Balloux, 2016). If time-calibrated phylogenies that are measurably evolving can be constructed, then sophisticated analyses of demographic histories can be inferred including the estimation of effective population sizes through time, implemented in coalescent-based algorithms such as BEAST (Drummond & Rambaut, 2007). Such analytical approaches have proven critical to understanding pandemics of viruses such as HIV (Faria et al., 2014) and the spread of bacterial pathogens (Croucher & Didelot, 2015). However, beyond the single example of our attempt to understand the date of *Bd*GPLs origin (Farrer et al., 2011), we are unaware of serious attempts to analyze EFPs using modern tip-calibrated approaches to estimating fungal molecular clocks with rigor.

3. EPIGENOMIC VARIATION WITHIN AND BETWEEN POPULATIONS OF EFPs

Phenotypic traits of EFPs are determined by their genomes, the environment, and their interactions. Epigenetics was a name given by Conrad Waddington to "the branch of biology which studies the causal interactions between genes and their products, which bring the phenotype into being" (Goldberg, Allis, & Bernstein, 2007). However, the term has since been used to describe a range of processes: for example, the temporal/spatial control of gene activity during animal development (Holliday, 1990), and changes in phenotype caused without alterations in the DNA sequence, that are either not necessarily heritable (Bernstein et al., 2010), or are exclusively heritable (Berger, Kouzarides, Shiekhattar, & Shilatifard, 2009). The latter definition (and the others listed) includes a wide range of processes ranging from base modifications such as cytosine methylation and cytosine hydroxymethylation, as well as histone posttranslational modifications, nucleosome positioning, and ncRNA regulating gene expression.

Epigenetic processes often culminate in differential expression, e.g., nucleosome occupancy negatively correlating with gene expression (Leach et al., 2016). Indeed, detecting expression values between conditions, or between isolates or even orthologous genes between species remains a key question for many EFPs and has been discussed in some detail in the previous sections. Many tools have been made available for detecting levels of expression and expression differences. A key normalized metric from RNAseq is the "reads per kilobase of transcript model per million reads" (RPKM). RPKM can be calculated by several tools such as EdgeR (Robinson, McCarthy, & Smyth, 2010) or Cufflinks (Trapnell et al., 2012). Alternatively, RPKM can be calculated simply by (1) counting the total number of reads in a sample divided by 1 million to give the "per million scaling factor" (PMSF), (2) dividing the number of reads aligned to a gene by the PMSF, and dividing that by the length of the gene in kilobases. A slightly updated metric is FPKM that looks at the number of fragments (the number of paired or individual reads that aligned). For single-end reads, FPKM equals RPKM. Finally, the transcripts per kilobase million (TPM) normalizes for the gene length first (rather than the scaling factor) and provides a relative abundance of transcripts. FPKM and RPKM can be further normalized using the trimmed mean of *M*-values (TMM) (Robinson & Oshlack, 2010), which includes additional scaling factors on the upper and lower expression values of the data, as is implemented in tools such as EdgeR (Robinson et al., 2010). Although each of these expression value metrics is designed to normalize RNAseq across samples or datasets, each may ultimately have a bias for longer or small gene families, library preparation or GC content, which should be identified during an analysis of differential expression.

Gene expression values (i.e., TMM, TPKM, or TPM) across multiple isolates or experiments are usually compared during differential expression analysis, which can require up to 12 biological replicates for the greatest accuracy rates (Schurch et al., 2016), although in practice usually only three are generated due to cost. Tools such as EdgeR (Robinson et al., 2010) and

Deseq2 (Love, Huber, & Anders, 2014) identify differentially expressed transcripts based on a generalized linear model for each gene assuming a negative binomial distribution and includes several other steps to eliminate bias in long genes or minimize "noisy" expression data. Other tools include DEGseq (Wang, Feng, Wang, Wang, & Zhang, 2010) which is also an R bioconductor package and assumes a poisson model which is appropriate for technical replicates but may overestimate expression differences between conditions. MMseq (multimapping RNA-seq analysis) (Turro, Astle, & Tavaré, 2014) to detect allele or isoform-specific expression and Cuffdiff (Cufflinks' method for estimating differential expression) (Trapnell et al., 2012) using quartile-based normalization are additional tools that may provide comparable or better results, and LOX to examine differential expression across multiple experiments, time points, or treatments (Zhang, López-Giráldez, & Townsend, 2010). Ultimately, studies usually have a defined cutoff, e.g., log fold changes between conditions, and/or FDR rates to identify genes that are changing most rapidly. Plots such as Volcano and MA plots can show the distribution of expression values for all genes, and those that are considered differentially expressed, thereby highlighting biases of those methods, e.g., bias of low average counts of reads/transcripts per million.

Numerous examples of differential expression have been discussed in the previous chapter, such as the secreted clade of G2M36 genes ($n = 57$) unique to *B. salamandrivorans*, which are mostly upregulated in salamander skin (Farrer et al., 2017). Notably, the study also generated a transcriptome of the Wenxian knobby newt (*T. wenxianensis*) to identify host genes that were differentially expressed during infection. Emerging fungal diseases are often nonmodel organisms, as is the case for *B. salamandrivorans*, and will themselves infect nonmodel organism hosts. To effectively study the genomics and epigenomics of these diseases, and their effect on the host, it is essential to move away from model-based systems and generate resources such as draft genome assemblies and gene sets for the growing repertoire of EFP's and their hosts.

The associations of mutations and changes in fitness, as well as transcriptional regulation, during pathogenicity are beginning to be characterized within a multitude of eukaryotic pathogens, e.g., *Cryptococcus* (Magditch, Liu, Xue, & Idnurm, 2012; Panepinto & Williamson, 2006). However, the modifications of both DNA and histones that play a key role in transcriptional regulation are to date largely uncharacterized in EFPs. In eukaryotes, histones assemble into octomers called nucleosomes, which wrap around approximately 147 base pairs of DNA (Stroud et al., 2012). While the position of each histone can be mapped independently by ChIP-seq, including variants of each type, a single type may be used as a proxy for nucleosome positions. Variation in histone-binding sites is found between isolates of fungal pathogens, as well as varying upon condition such as *C. albicans* during heat shock (Leach et al., 2016). Furthermore, nucleosome levels in *C. albicans* decrease near to the transcription factor-binding sites of key pathogenicity genes, allowing activation by transcription factors and RNA polymerase (Leach et al., 2016).

Histones undergo posttranslational modifications on their N-terminal tails that alter their interactions with the DNA and other proteins that they bind. Modifications can be made to any of the four types of histones at several amino acid sites and can include acetylation, phosphorylation, methylation, deamination/citrullination (arginine → citrulline), β-*N*-acetylglucosamination, ADP ribosylation, ubiquitination and small ubiquitin-like modifier (SUMO)-lyation, tail-clipping, and proline isomerization (Bannister & Kouzarides, 2011). These modifications ultimately alter the chromatin structure, which can manifest into changes in transcription, repair, replication, and recombination. For example, acetylation of lysine residues on H3 and H4 by protein complexes involving histone acetyltransferases (HATs) is associated with active transcription for several fungal pathogens (Jeon, Kwon, & Lee, 2014). Notably, the Rtt109 HAT is responsible for acetylation of H3K56 and contributes to pathogenicity of *C. albicans* in mouse macrophages (Lopes da Rosa, Boyartchuk, Zhu, & Kaufman, 2010). Another family of HATs are the Gcn5-related *N*-acetyltransferases (GNAT), including the GCN5 protein implicated in *C. neoformans* growth rates at high temperatures, capsule attachment, and tolerance of oxidative stress (O'Meara, Hay, Price, Giles, & Alspaugh, 2010).

DNA methylation is another important mechanism for epigenetic changes regulating gene expression and transposon silencing (Lister et al., 2009). Whole-genome bisulfite sequencing and methylated DNA immunoprecipitation are methods to profile the methylation of cytosine (carbon 5) to 5-methylcytosine (5-meC) in eukaryotes, generally within cytosine-rich genomic islands (CpG, CpHpG, and CpHpH) (Lou et al., 2014). DNA methylation is achieved via a number of DNA methyltransferase (DNMT), dependent on the species, resulting in 5-meC that can be heritable (e.g., via DNMT1 and UHRF1) (Law & Jacobsen, 2010). 5-meC is widespread in bacteria, plants, and mammalian cells, but differentially conserved across the fungal kingdom. Notably, 5-meC appears to be absent in a number of fungal genera including *Saccharomyces* and *Pichia* (Capuano, Mülleder, Kok, Blom, & Ralser, 2014). However, in *Neurospora* 5-meC within CpG islands is located in ex-transposons targeted by RIP mutations, where its presence is dependent on a single DNMT named DIM-2, directed by a histone H3 methyltransferase (Selker et al., 2003).

In *C. neoformans* isolate H99, Huff et al. have identified DNMT5 as a CG-specific DNMT and show that knockouts appear to completely remove 5-meC (Huff & Zilberman, 2014, p. 1). Separately, DNMT5 has been implicated in infection in mice (Liu et al., 2008), where knockouts show significantly reduced virulence. 5-meC in *Cryptococcus* is primarily associated with transposable elements, and the methylation directly disfavors nucleosome binding (Huff & Zilberman, 2014, p. 1) (determined using micrococcal nuclease (MNase) to digest chromatin followed by sequencing). Huff et al. show that 5-meC is negatively associated with nucleosome positions, but it remains to be shown how the patterns and associations with nucleosomes varies between isolates or during infection, and as suggested by Liu et al., it may reveal insights into the mechanisms of infection (Liu et al., 2008).

Epigenomics in fungal pathology remains an active of area of research that compliments the larger field of genomics (i.e., DNAseq) in identifying new genotypic features of EFPs and particularly dynamic changes associated with virulence traits. However, since most fungal pathogens remain unculturable, and some (such as *Microsporidia*) are obligate intracellular pathogens—obtaining high quality and sufficient depth of coverage for RNAseq, let alone ChIPseq or Methylseq remains an obstacle. An increase in sampling across fungal pathogens and their nonpathogenic relatives, especially for generating new high-quality genomes for comparison, but also transcriptomics is likely to improve our understanding of fungal pathogenesis. Sampling nonpathogenic relatives will require a move away from focusing solely on outbreak strains, and also looking for fungal relatives in host populations that are not experiencing population declines may yield novel-related isolates. The field of metagenomics also promises to identify new locations and relatives for EFPs.

ncRNA such as miRNA and siRNA of the RNAi pathways are prominent epigenetic components found throughout the fungal kingdom, where they function to silence or downregulate gene expression via complimentary sequences to mRNA targets (Pasquinelli, 2012) or gene promoters (Chu, Kalantari, Dodd, & Corey, 2012). RNAi is achieved via either microRNA (miRNA) derived from single-stranded RNA transcripts that fold to form ~70 nt hairpins, or small interfering RNAs (siRNAs; short interfering RNA; silencing RNA) that derive from longer regions of double-stranded RNA. siRNA ultimately cleaves sequence-specific mRNAs, compared with miRNA that has reduced specificity and therefore may target a wider range of mRNAs (Lam, Chow, Zhang, & Leung, 2015). Both miRNA and siRNA are cleaved by the RNase III endoribonuclease Dicer (Dicer-1 and Dicer-2, respectively) prior to being incorporated into either the cytoplasmic RNA-induced silencing complex (RISC) or the nuclear RNA-induced transcriptional silencing (RITS) complex, where the RNA (guide strand) binds target mRNA (such as miRNA response elements; MRE; found in 3′ UTRs), which is cleaved by the PIWI domain of a catalytic Argonaute protein (a major component of both the RISC and RITS) thereby causing degradation of the transcript. Another category of RNA silencing molecules is the Dicer-independent PIWI-associated/interacting RNAs (piRNAs), some of which are classified as repeat-associated small interfering RNA (rasiRNA)—however, both sets are thought to be absent in the fungal kingdom.

RNAi silencing machinery (in contrast to piRNA and rasiRNA) is prominent throughout the fungal kingdom, especially filamentous fungi, although is lost sporadically in some species of both yeasts and filamentous fungi (Dang, Yang, Xue, & Liu, 2011). Excitingly, exogenous/artificial (in addition to endogenous) miRNA and siRNA derived from double-stranded RNA or hairpin RNA with complementary sequence to target gene promoters (Chu et al., 2012) or mRNA targets (Pasquinelli, 2012) are being increasingly used for therapeutics against fungi that cause disease in plants (Duan, Wang, & Guo, 2012) and humans (Khatri & Rajam, 2007). For example, a synthetic 23-nucleotide siRNA was designed with complementary base pairs to the sequence of a key polyamine biosynthesis gene (ornithine decarboxylase) in *A. nidulans* required for normal growth, resulting in a reduction in mycelial growth, target mRNA titers, and cellular polyamine concentrations (Khatri & Rajam, 2007). Despite their important role in endogenous gene control (especially transposons), and their potential therapeutic role, endogenous miRNAs and siRNAs (and their respective targets) are not routinely predicted from the genome sequence, despite various in silico strategies existing (i.e., Bengert & Dandekar, 2005). Currently, the extent that RNAi has a role on gene regulation in many fungal pathogens including EFPs is unclear. However, as described later, the study of RNAi is a rapidly emerging field that holds great promise not only as a tool for understanding fungal virulence but also as a novel approach to disrupt fungal pathogenicity.

RNAi has been shown in numerous biological roles across the fungal kingdom. For example, *N. crassa* can initiate potent RNAi-mediated gene silencing to defend against viral and transposon invasion (Dang et al., 2011). Other functions of RNAi include sex-induced silencing in *C. neoformans*, which is mediated by RNAi via sequence-specific small RNAs (Wang, Hsueh, et al., 2010). Interestingly, one lineage of the related *C. gattii* (VGII) is missing PAZ, Piwi, and DUF1785 domains, all of which are components of the RNAi machinery. This loss of RNAi has been hypothesized to contribute to increased genome plasticity in this lineage that may have contributed to specific hypervirulent traits in VGII (D'Souza et al., 2011; Farrer et al., 2015; Wang, Hsueh, et al., 2010).

The discovery that communication between host and pathogen can occur through the transfer of extracellular microvesicles (ExMVs) has opened a new research field into the horizontal transfer of bioactive molecules in cell-to-cell communication (Ratajczak & Ratajczak, 2016). It has now been well documented that horizontal transfer of miRNAs occurs between fungal and host cells occurs via the action of ExMVs, that this transfer is bidirectional, and that the transfer of miRNAs can result in RNAi that induces host susceptibility to a pathogen (Wang et al., 2016). RNAi that is mediated via such "cross-kingdom" transfer of ExMVs has been show to occur in the aggressive pathogenic fungus *Blumeria cinerea*, where selective silencing of host plant immune genes occurs by the introduction of miRNA virulence effectors (Weiberg et al., 2013). The characterization of miRNAs in EFPs using high-throughput RNA sequencing approaches followed by identification of matching host sequences therefore offers an opportunity to identify potential RNA-based virulence effectors. Moreover, the recognition that virulence can be mediated epigenomically has opened up new opportunities to control fungal diseases using nonfungicide means. For instance, recent work has shown that in *B. cinerea*, the majority of miRNA effectors are derived from retrotransposon LTRs which, when miRNA production is knocked-down through deletion of the key component of the *B. cinerea* RNAi pathway Dicer, that virulence is abrogated *in planta* (Wang et al., 2016). This seminal result was then extended to show that engineering the host plant, in this case *Arabidopsis*, to express the anti-Dicer RNAi conferred resistance against *B. cinerea* demonstrating that host-induced gene silencing of the pathogen occurs. Finally, it was then demonstrated that the simple application of synthetic environmental anti-Dicer RNAi to the fungus, while in the act of infecting the host, resulted in the attenuation of virulence as the fungus took up the RNAi constructs via ExMVs. Studies such as these showing that pathogenic fungi can be epigenomically silenced through nonfungicide-based means, and by the simple application of a nontoxic and highly specific RNAi construct, are clearly a disruptive approach that has broad applicability to a broad span of the non-model EFPs that we have discussing here and shows much promise.

4. CONCLUDING REMARKS

Presently, phylogenomic, comparative genomic, and epigenomic methods are becoming the *modus operandi* for detection and characterization of virulence determinants and epidemiological parameters among EFPs (Hasman et al., 2014; Lecuit & Eloit, 2014), which are themselves increasingly taking center stage for contemporaneous epidemics of plants, humans, and other animals (Fisher et al., 2016). Testaments to the success of this approach are the many examples of traits underpinning EFP that have been identified using these methods. While the full scope and potential of these experimental techniques and resulting compendiums of data are being realized, many challenges remain. Importantly, the continuing adoption of best practices, repeatable protocols, standardizations, and data storage need to be developed to guide future studies working with these new data types and developing powerful new experimental designs. The rapidity of disease outbreaks far outpaces current systems for genomic/epigenomic data acquisition and distribution. Expeditious evaluation and disease mitigation require collaborative research groups that can contribute and coordinate the varied expertise and skills that are needed to tackle new outbreaks. Given the pace and scope of genomics and epigenomic techniques, these fields will likely continue to shape our understanding of pathogen evolution and provide additional approaches to combatting the increasing threat that EFPs pose to biodiversity and ecosystem health.

ACKNOWLEDGMENTS

R.A.F. is supported by an MIT/Wellcome Trust Fellowship. M.C.F. is supported by the UK Natural Research Council, the Leverhulme Trust, and the Morris Animal Foundation.

REFERENCES

Abbey, D., Hickman, M., Gresham, D., & Berman, J. (2011). High-resolution SNP/CGH microarrays reveal the accumulation of loss of heterozygosity in commonly used Candida albicans strains. *G3 (Bethesda, Md)*, *1*, 523–530. https://doi.org/10.1534/g3.111.000885.
Abdolrasouli, A., Rhodes, J., Beale, M. A., Hagen, F., Rogers, T. R., Chowdhary, A., et al. (2015). Genomic context of azole resistance mutations in Aspergillus fumigatus

determined using whole-genome sequencing. *mBio6*. e00536. https://doi.org/10.1128/mBio.00536-15.

Abramyan, J., & Stajich, J. E. (2012). Species-specific chitin-binding module 18 expansion in the amphibian pathogen Batrachochytrium dendrobatidis. *mBio 3*. e00150-112. https://doi.org/10.1128/mBio.00150-12.

Altschul, S. F., Gish, W., Miller, W., Myers, E. W., & Lipman, D. J. (1990). Basic local alignment search tool. *Journal of Molecular Biology, 215*, 403–410. https://doi.org/10.1016/S0022-2836(05)80360-2.

Amselem, J., Lebrun, M.-H., & Quesneville, H. (2015). Whole genome comparative analysis of transposable elements provides new insight into mechanisms of their inactivation in fungal genomes. *BMC Genomics, 16*, 141. https://doi.org/10.1186/s12864-015-1347-1.

Arnold, B., Bomblies, K., & Wakeley, J. (2012). Extending coalescent theory to autotetraploids. *Genetics, 192*, 195–204. https://doi.org/10.1534/genetics.112.140582.

Bairoch, A., & Apweiler, R. (2000). The SWISS-PROT protein sequence database and its supplement TrEMBL in 2000. *Nucleic Acids Research, 28*, 45–48.

Banfield, D. K. (2011). Mechanisms of protein retention in the Golgi. *Cold Spring Harbor Perspectives in Biology, 3*(8), a005264. https://doi.org/10.1101/cshperspect.a005264.

Bankevich, A., Nurk, S., Antipov, D., Gurevich, A. A., Dvorkin, M., Kulikov, A. S., et al. (2012). SPAdes: A new genome assembly algorithm and its applications to single-cell sequencing. *Journal of Computational Biology: A Journal of Computational Molecular Cell Biology, 19*, 455–477. https://doi.org/10.1089/cmb.2012.0021.

Bannister, A. J., & Kouzarides, T. (2011). Regulation of chromatin by histone modifications. *Cell Research, 21*, 381–395. https://doi.org/10.1038/cr.2011.22.

Bellaousov, S., & Mathews, D. H. (2010). ProbKnot: Fast prediction of RNA secondary structure including pseudoknots. *RNA, 16*, 1870–1880. https://doi.org/10.1261/rna.2125310.

Belle, E. M. S., Piganeau, G., Gardner, M., & Eyre-Walker, A. (2005). An investigation of the variation in the transition bias among various animal mitochondrial DNA. *Gene, 355*, 58–66. https://doi.org/10.1016/j.gene.2005.05.019.

Bendtsen, J. D., Jensen, L. J., Blom, N., Von Heijne, G., & Brunak, S. (2004). Feature-based prediction of non-classical and leaderless protein secretion. *Protein Engineering, Design & Selection: PEDS, 17*, 349–356. https://doi.org/10.1093/protein/gzh037.

Bengert, P., & Dandekar, T. (2005). Current efforts in the analysis of RNAi and RNAi target genes. *Briefings in Bioinformatics, 6*, 72–85.

Benjamini, Y., & Hochberg, Y. (1995). Controlling the false discovery rate: A practical and powerful approach to multiple testing. *Journal of the Royal Statistical Society: Series B: Methodological, 57*, 289–300.

Benson, G. (1999). Tandem repeats finder: A program to analyze DNA sequences. *Nucleic Acids Research, 27*, 573–580.

Benson, D. A., Karsch-Mizrachi, I., Clark, K., Lipman, D. J., Ostell, J., & Sayers, E. W. (2012). GenBank. *Nucleic Acids Research, 40*, D48–D53. https://doi.org/10.1093/nar/gkr1202.

Berger, S. L., Kouzarides, T., Shiekhattar, R., & Shilatifard, A. (2009). An operational definition of epigenetics. *Genes & Development, 23*, 781–783. https://doi.org/10.1101/gad.1787609.

Berger, L., Speare, R., Daszak, P., Green, D. E., Cunningham, A. A., Goggin, C. L., et al. (1998). Chytridiomycosis causes amphibian mortality associated with population declines in the rain forests of Australia and Central America. *Proceedings of the National Academy of Sciences of the United States of America, 95*, 9031–9036.

Bernstein, B. E., Stamatoyannopoulos, J. A., Costello, J. F., Ren, B., Milosavljevic, A., Meissner, A., et al. (2010). The NIH roadmap epigenomics mapping consortium. *Nature Biotechnology, 28*, 1045–1048. https://doi.org/10.1038/nbt1010-1045.

Besemer, J., & Borodovsky, M. (1999). Heuristic approach to deriving models for gene finding. *Nucleic Acids Research*, *27*, 3911–3920.

Bigirimana, V. d. P., Hua, G. K. H., Nyamangyoku, O. I., & Höfte, M. (2015). Rice sheath rot: An emerging ubiquitous destructive disease complex. *Frontiers in Plant Science*, *6*, 1066. https://doi.org/10.3389/fpls.2015.01066.

Birney, E., Clamp, M., & Durbin, R. (2004). GeneWise and genomewise. *Genome Research*, *14*, 988–995. https://doi.org/10.1101/gr.1865504.

Biswas, S., & Akey, J. M. (2006). Genomic insights into positive selection. *Trends in Genetics: TIG*, *22*, 437–446. https://doi.org/10.1016/j.tig.2006.06.005.

Blackwell, M. (2011). The fungi: 1, 2, 3 ... 5.1 million species? *American Journal of Botany*, *98*, 426–438. https://doi.org/10.3732/ajb.1000298.

Blanchette, M., Kent, W. J., Riemer, C., Elnitski, L., Smit, A. F. A., Roskin, K. M., et al. (2004). Aligning multiple genomic sequences with the threaded blockset aligner. *Genome Research*, *14*, 708–715. https://doi.org/10.1101/gr.1933104.

Blanco, E., Parra, G., & Guigó, R. (2007). Using geneid to identify genes. *Current Protocols in Bioinformatics* chapter 4, Unit 4.3https://doi.org/10.1002/0471250953.bi0403s18.

Blehert, D. S., Hicks, A. C., Behr, M., Meteyer, C. U., Berlowski-Zier, B. M., Buckles, E. L., et al. (2009). Bat white-nose syndrome: An emerging fungal pathogen? *Science*, *323*, 227. https://doi.org/10.1126/science.1163874.

Boetzer, M., & Pirovano, W. (2014). SSPACE-LongRead: Scaffolding bacterial draft genomes using long read sequence information. *BMC Bioinformatics*, *15*, 211. https://doi.org/10.1186/1471-2105-15-211.

Bos, J. I. B., Kanneganti, T.-D., Young, C., Cakir, C., Huitema, E., Win, J., et al. (2006). The C-terminal half of Phytophthora infestans RXLR effector AVR3a is sufficient to trigger R3a-mediated hypersensitivity and suppress INF1-induced cell death in Nicotiana benthamiana. *The Plant Journal: For Cell and Molecular Biology*, *48*, 165–176. https://doi.org/10.1111/j.1365-313X.2006.02866.x.

Bovers, M., Hagen, F., Kuramae, E. E., Diaz, M. R., Spanjaard, L., Dromer, F., et al. (2006). Unique hybrids between the fungal pathogens Cryptococcus neoformans and Cryptococcus gattii. *FEMS Yeast Research*, *6*, 599–607. https://doi.org/10.1111/j.1567-1364.2006.00082.x.

Boyce, K. J., Wang, Y., Verma, S., Shakya, V. P. S., Xue, C., & Idnurm, A. (2017). Mismatch repair of DNA replication errors contributes to microevolution in the pathogenic fungus Cryptococcus neoformans. *mBio*, *8*(3), pii: e00595-17. https://doi.org/10.1128/mBio.00595-17.

Bradnam, K. R., Fass, J. N., Alexandrov, A., Baranay, P., Bechner, M., Birol, I., et al. (2013). Assemblathon 2: Evaluating de novo methods of genome assembly in three vertebrate species. *GigaScience*, *2*, 10. https://doi.org/10.1186/2047-217X-2-10.

Butler, J., MacCallum, I., Kleber, M., Shlyakhter, I. A., Belmonte, M. K., Lander, E. S., et al. (2008). ALLPATHS: De novo assembly of whole-genome shotgun microreads. *Genome Research*, *18*, 810–820. https://doi.org/10.1101/gr.7337908.

Büttner, P., Koch, F., Voigt, K., Quidde, T., Risch, S., Blaich, R., et al. (1994). Variations in ploidy among isolates of Botrytis cinerea: Implications for genetic and molecular analyses. *Current Genetics*, *25*, 445–450.

Byrnes, E. J., III, Li, W., Lewit, Y., Ma, H., Voelz, K., Ren, P., et al. (2010). Emergence and pathogenicity of highly virulent Cryptococcus gattii genotypes in the Northwest United States. *PLoS Pathogens 6*. e1000850. https://doi.org/10.1371/journal.ppat.1000850.

Cantarel, B. L., Korf, I., Robb, S. M. C., Parra, G., Ross, E., Moore, B., et al. (2008). MAKER: An easy-to-use annotation pipeline designed for emerging model organism genomes. *Genome Research*, *18*, 188–196. https://doi.org/10.1101/gr.6743907.

Capella-Gutiérrez, S., Silla-Martínez, J. M., & Gabaldón, T. (2009). trimAl: A tool for automated alignment trimming in large-scale phylogenetic analyses. *Bioinformatics (Oxford, England)*, *25*, 1972–1973. https://doi.org/10.1093/bioinformatics/btp348.

Capuano, F., Mülleder, M., Kok, R., Blom, H. J., & Ralser, M. (2014). Cytosine DNA methylation is found in Drosophila melanogaster but absent in Saccharomyces cerevisiae, Schizosaccharomyces pombe, and other yeast species. *Analytical Chemistry*, *86*, 3697–3702. https://doi.org/10.1021/ac500447w.

Carr, J., & Shearer, G. (1998). Genome size, complexity, and ploidy of the pathogenic fungus Histoplasma capsulatum. *Journal of Bacteriology*, *180*, 6697–6703.

Casadevall, A., Freij, J. B., Hann-Soden, C., & Taylor, J. (2017). Continental drift and speciation of the *Cryptococcus neoformans* and *Cryptococcus gattii* species complexes. *mSphere2*. e00103-17. https://doi.org/10.1128/mSphere.00103-17.

Chen, J.-L., & Greider, C. W. (2005). Functional analysis of the pseudoknot structure in human telomerase RNA. *Proceedings of the National Academy of Sciences of the United States of America*, *102*, 8080–8085. https://doi.org/10.1073/pnas.0502259102.

Chen, X.-R., Xing, Y.-P., Li, Y.-P., Tong, Y.-H., & Xu, J.-Y. (2013). RNA-Seq reveals infection-related gene expression changes in Phytophthora capsici. *PLoS One8*. , https://doi.org/10.1371/journal.pone.0074588.

Chowdhary, A., Sharma, C., Hagen, F., & Meis, J. F. (2014). Exploring azole antifungal drug resistance in Aspergillus fumigatus with special reference to resistance mechanisms. *Future Microbiology*, *9*, 697–711. https://doi.org/10.2217/fmb.14.27.

Chowdhary, A., Sharma, C., & Meis, J. F. (2017). Candida auris: A rapidly emerging cause of hospital-acquired multidrug-resistant fungal infections globally. *PLoS Pathogens13*. e1006290. https://doi.org/10.1371/journal.ppat.1006290.

Chu, Y., Kalantari, R., Dodd, D. W., & Corey, D. R. (2012). Transcriptional silencing by hairpin RNAs complementary to a gene promoter. *Nucleic Acid Therapeutics*, *22*, 147–151. https://doi.org/10.1089/nat.2012.0360.

Cibulskis, K., Lawrence, M. S., Carter, S. L., Sivachenko, A., Jaffe, D., Sougnez, C., et al. (2013). Sensitive detection of somatic point mutations in impure and heterogeneous cancer samples. *Nature Biotechnology*, *31*, 213–219. https://doi.org/10.1038/nbt.2514.

Cole, J. R., Wang, Q., Fish, J. A., Chai, B., McGarrell, D. M., Sun, Y., et al. (2014). Ribosomal database project: Data and tools for high throughput rRNA analysis. *Nucleic Acids Research*, *42*, D633–642. https://doi.org/10.1093/nar/gkt1244.

Conesa, A., Götz, S., García-Gómez, J. M., Terol, J., Talón, M., & Robles, M. (2005). Blast2GO: A universal tool for annotation, visualization and analysis in functional genomics research. *Bioinformatics (Oxford, England)*, *21*, 3674–3676. https://doi.org/10.1093/bioinformatics/bti610.

Cools, H. J., & Fraaije, B. A. (2008). Are azole fungicides losing ground against Septoria wheat disease? Resistance mechanisms in Mycosphaerella graminicola. *Pest Management Science*, *64*, 681–684. https://doi.org/10.1002/ps.1568.

Cooper, D. N., Mort, M., Stenson, P. D., Ball, E. V., & Chuzhanova, N. A. (2010). Methylation-mediated deamination of 5-methylcytosine appears to give rise to mutations causing human inherited disease in CpNpG trinucleotides, as well as in CpG dinucleotides. *Human Genomics*, *4*, 406–410.

Croucher, N. J., & Didelot, X. (2015). The application of genomics to tracing bacterial pathogen transmission. *Current Opinion in Microbiology*, *23*, 62–67. https://doi.org/10.1016/j.mib.2014.11.004.

Cushion, M. T., & Stringer, J. R. (2010). Stealth and opportunism: Alternative lifestyles of species in the fungal genus Pneumocystis. *Annual Review of Microbiology*, *64*, 431–452. https://doi.org/10.1146/annurev.micro.112408.134335.

D'Souza, C. A., Kronstad, J. W., Taylor, G., Warren, R., Yuen, M., Hu, G., et al. (2011). Genome variation in Cryptococcus gattii, an emerging pathogen of immunocompetent hosts. *mBio 2*. e00342-310. https://doi.org/10.1128/mBio.00342-10.

Dang, Y., Yang, Q., Xue, Z., & Liu, Y. (2011). RNA interference in fungi: Pathways, functions, and applications. *Eukaryotic Cell*, *10*, 1148–1155. https://doi.org/10.1128/EC.05109-11.

de Roode, J. C., Pansini, R., Cheesman, S. J., Helinski, M. E. H., Huijben, S., Wargo, A. R., et al. (2005). Virulence and competitive ability in genetically diverse malaria infections. *Proceedings of the National Academy of Sciences of the United States of America*, *102*, 7624–7628. https://doi.org/10.1073/pnas.0500078102.

Dettman, J. R., Jacobson, D. J., & Taylor, J. W. (2003). A multilocus genealogical approach to phylogenetic species recognition in the model eukaryote Neurospora. *Evolution; International Journal of Organic Evolution*, *57*, 2703–2720.

Dewey, C. N. (2007). Aligning multiple whole genomes with Mercator and MAVID. *Methods in Molecular Biology (Clifton, NJ)*, *395*, 221–236.

Diaz-Guerra, T. M., Mellado, E., Cuenca-Estrella, M., & Rodriguez-Tudela, J. L. (2003). A point mutation in the 14alpha-sterol demethylase gene cyp51A contributes to itraconazole resistance in Aspergillus fumigatus. *Antimicrobial Agents and Chemotherapy*, *47*, 1120–1124.

Dobin, A., Davis, C. A., Schlesinger, F., Drenkow, J., Zaleski, C., Jha, S., et al. (2013). STAR: Ultrafast universal RNA-seq aligner. *Bioinformatics (Oxford, England)*, *29*, 15–21. https://doi.org/10.1093/bioinformatics/bts635.

Dong, S., Shen, Z., Xu, L., & Zhu, F. (2010). Sequence and phylogenetic analysis of SSU rRNA gene of five microsporidia. *Current Microbiology*, *60*, 30–37. https://doi.org/10.1007/s00284-009-9495-7.

Drummond, A. J., & Rambaut, A. (2007). BEAST: Bayesian evolutionary analysis by sampling trees. *BMC Evolutionary Biology*, 7, 214. https://doi.org/10.1186/1471-2148-7-214.

Duan, C.-G., Wang, C.-H., & Guo, H.-S. (2012). Application of RNA silencing to plant disease resistance. *Silence*, *3*, 5. https://doi.org/10.1186/1758-907X-3-5.

Dujon, B. A., & Louis, E. J. (2017). Genome diversity and evolution in the budding yeasts (Saccharomycotina). *Genetics*, *206*, 717–750. https://doi.org/10.1534/genetics.116.199216.

Dunn, O. J. (1959). Estimation of the medians for dependent variables. *Annals of Mathematical Statistics*, *30*, 192–197. https://doi.org/10.1214/aoms/1177706374.

Eddy, S. R. (2011). Accelerated profile HMM searches. *PLoS Computational Biology* 7. e1002195. https://doi.org/10.1371/journal.pcbi.1002195.

Eddy, S. R., & Durbin, R. (1994). RNA sequence analysis using covariance models. *Nucleic Acids Research*, *22*, 2079–2088.

Edgar, R. C. (2004a). MUSCLE: Multiple sequence alignment with high accuracy and high throughput. *Nucleic Acids Research*, *32*, 1792–1797. https://doi.org/10.1093/nar/gkh340.

Edgar, R. C. (2004b). MUSCLE: A multiple sequence alignment method with reduced time and space complexity. *BMC Bioinformatics*, *5*, 113. https://doi.org/10.1186/1471-2105-5-113.

Eilertson, K. E., Booth, J. G., & Bustamante, C. D. (2012). SnIPRE: Selection inference using a poisson random effects model. *PLoS Computational Biology 8*. e1002806. https://doi.org/10.1371/journal.pcbi.1002806.

Engelthaler, D. M., Hicks, N. D., Gillece, J. D., Roe, C. C., Schupp, J. M., Driebe, E. M., et al. (2014). Cryptococcus gattii in North American Pacific Northwest: Whole-population genome analysis provides insights into species evolution and dispersal. *mBio 5*. e01464-14. https://doi.org/10.1128/mBio.01464-14.

Enright, A. J., Van Dongen, S., & Ouzounis, C. A. (2002). An efficient algorithm for large-scale detection of protein families. *Nucleic Acids Research*, *30*, 1575–1584.

European Centre for Disease Prevention and Control. (2013). *Risk assessment on the impact of environmental usage of triazoles on the development and spread of resistance to medical triazoles in Aspergillus species*. Stockholm: European Centre for Disease Prevention and Control.

Faria, N. R., Rambaut, A., Suchard, M. A., Baele, G., Bedford, T., Ward, M. J., et al. (2014). HIV epidemiology. The early spread and epidemic ignition of HIV-1 in human populations. *Science*, *346*, 56–61. https://doi.org/10.1126/science.1256739.

Farrer, R. A., Desjardins, C. A., Sakthikumar, S., Gujja, S., Saif, S., Zeng, Q., et al. (2015). Genome evolution and innovation across the four major lineages of Cryptococcus gattii. *mBio*, *6*(5), e00868-15. https://doi.org/10.1128/mBio.00868-15.

Farrer, R. A., Henk, D. A., Garner, T. W. J., Balloux, F., Woodhams, D. C., & Fisher, M. C. (2013). Chromosomal copy number variation, selection and uneven rates of recombination reveal cryptic genome diversity linked to pathogenicity. *PLoS Genetics 9*. e1003703. https://doi.org/10.1371/journal.pgen.1003703.

Farrer, R. A., Henk, D. A., MacLean, D., Studholme, D. J., & Fisher, M. C. (2013). Using false discovery rates to benchmark SNP-callers in next-generation sequencing projects. *Scientific Reports*, *3*, 1512. https://doi.org/10.1038/srep01512.

Farrer, R. A., Martel, A., Verbrugghe, E., Abouelleil, A., Ducatelle, R., Longcore, J. E., et al. (2017). Genomic innovations linked to infection strategies across emerging pathogenic chytrid fungi. *Nature Communications 8*. 14742. https://doi.org/10.1038/ncomms14742.

Farrer, R. A., Voelz, K., Henk, D. A., Johnston, S. A., Fisher, M. C., May, R. C., et al. (2016). Microevolutionary traits and comparative population genomics of the emerging pathogenic fungus Cryptococcus gattii. *Philosophical Transactions of the Royal Society of London Series B, Biological Sciences 371*. 20160021. https://doi.org/10.1098/rstb.2016.0021.

Farrer, R. A., Weinert, L. A., Bielby, J., Garner, T. W. J., Balloux, F., Clare, F., et al. (2011). Multiple emergences of genetically diverse amphibian-infecting chytrids include a globalized hypervirulent recombinant lineage. *Proceedings of the National Academy of Sciences of the United States of America*, *108*, 18732–18736. https://doi.org/10.1073/pnas.1111915108.

Finn, R. D., Bateman, A., Clements, J., Coggill, P., Eberhardt, R. Y., Eddy, S. R., et al. (2014). Pfam: The protein families database. *Nucleic Acids Research*, *42*, D222–D230. https://doi.org/10.1093/nar/gkt1223.

Finn, R. D., Clements, J., & Eddy, S. R. (2011). HMMER web server: Interactive sequence similarity searching. *Nucleic Acids Research*, *39*, W29–W37. https://doi.org/10.1093/nar/gkr367.

Fishel, B., Amstutz, H., Baum, M., Carbon, J., & Clarke, L. (1988). Structural organization and functional analysis of centromeric DNA in the fission yeast Schizosaccharomyces pombe. *Molecular and Cellular Biology*, *8*, 754–763.

Fisher, M. C., Gow, N. A. R., & Gurr, S. J. (2016). Tackling emerging fungal threats to animal health, food security and ecosystem resilience. *Philosophical Transactions of the Royal Society of London Series B, Biological Sciences 371*. 20160332. https://doi.org/10.1098/rstb.2016.0332.

Fisher, M. C., Henk, D. A., Briggs, C. J., Brownstein, J. S., Madoff, L. C., McCraw, S. L., et al. (2012). Emerging fungal threats to animal, plant and ecosystem health. *Nature*, *484*, 186–194. https://doi.org/10.1038/nature10947.

Fones, H. N., Fisher, M. C., & Gurr, S. J. (2017). Emerging fungal threats to plants and animals challenge agriculture and ecosystem resilience. *Microbiology Spectrum*, *5*(2). https://doi.org/10.1128/microbiolspec.FUNK-0027-2016.

Forche, A., Alby, K., Schaefer, D., Johnson, A. D., Berman, J., & Bennett, R. J. (2008). The parasexual cycle in Candida albicans provides an alternative pathway to meiosis for the formation of recombinant strains. *PLoS Biology 6*. e110. https://doi.org/10.1371/journal.pbio.0060110.

Forche, A., Magee, P. T., Selmecki, A., Berman, J., & May, G. (2009). Evolution in Candida albicans populations during a single passage through a mouse host. *Genetics*, *182*, 799–811. https://doi.org/10.1534/genetics.109.103325.

Fraczek, M. G., Bromley, M., Buied, A., Moore, C. B., Rajendran, R., Rautemaa, R., et al. (2013). The cdr1B efflux transporter is associated with non-cyp51a-mediated itraconazole resistance in Aspergillus fumigatus. *The Journal of Antimicrobial Chemotherapy*, *68*, 1486–1496. https://doi.org/10.1093/jac/dkt075.

Fraser, J. A., Giles, S. S., Wenink, E. C., Geunes-Boyer, S. G., Wright, J. R., Diezmann, S., et al. (2005). Same-sex mating and the origin of the Vancouver Island Cryptococcus gattii outbreak. *Nature*, *437*, 1360–1364. https://doi.org/10.1038/nature04220.

Friesen, T. L., Stukenbrock, E. H., Liu, Z., Meinhardt, S., Ling, H., Faris, J. D., et al. (2006). Emergence of a new disease as a result of interspecific virulence gene transfer. *Nature Genetics*, *38*, 953–956. https://doi.org/10.1038/ng1839.

Frith, M. C., Saunders, N. F. W., Kobe, B., & Bailey, T. L. (2008). Discovering sequence motifs with arbitrary insertions and deletions. *PLoS Computational Biology*, *4*(5), e1000071. https://doi.org/10.1371/journal.pcbi.1000071.

Galagan, J. E., Henn, M. R., Ma, L.-J., Cuomo, C. A., & Birren, B. (2005). Genomics of the fungal kingdom: Insights into eukaryotic biology. *Genome Research*, *15*, 1620–1631. https://doi.org/10.1101/gr.3767105.

Gardiner, D. M., McDonald, M. C., Covarelli, L., Solomon, P. S., Rusu, A. G., Marshall, M., et al. (2012). Comparative pathogenomics reveals horizontally acquired novel virulence genes in fungi infecting cereal hosts. *PLoS Pathogens 8*. e1002952. https://doi.org/10.1371/journal.ppat.1002952.

Garner, T. W., Perkins, M. W., Govindarajulu, P., Seglie, D., Walker, S., Cunningham, A. A., et al. (2006). The emerging amphibian pathogen Batrachochytrium dendrobatidis globally infects introduced populations of the North American bullfrog, Rana catesbeiana. *Biology Letters*, *2*, 455–459. https://doi.org/10.1098/rsbl.2006.0494.

Garrison, E., & Marth, G. (2012). *Haplotype-based variant detection from short-read sequencing*. ArXiv12073907 Q-Bio.

Gauthier, C., Weber, S., Alarco, A.-M., Alqawi, O., Daoud, R., Georges, E., et al. (2003). Functional similarities and differences between Candida albicans Cdr1p and Cdr2p transporters. *Antimicrobial Agents and Chemotherapy*, *47*, 1543–1554. https://doi.org/10.1128/AAC.47.5.1543-1554.2003.

Gelfand, Y., Rodriguez, A., & Benson, G. (2007). TRDB—The tandem repeats database. *Nucleic Acids Research*, *35*, D80–D87. https://doi.org/10.1093/nar/gkl1013.

Giraud, T., Gladieux, P., & Gavrilets, S. (2010). Linking the emergence of fungal plant diseases with ecological speciation. *Trends in Ecology & Evolution*, *25*, 387–395. https://doi.org/10.1016/j.tree.2010.03.006.

Gnerre, S., Maccallum, I., Przybylski, D., Ribeiro, F. J., Burton, J. N., Walker, B. J., et al. (2011). High-quality draft assemblies of mammalian genomes from massively parallel sequence data. *Proceedings of the National Academy of Sciences of the United States of America*, *108*, 1513–1518. https://doi.org/10.1073/pnas.1017351108.

Goldberg, A. D., Allis, C. D., & Bernstein, E. (2007). Epigenetics: A landscape takes shape. *Cell*, *128*, 635–638. https://doi.org/10.1016/j.cell.2007.02.006.

Grice, C. M., Bertuzzi, M., & Bignell, E. M. (2013). Receptor-mediated signaling in Aspergillus fumigatus. *Frontiers in Microbiology*, *4*, 26. https://doi.org/10.3389/fmicb.2013.00026.

Griffiths-Jones, S., Bateman, A., Marshall, M., Khanna, A., & Eddy, S. R. (2003). Rfam: An RNA family database. *Nucleic Acids Research*, *31*, 439–441.

Haas, B. J., Kamoun, S., Zody, M. C., Jiang, R. H. Y., Handsaker, R. E., Cano, L. M., et al. (2009). Genome sequence and analysis of the Irish potato famine pathogen Phytophthora infestans. *Nature*, *461*, 393–398. https://doi.org/10.1038/nature08358.

Haas, B. J., Papanicolaou, A., Yassour, M., Grabherr, M., Blood, P. D., Bowden, J., et al. (2013). De novo transcript sequence reconstruction from RNA-Seq: Reference generation and analysis with Trinity. *Nature Protocols*, *8*, 1494–1512. https://doi.org/10.1038/nprot.2013.084.

Haas, B. J., Salzberg, S. L., Zhu, W., Pertea, M., Allen, J. E., Orvis, J., et al. (2008). Automated eukaryotic gene structure annotation using EVidenceModeler and the program to assemble spliced alignments. *Genome Biology, 9*, R7. https://doi.org/10.1186/gb-2008-9-1-r7.

Haas, B. J., Zeng, Q., Pearson, M. D., Cuomo, C. A., & Wortman, J. R. (2011). Approaches to fungal genome annotation. *Mycology, 2*, 118–141. https://doi.org/10.1080/21501203.2011.606851.

Haft, D. H., Selengut, J. D., & White, O. (2003). The TIGRFAMs database of protein families. *Nucleic Acids Research, 31*, 371–373.

Hagen, F., Ceresini, P. C., Polacheck, I., Ma, H., van Nieuwerburgh, F., Gabaldón, T., et al. (2013). Ancient dispersal of the human fungal pathogen Cryptococcus gattii from the Amazon rainforest. *PLoS One 8*. e71148. https://doi.org/10.1371/journal.pone.0071148.

Hagen, F., Khayhan, K., Theelen, B., Kolecka, A., Polacheck, I., Sionov, E., et al. (2015). Recognition of seven species in the Cryptococcus gattii/Cryptococcus neoformans species complex. *Fungal Genetics and Biology: FG & B, 78*, 16–48. https://doi.org/10.1016/j.fgb.2015.02.009.

Hartmann, F. E., Sánchez-Vallet, A., McDonald, B. A., & Croll, D. (2017). A fungal wheat pathogen evolved host specialization by extensive chromosomal rearrangements. *The ISME Journal, 11*, 1189–1204. https://doi.org/10.1038/ismej.2016.196.

Hasman, H., Saputra, D., Sicheritz-Ponten, T., Lund, O., Svendsen, C. A., Frimodt-Møller, N., et al. (2014). Rapid whole-genome sequencing for detection and characterization of microorganisms directly from clinical samples. *Journal of Clinical Microbiology, 52*, 139–146. https://doi.org/10.1128/JCM.02452-13.

Healey, K. R., Zhao, Y., Perez, W. B., Lockhart, S. R., Sobel, J. D., Farmakiotis, D., et al. (2016). Prevalent mutator genotype identified in fungal pathogen Candida glabrata promotes multi-drug resistance. *Nature Communications 7*. 11128. https://doi.org/10.1038/ncomms11128.

Hibbett, D. S., Binder, M., Bischoff, J. F., Blackwell, M., Cannon, P. F., Eriksson, O. E., et al. (2007). A higher-level phylogenetic classification of the Fungi. *Mycological Research, 111*, 509–547. https://doi.org/10.1016/j.mycres.2007.03.004.

Hittalmani, S., Mahesh, H. B., Mahadevaiah, C., & Prasannakumar, M. K. (2016). De novo genome assembly and annotation of rice sheath rot fungus Sarocladium oryzae reveals genes involved in Helvolic acid and Cerulenin biosynthesis pathways. *BMC Genomics, 17*, 271. https://doi.org/10.1186/s12864-016-2599-0.

Hogenhout, S. A., Van der Hoorn, R. A. L., Terauchi, R., & Kamoun, S. (2009). Emerging concepts in effector biology of plant-associated organisms. *Molecular Plant-Microbe Interactions: MPMI, 22*, 115–122. https://doi.org/10.1094/MPMI-22-2-0115.

Holliday, R. (1990). DNA methylation and epigenetic inheritance. *Philosophical Transactions of the Royal Society of London Series B, Biological Sciences, 326*, 329–338. https://doi.org/10.1098/rstb.1990.0015.

Hsueh, P.-R., Teng, L.-J., Hung, C.-C., Hsu, J.-H., Yang, P.-C., Ho, S.-W., et al. (2000). Molecular evidence for strain dissemination of Penicillium marneffei: An emerging pathogen in Taiwan. *The Journal of Infectious Diseases, 181*, 1706–1712. https://doi.org/10.1086/315432.

Hu, G., Wang, J., Choi, J., Jung, W. H., Liu, I., Litvintseva, A. P., et al. (2011). Variation in chromosome copy number influences the virulence of Cryptococcus neoformans and occurs in isolates from AIDS patients. *BMC Genomics, 12*, 526. https://doi.org/10.1186/1471-2164-12-526.

Huff, J. T., & Zilberman, D. (2014). Dnmt1-independent CG methylation contributes to nucleosome positioning in diverse eukaryotes. *Cell, 156*, 1286–1297. https://doi.org/10.1016/j.cell.2014.01.029.

Hunt, M., Kikuchi, T., Sanders, M., Newbold, C., Berriman, M., & Otto, T. D. (2013). REAPR: A universal tool for genome assembly evaluation. *Genome Biology, 14*, R47. https://doi.org/10.1186/gb-2013-14-5-r47.

Inderbitzin, P., Davis, R. M., Bostock, R. M., & Subbarao, K. V. (2011). The ascomycete Verticillium longisporum is a hybrid and a plant pathogen with an expanded host range. *PLoS One 6.* e18260. https://doi.org/10.1371/journal.pone.0018260.

Islam, M. T., Croll, D., Gladieux, P., Soanes, D. M., Persoons, A., Bhattacharjee, P., et al. (2016). Emergence of wheat blast in Bangladesh was caused by a South American lineage of Magnaporthe oryzae. *BMC Biology, 14,* 84. https://doi.org/10.1186/s12915-016-0309-7.

James, T. Y., Litvintseva, A. P., Vilgalys, R., Morgan, J. A. T., Taylor, J. W., Fisher, M. C., et al. (2009). Rapid global expansion of the fungal disease chytridiomycosis into declining and healthy amphibian populations. *PLoS Pathogens 5.* e1000458. https://doi.org/10.1371/journal.ppat.1000458.

Janbon, G., Ormerod, K. L., Paulet, D., Iii, E. J. B., Yadav, V., Chatterjee, G., et al. (2014). Analysis of the genome and transcriptome of Cryptococcus neoformans var. grubii reveals complex RNA expression and microevolution leading to virulence attenuation. *PLoS Genetics 10.* e1004261. https://doi.org/10.1371/journal.pgen.1004261.

Jeon, J., Choi, J., Lee, G.-W., Park, S.-Y., Huh, A., Dean, R. A., et al. (2015). Genome-wide profiling of DNA methylation provides insights into epigenetic regulation of fungal development in a plant pathogenic fungus, Magnaporthe oryzae. *Scientific Reports, 5,* 8567. https://doi.org/10.1038/srep08567.

Jeon, J., Kwon, S., & Lee, Y.-H. (2014). Histone acetylation in fungal pathogens of plants. *Plant Pathology Journal, 30,* 1–9. https://doi.org/10.5423/PPJ.RW.01.2014.0003.

Kajitani, R., Toshimoto, K., Noguchi, H., Toyoda, A., Ogura, Y., Okuno, M., et al. (2014). Efficient de novo assembly of highly heterozygous genomes from whole-genome shotgun short reads. *Genome Research, 24,* 1384–1395. https://doi.org/10.1101/gr.170720.113.

Kanehisa, M., & Goto, S. (2000). KEGG: Kyoto encyclopedia of genes and genomes. *Nucleic Acids Research, 28,* 27–30.

Karvonen, A., Rellstab, C., Louhi, K.-R., & Jokela, J. (2012). Synchronous attack is advantageous: Mixed genotype infections lead to higher infection success in trematode parasites. *Proceedings of the Biological Sciences, 279,* 171–176. https://doi.org/10.1098/rspb.2011.0879.

Kasuga, T., White, T. J., & Taylor, J. W. (2002). Estimation of nucleotide substitution rates in eurotiomycete fungi. *Molecular Biology and Evolution, 19*(12), 2318–2324.

Katoh, K., & Standley, D. M. (2013). MAFFT multiple sequence alignment software version 7: Improvements in performance and usability. *Molecular Biology and Evolution, 30,* 772–780. https://doi.org/10.1093/molbev/mst010.

Kent, W. J. (2002). BLAT—The BLAST-like alignment tool. *Genome Research, 12,* 656–664. https://doi.org/10.1101/gr.229202. Article published online before March 2002.

Khatri, M., & Rajam, M. V. (2007). Targeting polyamines of Aspergillus nidulans by siRNA specific to fungal ornithine decarboxylase gene. *Medical Mycology, 45,* 211–220. https://doi.org/10.1080/13693780601158779.

Kim, K.-H., Willger, S. D., Park, S.-W., Puttikamonkul, S., Grahl, N., Cho, Y., et al. (2009). TmpL, a transmembrane protein required for intracellular redox homeostasis and virulence in a plant and an animal fungal pathogen. *PLoS Pathogens 5.* e1000653. https://doi.org/10.1371/journal.ppat.1000653.

Kõljalg, U., Nilsson, R. H., Abarenkov, K., Tedersoo, L., Taylor, A. F. S., Bahram, M., et al. (2013). Towards a unified paradigm for sequence-based identification of fungi. *Molecular Ecology, 22,* 5271–5277. https://doi.org/10.1111/mec.12481.

Koren, S., Walenz, B. P., Berlin, K., Miller, J. R., Bergman, N. H., & Phillippy, A. M. (2017). Canu: Scalable and accurate long-read assembly via adaptive k-mer weighting and repeat separation. *Genome Research, 27*(5), 722–736. https://doi.org/10.1101/gr.215087.116.

Korf, I. (2004). Gene finding in novel genomes. *BMC Bioinformatics*, *5*, 59. https://doi.org/10.1186/1471-2105-5-59.

Krogh, A., Larsson, B., von Heijne, G., & Sonnhammer, E. L. (2001). Predicting transmembrane protein topology with a hidden Markov model: Application to complete genomes. *Journal of Molecular Biology*, *305*, 567–580. https://doi.org/10.1006/jmbi.2000.4315.

Kryazhimskiy, S., & Plotkin, J. B. (2008). The population genetics of dN/dS. *PLoS Genetics*, *4*(12), e1000304. https://doi.org/10.1371/journal.pgen.1000304.

Kuo, D., Tan, K., Zinman, G., Ravasi, T., Bar-Joseph, Z., & Ideker, T. (2010). Evolutionary divergence in the fungal response to fluconazole revealed by soft clustering. *Genome Biology*, *11*, R77. https://doi.org/10.1186/gb-2010-11-7-r77.

Kurtz, S., Phillippy, A., Delcher, A. L., Smoot, M., Shumway, M., Antonescu, C., et al. (2004). Versatile and open software for comparing large genomes. *Genome Biology*, *5*, R12. https://doi.org/10.1186/gb-2004-5-2-r12.

Kwon-Chung, K. J., Bennett, J. E., Wickes, B. L., Meyer, W., Cuomo, C. A., Wollenburg, K. R., et al. (2017). The case for adopting the "Species Complex" nomenclature for the etiologic agents of cryptococcosis. *mSphere*, *2*(1), pii: e00357-16. https://doi.org/10.1128/mSphere.00357-16.

Kwon-Chung, K. J., & Chang, Y. C. (2012). Aneuploidy and drug resistance in pathogenic fungi. *PLoS Pathogens*, *8*(11), e1003022. https://doi.org/10.1371/journal.ppat.1003022.

Lagesen, K., Hallin, P., Rødland, E. A., Staerfeldt, H.-H., Rognes, T., & Ussery, D. W. (2007). RNAmmer: Consistent and rapid annotation of ribosomal RNA genes. *Nucleic Acids Research*, *35*, 3100–3108. https://doi.org/10.1093/nar/gkm160.

Lam, J. K. W., Chow, M. Y. T., Zhang, Y., & Leung, S. W. S. (2015). siRNA versus miRNA as therapeutics for gene silencing. *Molecular Therapy Nucleic Acids*, *4*, e252. https://doi.org/10.1038/mtna.2015.23.

Langmead, B., & Salzberg, S. L. (2012). Fast gapped-read alignment with Bowtie 2. *Nature Methods*, *9*, 357–359. https://doi.org/10.1038/nmeth.1923.

Lassmann, T., & Sonnhammer, E. L. L. (2005). Automatic assessment of alignment quality. *Nucleic Acids Research*, *33*, 7120–7128. https://doi.org/10.1093/nar/gki1020.

Law, J. A., & Jacobsen, S. E. (2010). Establishing, maintaining and modifying DNA methylation patterns in plants and animals. *Nature Reviews Genetics*, *11*, 204–220. https://doi.org/10.1038/nrg2719.

Leach, M. D., Farrer, R. A., Tan, K., Miao, Z., Walker, L. A., Cuomo, C. A., et al. (2016). Hsf1 and Hsp90 orchestrate temperature-dependent global transcriptional remodelling and chromatin architecture in Candida albicans. *Nature Communications*7, 11704. https://doi.org/10.1038/ncomms11704.

Lecuit, M., & Eloit, M. (2014). The diagnosis of infectious diseases by whole genome next generation sequencing: A new era is opening. *Frontiers in Cellular and Infection Microbiology*, *4*, 25. https://doi.org/10.3389/fcimb.2014.00025.

Lengeler, K. B., Cox, G. M., & Heitman, J. (2001). Serotype AD strains of Cryptococcus neoformans are diploid or aneuploid and are heterozygous at the mating-type locus. *Infection and Immunity*, *69*, 115–122. https://doi.org/10.1128/IAI.69.1.115-122.2001.

Leopardi, S., Blake, D., & Puechmaille, S. J. (2015). White-nose syndrome fungus introduced from Europe to North America. *Current Biology: CB*, *25*, R217–219. https://doi.org/10.1016/j.cub.2015.01.047.

Li, H., & Durbin, R. (2010). Fast and accurate long-read alignment with Burrows-Wheeler transform. *Bioinformatics (Oxford, England)*, *26*, 589–595. https://doi.org/10.1093/bioinformatics/btp698.

Li, W., & Godzik, A. (2006). Cd-hit: A fast program for clustering and comparing large sets of protein or nucleotide sequences. *Bioinformatics (Oxford, England)*, *22*, 1658–1659. https://doi.org/10.1093/bioinformatics/btl158.

Li, R., Zhu, H., Ruan, J., Qian, W., Fang, X., Shi, Z., et al. (2010). De novo assembly of human genomes with massively parallel short read sequencing. *Genome Research, 20*, 265–272. https://doi.org/10.1101/gr.097261.109.

Lister, R., Pelizzola, M., Dowen, R. H., Hawkins, R. D., Hon, G., Tonti-Filippini, J., et al. (2009). Human DNA methylomes at base resolution show widespread epigenomic differences. *Nature, 462*, 315–322. https://doi.org/10.1038/nature08514.

Liti, G., Carter, D. M., Moses, A. M., Warringer, J., Parts, L., James, S. A., et al. (2009). Population genomics of domestic and wild yeasts. *Nature, 458*, 337–341. https://doi.org/10.1038/nature07743.

Litvintseva, A. P., Carbone, I., Rossouw, J., Thakur, R., Govender, N. P., & Mitchell, T. G. (2011). Evidence that the human pathogenic fungus Cryptococcus neoformans var. grubii may have evolved in Africa. *PLoS One 6*. e19688. https://doi.org/10.1371/journal.pone.0019688.

Liu, O. W., Chun, C. D., Chow, E. D., Chen, C., Madhani, H. D., & Noble, S. M. (2008). Systematic genetic analysis of virulence in the human fungal pathogen Cryptococcus neoformans. *Cell, 135*, 174–188. https://doi.org/10.1016/j.cell.2008.07.046.

Liu, P., & Stajich, J. E. (2015). Characterization of the carbohydrate binding module 18 gene family in the amphibian pathogen Batrachochytrium dendrobatidis. *Fungal Genetics and Biology: FG & B*, 77, 31–39. https://doi.org/10.1016/j.fgb.2015.03.003.

Liu, T., Ye, W., Ru, Y., Yang, X., Gu, B., Tao, K., et al. (2011). Two host cytoplasmic effectors are required for pathogenesis of Phytophthora sojae by suppression of host defenses. *Plant Physiology, 155*, 490–501. https://doi.org/10.1104/pp.110.166470.

Longo, A. V., Burrowes, P. A., & Zamudio, K. R. (2014). Genomic studies of disease-outcome in host–pathogen dynamics. *Integrative and Comparative Biology, 54*, 427–438. https://doi.org/10.1093/icb/icu073.

Lopes da Rosa, J., Boyartchuk, V. L., Zhu, L. J., & Kaufman, P. D. (2010). Histone acetyltransferase Rtt109 is required for Candida albicans pathogenesis. *Proceedings of the National Academy of Sciences of the United States of America, 107*, 1594–1599. https://doi.org/10.1073/pnas.0912427107.

Lou, S., Lee, H.-M., Qin, H., Li, J.-W., Gao, Z., Liu, X., et al. (2014). Whole-genome bisulfite sequencing of multiple individuals reveals complementary roles of promoter and gene body methylation in transcriptional regulation. *Genome Biology, 15*, 408. https://doi.org/10.1186/s13059-014-0408-0.

Love, M. I., Huber, W., & Anders, S. (2014). Moderated estimation of fold change and dispersion for RNA-seq data with DESeq2. *Genome Biology, 15*, 550. https://doi.org/10.1186/s13059-014-0550-8.

Love, R. R., Weisenfeld, N. I., Jaffe, D. B., Besansky, N. J., & Neafsey, D. E. (2016). Evaluation of DISCOVAR de novo using a mosquito sample for cost-effective short-read genome assembly. *BMC Genomics, 17*, 187. https://doi.org/10.1186/s12864-016-2531-7.

Lowe, T. M., & Eddy, S. R. (1997). tRNAscan-SE: A program for improved detection of transfer RNA genes in genomic sequence. *Nucleic Acids Research, 25*, 955–964.

Lukashin, A. V., & Borodovsky, M. (1998). GeneMark.hmm: New solutions for gene finding. *Nucleic Acids Research, 26*, 1107–1115.

Lyngsø, R. B., & Pedersen, C. N. (2000). RNA pseudoknot prediction in energy-based models. *Journal of Computational Biology: A Journal of Computational Molecular Cell Biology*, 7, 409–427. https://doi.org/10.1089/106652700750050862.

Magditch, D. A., Liu, T.-B., Xue, C., & Idnurm, A. (2012). DNA mutations mediate microevolution between host-adapted forms of the pathogenic fungus Cryptococcus neoformans. *PLoS Pathogens 8*. e1002936. https://doi.org/10.1371/journal.ppat.1002936.

Majoros, W. H., Pertea, M., & Salzberg, S. L. (2004). TigrScan and GlimmerHMM: Two open source ab initio eukaryotic gene-finders. *Bioinformatics (Oxford, England)*, *20*, 2878–2879. https://doi.org/10.1093/bioinformatics/bth315.

Martel, A., Blooi, M., Adriaensen, C., Van Rooij, P., Beukema, W., Fisher, M. C., et al. (2014). Wildlife disease. Recent introduction of a chytrid fungus endangers Western Palearctic salamanders. *Science*, *346*, 630–631. https://doi.org/10.1126/science.1258268.

Martel, A., Spitzen-van der Sluijs, A., Blooi, M., Bert, W., Ducatelle, R., Fisher, M. C., et al. (2013). Batrachochytrium salamandrivorans sp. nov. causes lethal chytridiomycosis in amphibians. *Proceedings of the National Academy of Sciences of the United States of America*, *110*, 15325–15329. https://doi.org/10.1073/pnas.1307356110.

McDonald, J. H., & Kreitman, M. (1991). Adaptive protein evolution at the Adh locus in Drosophila. *Nature*, *351*, 652–654. https://doi.org/10.1038/351652a0.

McDonald, B. A., & Stukenbrock, E. H. (2016). Rapid emergence of pathogens in agro-ecosystems: Global threats to agricultural sustainability and food security. *Philosophical Transactions of the Royal Society of London Series B, Biological Sciences*, *371*(1709), pii: 20160026. https://doi.org/10.1098/rstb.2016.0026.

McKenna, A., Hanna, M., Banks, E., Sivachenko, A., Cibulskis, K., Kernytsky, A., et al. (2010). The genome analysis toolkit: A MapReduce framework for analyzing next-generation DNA sequencing data. *Genome Research*, *20*, 1297–1303. https://doi.org/10.1101/gr.107524.110.

Meis, J. F., Chowdhary, A., Rhodes, J. L., Fisher, M. C., & Verweij, P. E. (2016). Clinical implications of globally emerging azole resistance in Aspergillus fumigatus. *Philosophical Transactions of the Royal Society of London Series B, Biological Sciences*, *371*(1709), pii: 20150460. https://doi.org/10.1098/rstb.2015.0460.

Menardo, F., Praz, C. R., Wyder, S., Ben-David, R., Bourras, S., Matsumae, H., et al. (2016). Hybridization of powdery mildew strains gives rise to pathogens on novel agricultural crop species. *Nature Genetics*, *48*, 201–205. https://doi.org/10.1038/ng.3485.

Meneau, I., Coste, A. T., & Sanglard, D. (2016). Identification of Aspergillus fumigatus multidrug transporter genes and their potential involvement in antifungal resistance. *Medical Mycology*, *54*, 616–627. https://doi.org/10.1093/mmy/myw005.

Mohammadi, R., Badiee, P., Badali, H., Abastabar, M., Safa, A. H., Hadipour, M., et al. (2015). Use of restriction fragment length polymorphism to identify Candida species, related to onychomycosis. *Advanced Biomedical Research*, *4*, 95. https://doi.org/10.4103/2277-9175.156659.

Morse, S. S. (1995). Factors in the emergence of infectious diseases. *Emerging Infectious Diseases*, *1*, 7–15. https://doi.org/10.3201/eid0101.950102.

Muñoz, J. F., Farrer, R. A., Desjardins, C. A., Gallo, J. E., Sykes, S., Sakthikumar, S., et al. (2016). Genome diversity, recombination, and virulence across the major lineages of paracoccidioides. *mSphere 1*. e00213-16. https://doi.org/10.1128/mSphere.00213-16.

Nakaune, R., Hamamoto, H., Imada, J., Akutsu, K., & Hibi, T. (2002). A novel ABC transporter gene, PMR5, is involved in multidrug resistance in the phytopathogenic fungus Penicillium digitatum. *Molecular Genetics and Genomics*, *267*, 179–185. https://doi.org/10.1007/s00438-002-0649-6.

Nawrocki, E. P., & Eddy, S. R. (2013). Infernal 1.1: 100-fold faster RNA homology searches. *Bioinformatics*, *29*, 2933–2935. https://doi.org/10.1093/bioinformatics/btt509.

Nickel, W., & Seedorf, M. (2008). Unconventional mechanisms of protein transport to the cell surface of eukaryotic cells. *Annual Review of Cell and Developmental Biology*, *24*, 287–308. https://doi.org/10.1146/annurev.cellbio.24.110707.175320.

Nucci, M., & Marr, K. A. (2005). Emerging fungal diseases. *Clinical Infectious Diseases*, *41*, 521–526. https://doi.org/10.1086/432060.

O'Meara, T. R., Hay, C., Price, M. S., Giles, S., & Alspaugh, J. A. (2010). Cryptococcus neoformans histone acetyltransferase Gcn5 regulates fungal adaptation to the host. *Eukaryotic Cell*, *9*, 1193–1202. https://doi.org/10.1128/EC.00098-10.

Ondov, B. D., Treangen, T. J., Melsted, P., Mallonee, A. B., Bergman, N. H., Koren, S., et al. (2016). Mash: Fast genome and metagenome distance estimation using MinHash. *Genome Biology*, *17*, 132.

Osawa, S., Jukes, T. H., Watanabe, K., & Muto, A. (1992). Recent evidence for evolution of the genetic code. *Microbiological Reviews*, *56*, 229–264.

Palmer, J. M., Kubatova, A., Novakova, A., Minnis, A. M., Kolarik, M., & Lindner, D. L. (2014). Molecular characterization of a heterothallic mating system in Pseudogymnoascus destructans, the Fungus causing white-nose syndrome of bats. *G3 (Bethesda, Md)*, *4*, 1755–1763. https://doi.org/10.1534/g3.114.012641.

Panepinto, J. C., & Williamson, P. R. (2006). Intersection of fungal fitness and virulence in Cryptococcus neoformans. *FEMS Yeast Research*, *6*, 489–498. https://doi.org/10.1111/j.1567-1364.2006.00078.x.

Parra, G., Bradnam, K., & Korf, I. (2007). CEGMA: A pipeline to accurately annotate core genes in eukaryotic genomes. *Bioinformatics (Oxford, England)*, *23*, 1061–1067. https://doi.org/10.1093/bioinformatics/btm071.

Pasquinelli, A. E. (2012). MicroRNAs and their targets: Recognition, regulation and an emerging reciprocal relationship. *Nature Reviews Genetics*, *13*, 271–282. https://doi.org/10.1038/nrg3162.

Pawlowska, T. E., & Taylor, J. W. (2004). Organization of genetic variation in individuals of arbuscular mycorrhizal fungi. *Nature*, *427*, 733–737. https://doi.org/10.1038/nature02290.

Pertea, M., Kim, D., Pertea, G. M., Leek, J. T., & Salzberg, S. L. (2016). Transcript-level expression analysis of RNA-seq experiments with HISAT, StringTie and Ballgown. *Nature Protocols*, *11*, 1650–1667. https://doi.org/10.1038/nprot.2016.095.

Petersen, T. N., Brunak, S., von Heijne, G., & Nielsen, H. (2011). SignalP 4.0: Discriminating signal peptides from transmembrane regions. *Nature Methods*, *8*, 785–786. https://doi.org/10.1038/nmeth.1701.

Plissonneau, C., Benevenuto, J., Mohd-Assaad, N., Fouché, S., Hartmann, F. E., & Croll, D. (2017). Using population and comparative genomics to understand the genetic basis of effector-driven fungal pathogen evolution. *Frontiers in Plant Science*, *8*, 119. https://doi.org/10.3389/fpls.2017.00119.

Quiroz Velasquez, P. F., Abiff, S. K., Fins, K. C., Conway, Q. B., Salazar, N. C., Delgado, A. P., et al. (2014). Transcriptome analysis of the entomopathogenic Oomycete Lagenidium giganteum reveals putative virulence factors. *Applied and Environmental Microbiology*, *80*, 6427–6436. https://doi.org/10.1128/AEM.02060-14.

Raffaele, S., Farrer, R. A., Cano, L. M., Studholme, D. J., MacLean, D., Thines, M., et al. (2010). Genome evolution following host jumps in the Irish potato famine pathogen lineage. *Science*, *330*, 1540–1543. https://doi.org/10.1126/science.1193070.

Ratajczak, M. Z., & Ratajczak, J. (2016). Horizontal transfer of RNA and proteins between cells by extracellular microvesicles: 14 years later. *Clinical and Translational Medicine*, *5*, 7. https://doi.org/10.1186/s40169-016-0087-4.

Rawlings, N. D., Barrett, A. J., & Finn, R. (2016). Twenty years of the MEROPS database of proteolytic enzymes, their substrates and inhibitors. *Nucleic Acids Research*, *44*, D343–350. https://doi.org/10.1093/nar/gkv1118.

Reedy, J. L., Floyd, A. M., & Heitman, J. (2009). Mechanistic plasticity of sexual reproduction and meiosis in the Candida pathogenic species complex. *Current Biology: CB*, *19*, 891–899. https://doi.org/10.1016/j.cub.2009.04.058.

Ren, J., Rastegari, B., Condon, A., & Hoos, H. H. (2005). HotKnots: Heuristic prediction of RNA secondary structures including pseudoknots. *RNA*, *11*, 1494–1504. https://doi.org/10.1261/rna.7284905.

Rhodes, J., Abdolrasouli, A., Farrer, R. A., Cuomo, C. A., Aanensen, D. M., Armstrong-James, D., et al. (2017). Rapid genome sequencing for outbreak analysis of the emerging human fungal pathogen. *Candida auris bioRxiv*. https://doi.org/10.1101/201343.

Rhodes, J., Beale, M. A., Vanhove, M., Jarvis, J. N., Kannambath, S., Simpson, J. A., et al. (2017). A population genomics approach to assessing the genetic basis of within-host microevolution underlying recurrent Cryptococcal meningitis infection. *G3 (Bethesda, Md)*, 7(4), 1165–1176. https://doi.org/10.1534/g3.116.037499.

Rhodes, J., Desjardins, C. A., Sykes, S. M., Beale, M. A., Vanhove, M., Sakthikumar, S., et al. (2017). Tracing genetic exchange and biogeography of Cryptococcus neoformans var. grubii at the global population level. *Genetics*, *207*, 327–346. https://doi.org/10.1534/genetics.117.203836.

Rieux, A., & Balloux, F. (2016). Inferences from tip-calibrated phylogenies: A review and a practical guide. *Molecular Ecology*, *25*, 1911–1924. https://doi.org/10.1111/mec.13586.

Robinson, M. D., McCarthy, D. J., & Smyth, G. K. (2010). edgeR: A Bioconductor package for differential expression analysis of digital gene expression data. *Bioinformatics*, *26*, 139–140. https://doi.org/10.1093/bioinformatics/btp616.

Robinson, M. D., & Oshlack, A. (2010). A scaling normalization method for differential expression analysis of RNA-seq data. *Genome Biology*, *11*, R25. https://doi.org/10.1186/gb-2010-11-3-r25.

Rosenblum, E. B., James, T. Y., Zamudio, K. R., Poorten, T. J., Ilut, D., Rodriguez, D., et al. (2013). Complex history of the amphibian-killing chytrid fungus revealed with genome resequencing data. *Proceedings of the National Academy of Sciences of the United States of America*, *110*, 9385–9390. https://doi.org/10.1073/pnas.1300130110.

Rosenblum, E. B., Stajich, J. E., Maddox, N., & Eisen, M. B. (2008). Global gene expression profiles for life stages of the deadly amphibian pathogen Batrachochytrium dendrobatidis. *Proceedings of the National Academy of Sciences of the United States of America*, *105*, 17034–17039. https://doi.org/10.1073/pnas.0804173105.

Salamov, A. A., & Solovyev, V. V. (2000). Ab initio gene finding in Drosophila genomic DNA. *Genome Research*, *10*, 516–522.

Salzberg, S. L., Phillippy, A. M., Zimin, A., Puiu, D., Magoc, T., Koren, S., et al. (2012). GAGE: A critical evaluation of genome assemblies and assembly algorithms. *Genome Research*, *22*, 557–567. https://doi.org/10.1101/gr.131383.111.

Sanguinetti, M., Posteraro, B., La Sorda, M., Torelli, R., Fiori, B., Santangelo, R., et al. (2006). Role of AFR1, an ABC transporter-encoding gene, in the in vivo response to fluconazole and virulence of Cryptococcus neoformans. *Infection and Immunity*, *74*, 1352–1359. https://doi.org/10.1128/IAI.74.2.1352-1359.2006.

Santos, M. A., Keith, G., & Tuite, M. F. (1993). Non-standard translational events in Candida albicans mediated by an unusual seryl-tRNA with a 5'-CAG-3' (leucine) anticodon. *The EMBO Journal*, *12*, 607–616.

Sanyal, K., Baum, M., & Carbon, J. (2004). Centromeric DNA sequences in the pathogenic yeast Candida albicans are all different and unique. *Proceedings of the National Academy of Sciences of the United States of America*, *101*, 11374–11379. https://doi.org/10.1073/pnas.0404318101.

Schoch, C. L., Seifert, K. A., Huhndorf, S., Robert, V., Spouge, J. L., Levesque, C. A., et al. (2012). Nuclear ribosomal internal transcribed spacer (ITS) region as a universal DNA barcode marker for Fungi. *Proceedings of the National Academy of Sciences of the United States of America*, *109*, 6241–6246. https://doi.org/10.1073/pnas.1117018109.

Schornack, S., van Damme, M., Bozkurt, T. O., Cano, L. M., Smoker, M., Thines, M., et al. (2010). Ancient class of translocated oomycete effectors targets the host nucleus. *Proceedings of the National Academy of Sciences of the United States of America*, *107*, 17421–17426. https://doi.org/10.1073/pnas.1008491107.

Schoustra, S. E., Debets, A. J. M., Slakhorst, M., & Hoekstra, R. F. (2007). Mitotic recombination accelerates adaptation in the fungus Aspergillus nidulans. *PLoS Genetics*, *3*(4), e68. https://doi.org/10.1371/journal.pgen.0030068.

Schurch, N. J., Schofield, P., Gierliński, M., Cole, C., Sherstnev, A., Singh, V., et al. (2016). How many biological replicates are needed in an RNA-seq experiment and which differential expression tool should you use? *RNA*, *22*, 839–851. https://doi.org/10.1261/rna.053959.115.

Selker, E. U., Tountas, N. A., Cross, S. H., Margolin, B. S., Murphy, J. G., Bird, A. P., et al. (2003). The methylated component of the Neurospora crassa genome. *Nature*, *422*, 893–897. https://doi.org/10.1038/nature01564.

Sharma, C., Kumar, N., Meis, J. F., Pandey, R., & Chowdhary, A. (2015). Draft genome sequence of a fluconazole-resistant candida auris strain from a Candidemia patient in India. *Genome Announcements*, *3*(4), pii: e00722-15. https://doi.org/10.1128/genomeA.00722-15.

Sharpton, T. J., Neafsey, D. E., Galagan, J. E., & Taylor, J. W. (2008). Mechanisms of intron gain and loss in Cryptococcus. *Genome Biology*, *9*(1), R24.

Sheltzer, J. M., Blank, H. M., Pfau, S. J., Tange, Y., George, B. M., Humpton, T. J., et al. (2011). Aneuploidy drives genomic instability in yeast. *Science*, *333*, 1026–1030. https://doi.org/10.1126/science.1206412.

Simão, F. A., Waterhouse, R. M., Ioannidis, P., Kriventseva, E. V., & Zdobnov, E. M. (2015). BUSCO: Assessing genome assembly and annotation completeness with single-copy orthologs. *Bioinformatics (Oxford, England)*, *31*, 3210–3212. https://doi.org/10.1093/bioinformatics/btv351.

Simpson, J. T., & Durbin, R. (2012). Efficient de novo assembly of large genomes using compressed data structures. *Genome Research*, *22*, 549–556. https://doi.org/10.1101/gr.126953.111.

Singh, R. P., Hodson, D. P., Huerta-Espino, J., Jin, Y., Bhavani, S., Njau, P., et al. (2011). The emergence of Ug99 races of the stem rust fungus is a threat to World wheat production. *Annual Review of Phytopathology*, *49*, 465–481. https://doi.org/10.1146/annurev-phyto-072910-095423.

Sionov, E., Lee, H., Chang, Y. C., & Kwon-Chung, K. J. (2010). Cryptococcus neoformans overcomes stress of azole drugs by formation of disomy in specific multiple chromosomes. *PLoS Pathogens 6*. e1000848. https://doi.org/10.1371/journal.ppat.1000848.

Smith, K. M., Galazka, J. M., Phatale, P. A., Connolly, L. R., & Freitag, M. (2012). Centromeres of filamentous fungi. *Chromosome Research*, *20*, 635–656. https://doi.org/10.1007/s10577-012-9290-3.

Song, G., Dickins, B. J. A., Demeter, J., Engel, S., Dunn, B., & Cherry, J. M. (2015). AGAPE (Automated Genome Analysis PipelinE) for pan-genome analysis of Saccharomyces cerevisiae. *PLoS One 10*. e0120671. https://doi.org/10.1371/journal.pone.0120671.

Spanu, P. D., Abbott, J. C., Amselem, J., Burgis, T. A., Soanes, D. M., Stüber, K., et al. (2010). Genome expansion and gene loss in powdery mildew fungi reveal tradeoffs in extreme parasitism. *Science*, *330*, 1543–1546. https://doi.org/10.1126/science.1194573.

Spatafora, J. W., Chang, Y., Benny, G. L., Lazarus, K., Smith, M. E., Berbee, M. L., et al. (2016). A phylum-level phylogenetic classification of zygomycete fungi based on genome-scale data. *Mycologia*, *108*, 1028–1046. https://doi.org/10.3852/16-042.

Stajich, J. E., Harris, T., Brunk, B. P., Brestelli, J., Fischer, S., Harb, O. S., et al. (2012). FungiDB: An integrated functional genomics database for fungi. *Nucleic Acids Research*, *40*(D1), D675–D681.

Stam, R., Jupe, J., Howden, A. J. M., Morris, J. A., Boevink, P. C., Hedley, P. E., et al. (2013). Identification and characterisation CRN Effectors in Phytophthora capsici Shows modularity and functional diversity. *PLoS One 8*. e59517. https://doi.org/10.1371/journal.pone.0059517.

Stanke, M., Keller, O., Gunduz, I., Hayes, A., Waack, S., & Morgenstern, B. (2006). AUGUSTUS: Ab initio prediction of alternative transcripts. *Nucleic Acids Research, 34*, W435–W439. https://doi.org/10.1093/nar/gkl200.

Stegen, G., Pasmans, F., Schmidt, B. R., Rouffaer, L. O., Van Praet, S., Schaub, M., et al. (2017). Drivers of salamander extirpation mediated by Batrachochytrium salamandrivorans. *Nature, 544*, 353–356. https://doi.org/10.1038/nature22059.

Stoletzki, N., & Eyre-Walker, A. (2011). Estimation of the neutrality index. *Molecular Biology and Evolution, 28*, 63–70. https://doi.org/10.1093/molbev/msq249.

Storey, J. D., & Tibshirani, R. (2003). Statistical significance for genomewide studies. *Proceedings of the National Academy of Sciences of the United States of America, 100*, 9440–9445. https://doi.org/10.1073/pnas.1530509100.

Stornaiuolo, M., Lotti, L. V., Borgese, N., Torrisi, M.-R., Mottola, G., Martire, G., et al. (2003). KDEL and KKXX retrieval signals appended to the same reporter protein determine different trafficking between endoplasmic reticulum, intermediate compartment, and Golgi complex. *Molecular Biology of the Cell, 14*, 889–902. https://doi.org/10.1091/mbc.E02-08-0468.

Stroud, H., Otero, S., Desvoyes, B., Ramírez-Parra, E., Jacobsen, S. E., & Gutierrez, C. (2012). Genome-wide analysis of histone H3.1 and H3.3 variants in Arabidopsis thaliana. *Proceedings of the National Academy of Sciences of the United States of America, 109*, 5370–5375. https://doi.org/10.1073/pnas.1203145109.

Studholme, D. J. (2016). Genome update. Let the consumer beware: Streptomyces genome sequence quality. *Microbial Biotechnology, 9*, 3–7. https://doi.org/10.1111/1751-7915.12344.

Stukenbrock, E. H., & McDonald, B. A. (2008). The origins of plant pathogens in agroecosystems. *Annual Review of Phytopathology, 46*, 75–100. https://doi.org/10.1146/annurev.phyto.010708.154114.

Tajima, F. (1989). Statistical method for testing the neutral mutation hypothesis by DNA polymorphism. *Genetics, 123*, 585–595.

Taylor, J. W., Jacobson, D. J., & Fisher, M. C. (1999). THE EVOLUTION OF ASEXUAL FUNGI: Reproduction, speciation and classification. *Annual Review of Phytopathology, 37*, 197–246. https://doi.org/10.1146/annurev.phyto.37.1.197.

Taylor, J. W., Jacobson, D. J., Kroken, S., Kasuga, T., Geiser, D. M., Hibbett, D. S., et al. (2000). Phylogenetic species recognition and species concepts in fungi. *Fungal Genetics and Biology: FG & B, 31*, 21–32. https://doi.org/10.1006/fgbi.2000.1228.

Thiel, T., Michalek, W., Varshney, R. K., & Graner, A. (2003). Exploiting EST databases for the development and characterization of gene-derived SSR-markers in barley (Hordeum vulgare L.). *TAG Theoretical and Applied Genetics Theoretische und Angewandte Genetik, 106*, 411–422. https://doi.org/10.1007/s00122-002-1031-0.

Trapnell, C., Roberts, A., Goff, L., Pertea, G., Kim, D., Kelley, D. R., et al. (2012). Differential gene and transcript expression analysis of RNA-seq experiments with TopHat and Cufflinks. *Nature Protocols*, 7, 562–578. https://doi.org/10.1038/nprot.2012.016.

Trivedi, J., Lachapelle, J., Vanderwolf, K., Misra, V., Willis, C. K. R., Ratcliffe, J., et al. (2017). Fungus causing white-nose syndrome in bats accumulates genetic variability in North America with no sign of recombination. *mSphere*, *2*(4), e00271–17. https://doi.org/10.1128/mSphereDirect.00271-17.

Turner, E., Jacobson, D. J., & Taylor, J. W. (2011). Genetic architecture of a reinforced, postmating, reproductive isolation barrier between Neurospora species indicates evolution via natural selection. *PLoS Genetics* 7. e1002204. https://doi.org/10.1371/journal.pgen.1002204.

Turro, E., Astle, W. J., & Tavaré, S. (2014). Flexible analysis of RNA-seq data using mixed effects models. *Bioinformatics (Oxford, England), 30*, 180–188. https://doi.org/10.1093/bioinformatics/btt624.

Tuteja, R. (2005). Type I signal peptidase: An overview. *Archives of Biochemistry and Biophysics*, *441*, 107–111. https://doi.org/10.1016/j.abb.2005.07.013.
Van Der Linden, J. W. M., Warris, A., & Verweij, P. E. (2011). Aspergillus species intrinsically resistant to antifungal agents. *Medical Mycology*, *49*(Suppl. 1), S82–89. https://doi.org/10.3109/13693786.2010.499916.
Van Rooij, P., Martel, A., D'Herde, K., Brutyn, M., Croubels, S., Ducatelle, R., et al. (2012). Germ tube mediated invasion of Batrachochytrium dendrobatidis in amphibian skin is host dependent. *PLoS One* 7. e41481. https://doi.org/10.1371/journal.pone.0041481.
Verweij, P. E., Chowdhary, A., Melchers, W. J. G., & Meis, J. F. (2016). Azole resistance in Aspergillus fumigatus: Can we retain the clinical use of mold-active antifungal azoles? *Clinical Infectious Diseases*, *62*, 362–368. https://doi.org/10.1093/cid/civ885.
Walker, B. J., Abeel, T., Shea, T., Priest, M., Abouelliel, A., Sakthikumar, S., et al. (2014). Pilon: An integrated tool for comprehensive microbial variant detection and genome assembly improvement. *PLoS One 9*. e112963. https://doi.org/10.1371/journal.pone.0112963.
Walker, S. F., Bosch, J., James, T. Y., Litvintseva, A. P., Oliver Valls, J. A., Piña, S., et al. (2008). Invasive pathogens threaten species recovery programs. *Current Biology: CB*, *18*, R853–854. https://doi.org/10.1016/j.cub.2008.07.033.
Wang, L., Feng, Z., Wang, X., Wang, X., & Zhang, X. (2010). DEGseq: An R package for identifying differentially expressed genes from RNA-seq data. *Bioinformatics (Oxford, England)*, *26*, 136–138. https://doi.org/10.1093/bioinformatics/btp612.
Wang, X., Hsueh, Y.-P., Li, W., Floyd, A., Skalsky, R., & Heitman, J. (2010). Sex-induced silencing defends the genome of Cryptococcus neoformans via RNAi. *Genes & Development*, *24*, 2566–2582. https://doi.org/10.1101/gad.1970910.
Wang, D. Y., Kumar, S., & Hedges, S. B. (1999). Divergence time estimates for the early history of animal phyla and the origin of plants, animals and fungi. *Proceedings of the Biological Sciences*, *266*, 163–171. https://doi.org/10.1098/rspb.1999.0617.
Wang, Q.-M., Liu, W.-Q., Liti, G., Wang, S.-A., & Bai, F.-Y. (2012). Surprisingly diverged populations of Saccharomyces cerevisiae in natural environments remote from human activity. *Molecular Ecology*, *21*, 5404–5417. https://doi.org/10.1111/j.1365-294X.2012.05732.x.
Wang, M., Weiberg, A., Lin, F.-M., Thomma, B. P. H. J., Huang, H.-D., & Jin, H. (2016). Bidirectional cross-kingdom RNAi and fungal uptake of external RNAs confer plant protection. *Nature Plants*, *2*, 16151, https://doi.org/10.1038/nplants.2016.151.
Weiberg, A., Wang, M., Lin, F.-M., Zhao, H., Zhang, Z., Kaloshian, I., et al. (2013). Fungal small RNAs suppress plant immunity by hijacking host RNA interference pathways. *Science*, *342*, 118–123. https://doi.org/10.1126/science.1239705.
Wiley, E. O. (1978). The evolutionary species concept reconsidered. *Systematic Zoology*, *27*, 17–26. https://doi.org/10.2307/2412809.
Wu, C. H., Apweiler, R., Bairoch, A., Natale, D. A., Barker, W. C., Boeckmann, B., et al. (2006). The universal protein resource (UniProt): An expanding universe of protein information. *Nucleic Acids Research*, *34*, D187–191. https://doi.org/10.1093/nar/gkj161.
Wu, T. D., & Watanabe, C. K. (2005). GMAP: A genomic mapping and alignment program for mRNA and EST sequences. *Bioinformatics (Oxford, England)*, *21*, 1859–1875. https://doi.org/10.1093/bioinformatics/bti310.
Yang, Z. (2007). PAML 4: Phylogenetic analysis by maximum likelihood. *Molecular Biology and Evolution*, *24*, 1586–1591. https://doi.org/10.1093/molbev/msm088.
Yike, I. (2011). Fungal proteases and their pathophysiological effects. *Mycopathologia*, *171*, 299 323. https://doi.org/10.1007/s11046-010-9386-2.

Zerbino, D. R., & Birney, E. (2008). Velvet: Algorithms for de novo short read assembly using de Bruijn graphs. *Genome Research*, *18*, 821–829. https://doi.org/10.1101/gr.074492.107.

Zhang, Z., López-Giráldez, F., & Townsend, J. P. (2010). LOX: Inferring Level Of eXpression from diverse methods of census sequencing. *Bioinformatics*, *26*, 1918–1919. https://doi.org/10.1093/bioinformatics/btq303.

Zhang, X., Wang, Y., Chi, W., Shi, Y., Chen, S., Lin, D., et al. (2014). Metalloprotease genes of Trichophyton mentagrophytes are important for pathogenicity. *Medical Mycology*, *52*, 36–45. https://doi.org/10.3109/13693786.2013.811552.

Zhao, Z.-M., Campbell, M. C., Li, N., Lee, D. S. W., Zhang, Z., & Townsend, J. P. (2017). Detection of regional variation in selection intensity within protein-coding genes using DNA sequence polymorphism and divergence. *Molecular Biology and Evolution*. https://doi.org/10.1093/molbev/msx213.

CHAPTER FOUR

Fungal Gene Cluster Diversity and Evolution

Jason C. Slot[1]
The Ohio State University, Columbus, OH, United States
[1]Corresponding author: e-mail address: slot.1@osu.edu

Contents

Abstract

Metabolic gene clusters (MGCs) have provided some of the earliest glimpses at the biochemical machinery of yeast and filamentous fungi. MGCs encode diverse genetic mechanisms for nutrient acquisition and the synthesis/degradation of essential and adaptive metabolites. Beyond encoding the enzymes performing these discrete anabolic or catabolic processes, MGCs may encode a range of mechanisms that enable their persistence as genetic consortia; these include enzymatic mechanisms to protect their host fungi from their inherent toxicities, and integrated regulatory machinery. This modular, self-contained nature of MGCs contributes to the metabolic and ecological adaptability of fungi. The phylogenetic and ecological patterns of MGC distribution reflect the

Advances in Genetics, Volume 100
ISSN 0065-2660
https://doi.org/10.1016/bs.adgen.2017.09.005

broad diversity of fungal life cycles and nutritional modes. While the origins of most gene clusters are enigmatic, MGCs are thought to be born into a genome through gene duplication, relocation, or horizontal transfer, and analyzing the death and decay of gene clusters provides clues about the mechanisms selecting for their assembly. Gene clustering may provide inherent fitness advantages through metabolic efficiency and specialization, but experimental evidence for this is currently limited. The identification and characterization of gene clusters will continue to be powerful tools for elucidating fungal metabolism as well as understanding the physiology and ecology of fungi.

1. INTRODUCTION

Fungi are among the most phylogenetically and functionally diverse organisms on earth, occupying a large variety of saprotrophic, biotrophic, and pathogenic niches. The fungal lifestyle of absorptive nutrition and hyphal growth is fundamentally versatile and largely built upon a foundation of metabolism and membrane transport. Fungi are notable among eukaryotes for their easily recognizable genomic structure in the form of spatial clustering of metabolically related genes. Metabolic gene clusters (MGCs) encode discrete pathways involved in nutrient acquisition, synthesis of vitamins, and the production and degradation of secondary metabolites (SMs). Fungal lineages vary in the degree of gene clustering they display, and several MGCs are found to have an ecological distribution. Both neutral and selective processes are involved in the formation of MGCs, and they are dispersed through both vertical and horizontal transfer. Recent work suggests there are specific fitness advantages linked to MGCs, and highlights the physicochemical constraints that shape their evolution. The discovery and characterization of MGCs promises to be a useful tool for deciphering fungal ecology in the coming years.

2. DIVERSITY AND DISTRIBUTION OF FUNGAL MGCs

2.1 The Types of Metabolic Processes in MGCs, an Expanding List

Fungal MGCs are loci that contain multiple genes from different gene families, which contribute to a discrete metabolic phenotype. To date, most of the metabolic phenotypes found to be encoded by MGCs participate in nutrient acquisition, or the biosynthesis/degradation of amino acids, cofactors, and SMs. The first fungal MGC to be identified was the *Saccharomyces cerevisiae* galactose utilization cluster (GAL; Douglas & Hawthorne, 1966). Subsequent nutrient acquisition MGC discoveries include the quinic acid catabolism cluster in *Neurospora crassa* (QA; Giles, Case, & Jacobson, 1973), the proline catabolism (PRO; Arst & MacDonald, 1975), and the nitrate assimilation (HANT; Johnstone et al., 1990) clusters in *Aspergillus nidulans*, all of which enjoyed extensive in-depth study prior to the era of whole-genome sequencing. The functional and evolutionary mechanisms driving the clustering of these nutritional genes quickly became the subject of speculation and debate, as it became clear from their order and regulation that MGCs are not simply eukaryotic versions of bacterial operons (Giles, 1978; Keller & Hohn, 1997; Walton, 2000). More recently, nutrient utilization clusters have been identified in partial and complete genome sequences through manual annotation of gene functions and by identification of fungal analogs of bacterial operons (Jeffries & Van Vleet, 2009; Marcet-Houben & Gabaldón, 2010; Yu, Chang, Bhatnagar, & Cleveland, 2000). These include MGCs involved in utilization of sugars (e.g., rhamnose and *N*-acetylglucosamine), amino acid catabolism, and iron metabolism.

MGCs are also involved in basic intracellular metabolism by participating in the synthesis of a number of vitamins and amino acids and other essential metabolites. Vitamins may be thought of as unique or complex metabolites, which certain essential metabolic pathways depend upon in minute quantities, but that are often acquired rather than synthesized endogenously. Similarly, rare but essential amino acids may be acquired from other organisms or synthesized. Several of the pathways for vitamin/amino acid synthesis are related by the tendency of the genes for their metabolism to cluster in both prokaryotes and (occasionally) fungi. While most vitamin biosynthetic pathways have been found clustered in prokaryotes, only a couple of fungal vitamin clusters have been identified, including the biotin cluster (Hall & Dietrich, 2007) and the yet-to-be functionally characterized pyridoxine cluster (Wightman, 2001) in *Saccharomyces*. Clustering is not exclusively a feature of niche-specific or "optional" pathways, however, although that is certainly the greater trend (Wisecaver, Slot, & Rokas, 2014). The AROM pathway for the synthesis of aromatic amino acids is the most universally conserved pathway among fungi. AROM was initially thought to be a gene cluster or operon in fungi like it is in bacteria, but was later found to be a pentafunctional peptide that resulted from the merging of monofunctional ancestral genes (Giles, 1978). Even the metabolism of the most essential molecules like pyrimidines is also found clustered in fungi.

While the roles of nutritional MGCs and those for essential metabolisms are readily inferred, the ecological roles of SM MGCs are not often apparent. Due to their often-demonstrated biological activity, it is generally understood that SMs as a compound class are important for defense, signaling, and competition (Raguso et al., 2015). Much of the massive diversity of SMs may not be under selection for precise ecological functions, but rather exists as part of a store of potentially active chemodiversity (Firn & Jones, 2003). Hundreds of fungal SM MGCs have been identified through targeted searches for genetic mechanisms and bioinformatic predictions (Blin et al., 2013; Brown et al., 1996; Hoffmeister & Keller, 2007). Driven primarily by concerns over SM toxicity and pathological effects against plants and animals from the beginning, most MGC research over the past 25 years has focused on SM clusters of agricultural and pharmaceutical interest. Among the MGC-encoded SMs that provide clear selective advantages in a defined context are host-selective toxins (Ahn & Walton, 1996; Wight, Labuda, & Walton, 2013), plant hormones (Siewers, 2006; Tudzynski & Hölter, 1998), feeding deterrents (Spiering, 2004), and β-lactam antibiotics (Díez et al., 1990). Neurotoxins produced by fungal endophyte SM MGCs benefit host plants through the reduction of herbivory, and endophyte-derived SMs may be more broadly involved in multitrophic interactions among endophytic communities, plants, and herbivores (Kusari, Singh, & Jayabaskaran, 2014; Panaccione et al., 2006). The encoding of SM pathways in tightly regulated gene clusters is a strong indication that there has been ecological selection for natural products which were identified to be bioactive in pharmaceutical screens for specific enzyme and cell inhibition, but these roles are open to speculation (Bushley et al., 2013; Kennedy et al., 1999).

Recently a new category of fungal MGCs that neutralize or degrade plant SMs has been reported (Glenn et al., 2016; Greene, McGary, Rokas, & Slot, 2014). The ability to degrade plant defense compounds facilitates both saprotrophic and biotrophic nutritional modes (Floudas et al., 2012; Hammerbacher et al., 2013), and is a critical limitation in the conversion of lignocellulosic biomass to fuels (Jönsson, Alriksson, & Nilvebrant, 2013). Due to the high diversity of plant metabolites and plant–fungal interactions, SM degradation MGCs may be found to be quite widespread. Identifying the MGCs for degrading plant defense molecules will be an important step in understanding the assembly of protective foliar phytobiomes (Van Bael, Estrada, & Arnold, 2017), and will aid in synthetic biology for biomass conversion.

2.2 Phylogenetic, Ecological, and Functional Diversity of MGCs

With steady improvements in genome and metabolism database infrastructure, the global identification of MGCs across fungi has become increasingly tractable in the past decade. Case studies of specific MGC families have indicated complex patterns of inheritance consistent with the previously characterized distribution of their functions (Khaldi, Collemare, Lebrun, & Wolfe, 2008; Patron et al., 2007; Slot & Hibbett, 2007). Subsequent systematic searches for fungal MGCs have used the inferred metabolic relationships among genes to identify clusters of functionally related genes in genomic databases. For example, Wisecaver et al. (2014) identified nearest metabolic neighbor enzyme-coding gene relationships as curated by the Kyoto Encyclopedia of Genes and Genomes (KEGG) (Kanehisa & Goto, 2000) that coincided with tight gene linkage in 208 fungal genomes. For the more rapidly evolving and diverse SM pathways, less restrictive associations among generally characterized gene functions have been searched for colocation (Khaldi et al., 2010), and probabilistic models of gene relationships have been built into SM cluster prediction pipelines (Blin et al., 2013). Comparative methods have further been used as rationale for predicting the core structures of biosynthetic MGC metabolic products (Brown & Proctor, 2016; Throckmorton, Wiemann, & Keller, 2015). However, methods to detect eukaryotic gene clusters without prior knowledge of gene function, and thereby discover truly novel metabolisms, though well developed for prokaryotic genomes (Langille, Hsiao, & Brinkman, 2010), have lagged behind for fungi and other eukaryotes.

To date, the study of fungal MGCs has been strongly biased toward a limited set of previously identified gene clusters (Li et al., 2016). By using broad search strategies, it is now becoming possible to not only characterize the true diversity, distribution, and ecological associations of MGCs but also to identify the recurring targets of natural selection on fungal metabolism. The systematic search for KEGG MGCs by Wisecaver et al. (2014) identified the metabolic classes and phylogenetic groups that are most affected by gene clustering. The most clustered metabolic pathways throughout fungi are related to environmental growth, particularly secondary metabolism, carbohydrate metabolism, and amino acid metabolism. A large portion of these processes might be considered "dispensable" because fungi would still be viable, though maybe not as competitive in all environments, without them. For example, SMs are associated with competition and defense mechanisms that are under constantly shifting selection pressures. Carbohydrate form and amino acid availability also vary with available substrates and community composition. The pathway classes least clustered in these searches, by contrast, are more often considered "essential," implying lethality in their absence, including processes in nucleotide metabolism, energy metabolism, and glycan metabolism.

Trends in MGC diversity are driven by those that are identified in Ascomycota, which are the most highly clustered among fungal genomes. KEGG enzyme genes from genomes in several classes of Pezizomycotina were found in clusters 3.6% of the time on average, while 3.7% of Saccharomycotina enzyme genes were clustered. Agaricomycetes enzymes were comparatively lacking in clustering at 1.6% clustered. These estimates of clustering are likely to be somewhat influenced by the sample of genomes within each lineage; genome sequencing has been largely focused on fungi of economic and societal importance, and the initiation and success of genome projects is related to the ease with which fungi are identified and cultured. However, it is not clear how these factors would influence estimates of clustering. It should also be noted that at the time of analyses, secondary metabolism was not well represented in KEGG maps, and secondary metabolic pathway neighbors are difficult to infer from gene homology. The percentages listed should be considered estimates of intermediary metabolic clustering, while overall clustering will be found to be much higher in future studies.

There are major differences in the type and extent of clustering among the various fungal lineages (Wisecaver et al., 2014). For example, while sequential steps in metabolic pathways are not often found in MGCs in Agaricomycetes, arrays of paralogous genes for ecology-specific functions (not usually considered MGCs) have been identified (Floudas et al., 2012). SM MGCs are rarely reported in Agaricomycetes (Quin, Flynn, & Schmidt-Dannert, 2014; Wawrzyn, Quin, Choudhary, López-Gallego, & Schmidt-Dannert, 2012), but it is not clear whether this difference in cluster discovery is a result of a bias toward research in the ecological niches dominated by Pezizomycotina (Bills, Gloer, & An, 2013; Strobel & Daisy, 2003), or whether there is a fundamental difference in genome structure between the lineages. The larger genome size and proliferation of noncoding DNA in Basidiomycetes is consistent with these fungi having smaller effective population sizes, which might not be sufficient to enable natural selection to fix clustered states with only weak fitness benefits (Lynch, 2006). Conversely, when clusters are found in these species, especially strong selection on these pathways might be inferred. Higher levels of clustering in Saccharomycotina and Microsporidia may be driven by the same effective population size environments that drive the streamlining of their genomes (Lynch, 2006).

It is difficult to infer whether ecological trends in gene clustering exist, because fungi tend to be opportunistic and adapt to ecological roles rapidly. However, extensive gene clustering in Eurotiomycetes, which includes both *Aspergillus* spp. (De Vries et al., 2017) and many of the microcolonial black yeasts (Teixeira et al., 2017), could suggest MGCs are favored in the soil saprotroph lifestyle where competition with diverse microorganisms is most intense, and metagenomes are very large. With the explosion of new fungal genome data and improvements in MGC detection algorithms, opportunities to study broader patterns of fungal SM MGC diversity are greater than they have ever been. To date, efforts to sample fungal genomes have prioritized characterizing phylogenetic diversity at higher levels or within genera or species (1000.fungalgenomes.org), and a few have looked in some depth at ecological guilds (De Vries et al., 2017; Kohler et al., 2015; Teixeira et al., 2017). Future efforts to sample genomes deeply within defined fungal communities will provide greater power to infer direct associations between MGC-encoded functions and fungal ecology.

2.3 Highly Clustered Enzyme Gene Pairs

To infer MGCs, pairs of genes that are near neighbors both on chromosomes and in metabolic pathways may be identified, and then overlapping pairs combined. Such clustered pairs are often more common than any of the clusters in which they participate, which may indicate more precise selection pressures on these pathway steps. The study of tyrosine metabolism gene clusters by Greene et al. (2014) was undertaken after it was discovered that the most highly clustered gene pair occurs in this pathway (Table 1). Specifically, gentisate 1,2-dioxygenase (EC:1.13.11.4) and acylpyruvate hydrolase (EC:3.7.1.5) are clustered in 76% of fungal genomes. This search identified two classes of MGCs: those degrading phenolics via a gentisate ring-cleavage route and those utilizing a homogentisate ring-cleavage route, e.g., in Schmaler-Ripcke et al. (2009). These phenolic compound degradation MGCs were found to have distributions with ecological signatures, with gentisate cleavage clusters sparsely distributed, but mostly found in grass-associated fungi, and three differently composed homogentisate cleavage clusters distributed to saprotrophs, plant pathogens, and melanized extremophiles, respectively. The phenolic compounds that are degraded in tyrosine metabolism represent a major and diverse component of plant

Table 1 Highly Clustered Metabolic Neighbor Gene Pairs in Fungi

Pathway (KEGG)	Enzyme Pair	%Genomes With Clusters (Wisecaver et al., 2014)
Tyrosine metabolism (00350)	Acylpyruvate hydrolase (EC:3.7.1.5); gentisate 1,2-dioxygenase (EC:1.13.11.4)	76
Valine, leucine, and isoleucine degradation (00280)	Methylcrotonyl-CoA carboxylase (EC:6.4.1.4); isovaleryl-CoA dehydrogenase (EC:1.3.8.4)	70
Amino sugar and nucleotide sugar metabolism (00520)	*N*-acetylglucosamine-6-phosphate deacetylase (EC:3.5.1.25); glucosamine phosphate deaminase (EC:3.5.99.6)	63
Amino sugar and nucleotide sugar metabolism (00520)	Glucosamine kinase (EC:2.7.1.1); glucosamine phosphate deaminase (EC:3.5.99.6)	60
Nitrogen metabolism (00910)	Nitrate reductase (EC:1.7.1.3); nitrite reductase (EC:1.7.1.4)	50
Biotin metabolism (00780)	7,8-diaminonanoate transaminase (EC:2.6.1.62); dethiobiotin synthase (EC:6.3.3.3)	50
Biotin metabolism (00780)	biotin synthase (EC:2.8.1.6); 7,8-diaminonanoate transaminase (EC:2.6.1.62)	45
Galactose metabolism (00052)	Galactokinase (EC:2.7.1.6); UDP-glucose 4-epimerase (EC:5.1.3.2)	42
Galactose metabolism (00052)	Uridyl transferase (EC:2.7.7.12); UDP-glucose 4-epimerase (EC:5.1.3.2)	40

chemical defenses against pathogens and pests, and are expected to impose intensive selection pressures on fungi (Lattanzio, Lattanzio, & Cardinali, 2006). Homogentisate degradation MGCs are additionally responsible for the production of pyomelanin, which has an uncertain role in fungal pathogenesis of plants and animals. Interestingly, homogentisate MGCs in plant pathogenic Dothideomycetes are limited to only the genes thought to be responsible for pyomelanin formation, and may be a mechanism for reduction of oxidative stress (Ahmad et al., 2016).

Other highly clustered gene pairs are involved in nitrate assimilation, galactose utilization, biotin biosynthesis, and *N*-acetylglucosamine degradation. It is not surprising in retrospect that many early cluster-related discoveries were made in these very pathways. Chitin degradation clusters are found broadly in Pezizomycotina and often contain a RON1 (regulator of *N*-acetylglucosamine catabolism) transcription factor, which highly upregulates MGC activity during growth on chitin, suggesting roles in both nutrition and developmental processes (Kappel, Gaderer, Flipphi, & Seidl-Seiboth, 2015). Chitin synthesis MGCs have also been identified in some fungi, further highlighting the demands on cell wall metabolism in fungi (Pacheco-Arjona & Ramirez-Prado, 2014). The genes for nitrate reductase (EC:1.7.1.3) and nitrite reductase (EC:1.7.1.4) are highly correlated in their presence in genomes, and are usually clustered with each other and at least one more gene (especially the high-affinity nitrate transporter, NRT2) when found (HANT), but clustering has been lost in Leotiomycetes and Sordariomycetes (Slot & Hibbett, 2007). The role of nitrate assimilation MGCs in fungi may extend beyond basic nutritional requirements under nitrogen limitation, as nitrate is found to be important in plant innate immunity to pathogens (Abrahamian, Ah Fong, Davis, Andreeva, & Judelson, 2016). In contrast, the galactose utilization genes, galactokinase (EC:2.7.1.6), UDP-glucose 4-epimerase (EC:5.1.3.2), and uridyl transferase (EC:2.7.7.12), are widespread in fungi, but are only clustered in yeasts. Interestingly, GAL clusters are found in yeasts from three different subphyla (Saccharomycotina, Taphrinomycotina, and Agaricomycotina), but not in their filamentous relatives (Slot & Rokas, 2010). Yeasts are also the only lineages with appreciable rates of loss of GAL genes, which may relate to the differential evolvability of clustered pathways (see below).

3. THE PATTERNS OF MGC EVOLUTION IN FUNGI

Several studies have sought to characterize the birth, life, and death of fungal MGCs, based on reconstructions of evolutionary events leading to the current distributions of clustered or unclustered gene states (Fig. 1). To accomplish this task, several recent studies performed exhaustive phylogenetic analyses of all constituent genes. Below are discussed some of the overarching patterns that have emerged to suggest evolutionary mechanisms affecting MGCs.

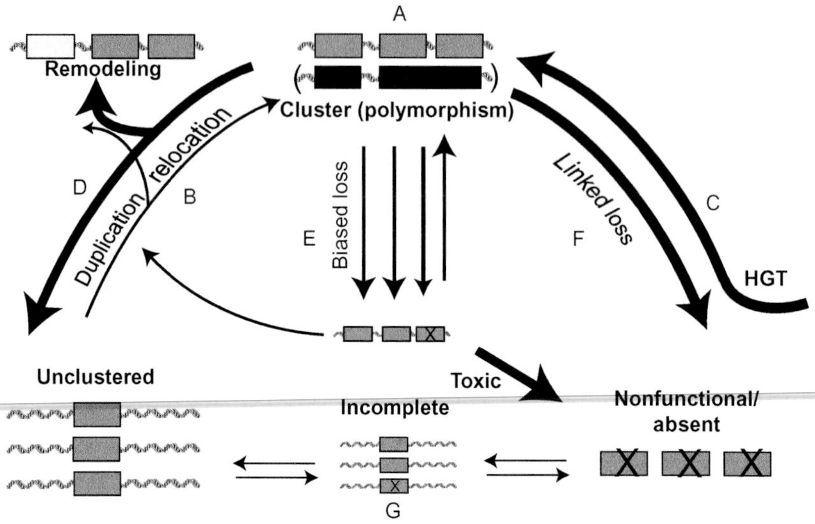

Fig. 1 Birth, remodeling, and death of metabolic gene clusters. Rates indicated by the thickness of *transition arrows* consider evidence of the roles of clustering in different evolutionary processes. (A) MGCs are genes simultaneously linked by position in a metabolic pathway and physical proximity in the genome. MGCs have been found/inferred to be polymorphic in recombining fungal populations. Such polymorphic loci could serve as entry points for novel MGCs constructed de novo (B) through duplication or relocation of endogenous genes to a common locus, or (C) by horizontal transfer of appropriately sized MGCs. (D) Duplication and relocation of genes may also contribute to MGC remodeling. (E) The loss of clustering might occur as unclustering of genes when the individual genes remain under purifying selection, or by the pseudogenization and loss of one or more genes in the MGC. Differentially retained MGC gene functions suggest there are physicochemical or other constraints on the evolution of the pathways they encode. These constraints may also drive the occurrence of whole cluster loss (F) observed for small, highly clustered pathways like GAL, for instance, incomplete or poorly coordinated pathways (G) may contribute to fitness defects that arise from the accumulation of toxic metabolic intermediate compounds.

3.1 How Do Fungal MGCs Originate?

The birth of MGCs is largely hidden in the phylogenetic record, with MGCs often appearing to be spontaneous occurrences with little evidence of intermediate states that are not alternatively explained by later remodeling (Bradshaw et al., 2013; Reynolds et al., 2017; Slot & Rokas, 2010). It is possible that the clustering of pathways is quite rapid after, for example, duplication or horizontal gene transfer (HGT) of chromosome arms or supernumerary chromosomes followed by selective retention of fitness-enhancing genes (Lawrence & Roth, 1996; Novo et al., 2009). There is also evidence that the origin and growth of MGCs has occurred by duplication and relocation of individual genes to a common locus by both vertical and horizontal transmission (Proctor, McCormick, Alexander, & Desjardins, 2009). For example, *Saccharomyces cerevisiae* MGCs encoding biotin synthesis (which are widespread in Pezizomycotina, but rare in Saccharomycotina) may have been assembled by HGT from multiple bacterial lineages and duplication of endogenous genes to common loci (Hall & Dietrich, 2007). Alternatively, the birth of the allantoin utilization (DAL) cluster appears to have formed by relocation of individual gene paralogs following the allopolyploidization of two *Saccharomyces* species (Marcet-Houben & Gabaldón, 2015; Wong & Wolfe, 2005). The convergent birth of a GAL cluster in *Cryptococcus neoformans* may have involved multiple duplications to a common locus, but can also be explained by HGT of a larger chromatin region from a close relative (Slot & Rokas, 2010). In another interesting convergence, both the *Torulaspora delbrueckii* (Wolfe et al., 2015) and the *Schizosaccharomyces japonicus* (Schja1|02949–Schja1|02954) GAL MGCs have grown from a core cluster to include upstream depolymerization of melibiose. The *T. delbrueckii* GAL cluster has more specifically acquired multiple additional genes suggesting it now encodes the complete shuttling of the malt disaccharide melibiose into glycolysis, a pathway that distinguishes lager yeast from ale yeast. Finally, cyanate detoxification (CCA) MGCs originated on at least three occasions by gene duplications to a common locus; such two-gene MGCs may be expected to arise more frequently by random rearrangements before subsequent selection than clusters of three or more genes.

3.2 Remodeling and Unclustering of MGCs

MGCs are often for convenience treated as defined genetic units when comparing homologous assemblages across fungi, but the clustered state requires continual maintenance in a background of rapidly rearranging genomes (Croll, Zala, & Mcdonald, 2013; Hane et al., 2011), and cluster remodeling has been frequently identified among fungi (Khaldi et al., 2008). For example, MGCs encoding the production of dothistromin, sterigmatocystin, and aflatoxins share a common ancestor, which is inferred to be similar to the present-day sterigmatocystin cluster based on synteny analysis (Bradshaw et al., 2013). Specifically, colocated paralogs of two enzymes in aflatoxin and sterigmatocystin MGCs are located in alternate miniclusters of the dothistromin MGC, strongly suggesting that the miniclusters were ancestrally linked. Descendants of the ancestral cluster subsequently became fragmented in Dothideomycetes, and the different lineages each recruited additional genes by duplication and relocation. It is worth noting that the same study identified reduced recombination within cluster fragments of different size in population analyses of *Dothistroma septosporum*. Cluster interaction or fusion may be another mechanism of metabolic pathway evolution. In one case, two overlapping clusters for the production of fumagillin and pseurotin share an embedded local regulator in *Aspergillus fumigatus* (Wiemann et al., 2013). Such cluster fusions may provide a mechanism of integrating novel metabolism into existing expression networks.

MGC remodeling can also be observed as the relocation of a small component of a cluster (e.g., only a module consisting of a transporter and regulator) into a different MGC, which could suggest repurposing to a new pathway. For example, by identifying gene clusters according to unexpected levels of synteny between loci in the divergent species, *Cadophora* sp. and *Chaetomium globosum*, alternate roles of the depudecin (DEP) transporter, *cis*-transcription factor, and polyketide synthase in a new MGC were suggested (Reynolds et al., 2017). The widespread gentisate catabolism cluster is clustered with alternative phenolic decomposition enzymes, suggesting it is part of a recurring point of metabolic stress (see below) in fungal decomposition of plants (Gluck-Thaler E. & Slot J.C., unpublished). Repurposing of subclusters can be observed at broader scales across prokaryotic gene clusters (Cimermancic et al., 2014), indicating that clustering occurs in both simple and more complex functional associations among genes. The clustered state of a metabolic pathway is not an exclusive indicator of coordinated function among genes, as metabolic pathways may become weakly linked or unclustered while maintaining their functions (Kemppainen, Alvarez Crespo, & Pardo, 2010; Slot et al., 2010). For a discrete pathway with conserved functions, like HANT, gradual loss of clustering since the origin of a cluster early in the radiation of filamentous fungi is a parsimonious explanation for unclustered pathways in Leotiomycetes and Sordariomycetes (Slot & Hibbett, 2007).

3.3 Tales From the MGC Crypt

Similar to the birth of MGCs, their deaths often occur over time windows that prevent the reconstruction of events given the current depth of sampling of extant genomes. For example, GAL pathways are most often found completely intact, completely absent, or completely composed of pseudogenes (Slot & Rokas, 2010). These observations may be partly due to the clustered state facilitating the accelerated removal of pathways. It was shown that the rate of transition from a clustered to a nonfunctional state (absence or pseudogene) is significantly greater than transitions from an unclustered to nonfunctional state among GAL genes (Slot & Rokas, 2010). Parallel gene loss in MGCs can be simply explained by the reduced scale and/or number of chromatin deletion events required for cluster loss. However, the parallel origins of pseudogenes, which are often found as minimal sense-disrupting nucleotide transitions or indels, require more complex mechanisms to be invoked. One explanation may be that, similar to parallel deletion due to colocation of genes, clustered genes may be under simultaneous relaxed selection due to simultaneous regulatory repression, or regional chromatin modifications (Pophaly & Tellier, 2015). These mechanisms assume that selection on each gene is independent of selection on others; however, it is more likely that selection on a gene is directly influenced by the loss of function of coordinately regulated genes.

Although fungal MGC death is rapid, the rate of genome change is significantly lower in fungi than in prokaryotes in general, so as the availability of complete genomes increases, it becomes possible to reconstruct and identify trends in the degradation of gene clusters using intermediate genotypes. Pseudogenes and their remnants are a key part of reconstructing the death of gene clusters in horizontal and vertical inheritance. Detailed reconstructions of gene cluster "taphonomy" (the processes of decay) have been performed for the bikaverin (BIK) and DEP MGCs (Campbell, Staats, van Kan, & Rokas, 2013; Reynolds et al., 2017). From these studies, common patterns have emerged. In both cases, degeneration was found to be quite rapid following MGC origin in a lineage, with both functional and highly degenerate clusters found in the same species or species complex (Reynolds et al., 2017; Schumacher et al., 2013)—but note the possibility of long-term retention of nonfunctional polymorphisms, such as in the *Saccharomyces kudriavzevii* GAL pathway (Hittinger et al., 2010). Another intriguing observation is the strong bias in the rates of degeneration among individual genes, independent of gene position in the cluster or size. In both BIK and DEP, regulatory and transport genes are retained longer and more often than those for catalytic enzymes, and are less likely to be found as pseudogenes. Differential selection by gene function may be explained by the relative modularity of the different functional classes, for example, transporters and regulators may be more readily adapted to new metabolic roles, while enzymes (particularly accessory enzymes in rapidly evolving SM pathways) may be more specialized in their functions and less likely to be directly repurposed (Nam et al., 2012). Evidence of transporter/regulator module repurposing has been discussed, and evidence for high modularity of fungal transporters is accumulating (Marsit & Dequin, 2015); however, other explanations for differential gene retention remain to be tested.

One possible explanation for differential gene retention is that gene functions within a pathway are unevenly distributed among members of recombining populations. In the case of filamentous Ascomycota, the formation of heterokaryons enables the distribution of the metabolic functions of a single mycelium among two or more nuclei in variable ratios that reflect environmental conditions (Samils, Oliva, & Johannesson, 2014). This process may maintain genetic diversity, while relieving the replication and regulatory burdens of pan-genomes on individual haplotypes. Cohabitation in a heterokaryon of producers and nonproducers of an SM may maintain selection to export or compartmentalize the metabolite from autotoxicity. Alternatively, similar selection pressures could arise among populations or communities of mycelia in a "public goods" context (Richards & Talbot, 2013). Not all members of a population may be required to produce a defense metabolite in order to realize a general benefit; however, they all must be able to tolerate the metabolite.

It is also intriguing to note that an association between the order of gene expression and the rates of gene loss was identified in the BIK MGC (Brown, Butchko, Busman, & Proctor, 2012; Campbell et al., 2013), first characterized in *Fusarium fujikuroi* (Wiemann et al., 2009). Namely, the genes expressed latest in *Fusarium* are the first and most frequently lost in *Botrytis*. This pattern could be a direct result of the feedback of gene expression on purifying selection (Pophaly & Tellier, 2015); however, the exact order of expression by function suggests there are evolutionary and physiological constraints involved. In the *F. verticillioides* BIK MGC, the transporter, BIK6, is the first structural gene expressed after the induction of the NMR-like regulator protein, BIK4, and the activating transcription factor, BIK. Catalytic enzymes are then expressed in reverse metabolic order (Arndt et al., 2015). This precise order of expression may enable complete control over the production of potentially toxic metabolites, by assembling all components of regulation, efflux, and intermediate metabolism prior to the formation of metabolic precursors for biosynthesis; while no obvious defects were observed in cultured strains expressing intermediates in bikaverin synthesis (Arndt et al., 2015; Wiemann et al., 2009), subtle differences in fitness may still be selected depending on effective population sizes and levels of stress in nature (Kraemer, Morgan, Ness, Keightley, & Colegrave, 2016; Lynch, 2006). It is thus interesting that in plants, there is limited correspondence between MGCs predicted by functional associations and those predicted by coordinated expression (Wisecaver et al., 2017); gene clusters may have coordinated expression, but not necessarily simultaneous expression. Reproduction of these patterns in other gene clusters would suggest that intergenic regulatory space is an integrated component of fungal SM MGCs that should be considered simultaneously with coding genes.

3.4 MGC Polymorphism

In addition to influencing macroevolutionary trajectories of metabolic pathways in fungal lineages, MGCs may also directly participate in population-level evolutionary processes. Loci that are polymorphic for entire gene clusters have been identified or inferred in multiple fungal lineages. Zhang, Rokas, and Slot (2012) reconstructed the simultaneous ancestral state of two complete SM MGCs of approximately the same size in a variable locus in Arthrodermataceae dermatophytes. In another example, Gibbons et al. (2012) identified two different SM MGCs sharing a terpene cyclase and a common locus in the *Aspergillus flavus* species complex. One of these MGC alleles, a homolog of which produces a sesquiterpene in *Trichoderma* spp., appears to be fixed in *A. flavus* strains (*A. oryzae*) that were domesticated in traditional fermented foods in Asia (Crutcher et al., 2013), and which the authors speculated could contribute to the flavor profiles of sake (Gibbons et al., 2012). While loci of clustered coadapted genes can lead to speciation due to hybrid incompatibility, "supergenes" are found to preserve locally adaptive phenotypes in recombining populations of plants and animals in a heterogenous landscape (Kokko et al., 2017; Purcell, Brelsford, Wurm, Perrin, & Chapuisat, 2014; Thompson & Jiggins, 2014). Polymorphisms covering MGC-sized regions may be part of a broader adaptive mechanism in some fungal populations, but may have been largely missed because de novo genome assembly has not been standard in genome resequencing (Plissonneau, Stürchler, & Croll, 2016).

3.5 HGT as Cause and/or Effect of Gene Clustering

A causal association between gene clustering and HGT, first proposed as "selfish clustering" in bacteria (Lawrence & Roth, 1996) and then fungi (Walton, 2000), has seen a steady trickle of support in the past two decades.

This hypothesis proposes that the state of being clustered with other genes with linked functions gives an individual gene increased fitness due to the added selective benefit to novel hosts following HGT. At the time of the hypothesis, there was little direct genomic evidence of HGT of MGCs in fungi (Rosewich & Kistler, 2000), but the discontinuous distribution of SMs among fungi questioned the strict vertical inheritance of SM MGCs. The first convincing cases of discrete horizontal MGC transfers into fungi involved the HANT MGC for nitrate assimilation, which was inferred to have been transferred from relatives of Ustilaginales (Basidiomycota) to Hypocreales (Ascomycota) when only a small sample of genomes was available (Slot & Hibbett, 2007). The fungal HANT MGC was further inferred to have been transferred from a younger osmotrophic lineage, Oomycetes (Heterokonta), to an ancestor of terrestrial fungi. Although the "selfish cluster hypothesis" was originally intended to explain the distribution of SM MGCs, the rapid diversification and discontinuous distributions of SM pathways made it difficult to identify and confirm whole MGC transfer events with a small genome sample (Khaldi et al., 2008; Patron et al., 2007). However, evidence of HGT of SM MGCs (Table 2) has steadily accumulated along with the increasing availability of fungal genomes (Campbell et al., 2012; Dhillon et al., 2015; Gibbons et al., 2012; Marcet-Houben & Gabaldón, 2016; Proctor et al., 2013; Reynolds et al., 2017; Slot & Rokas, 2011). Horizontal transfers across large phylogenetic divides were also involved in the evolution of degradative MGCs including GAL and specialized phenolic compound degradation (Greene et al., 2014; Slot & Rokas, 2010).

In order to test the hypothesis that clustered genes have increased dispersal by HGT, KEGG-encoded enzymes were investigated for evidence of HGT utilizing an automated gene tree–species tree reconciliation approach (Wisecaver et al., 2014). This search was heavily biased toward core and nutritional metabolic pathways which were most represented in KEGG at the time. By the assumptions made in the study, HGT was found to impact an average of 2.8% of enzymes, but the HGT frequency for clustered genes in the set was 4.8%. The increased rate of HGT among clustered genes was primarily driven by genomes in Pezizomycotina, which were also the most clustered among the genomes sampled. HGT was also most frequent in the KEGG metabolic categories that contained the most clusters in Pezizomycotina, including biosynthesis of SMs and metabolism of carbohydrates and amino acids. However, clustering was inversely related to HGT in Saccharomycotina, which may reflect the genome reduction trends in Saccharomycotina coupled with the increased rates of loss of clustered genes.

Table 2 Examples of HGT of Complete Fungal MGCs (Note Hypothetical Transfers Incidental to Genome Projects Not Included)

Cluster		Phenotype (Symbol for Confirmed or Presumed)	Fitness Benefit (Evidence or Not)	(Hypothetical) Ecological Context of Transfer (Exact Donor/Recipient Rarely Evident)	Ref.
Nutritional					
HANT	3	Nitrate assimilation	Resource availability	Invasion of land phytobiome community interactions	Slot and Hibbett (2007)
GAL	4	Galactose utilization	Resource availability	Fruit/grain fermentation microbiome interactions	Slot and Rokas (2010)
Degradation					
STILB	6	Stilbene degradation/ utilization	Substrate detoxification	Grass phytobiome community interactions	Greene et al. (2014)
TD-BE	8	Tyrosine degradation/ utilization, pyomelanin synthesis	Abiotic stress reduction	Extreme microenvironment community interactions	Greene et al. (2014)
CCA	2	Cyanate degradation	Substrate detoxification	Endophyte/pathogen community interactions	Elmore et al. (2015)
FDB1, FDB2	8, 14	BOA detoxification	Substrate detoxification/ predation	Endophytobiome community interactions	Glenn et al. (2016)
Secondary metabolism					
ST	23	Sterigmatocystin biosynthesis	Defense against invertebrates/ microbes	Soil community interactions, mycoparasitism	Matasyoh, Dittrich, Schueffler, and Laatsch (2010) and Slot and Rokas (2011)

Continued

Table 2 Examples of HGT of Complete Fungal MGCs (Note Hypothetical Transfers Incidental to Genome Projects Not Included)—cont'd

Cluster		Phenotype (Symbol for Confirmed or Presumed)	Fitness Benefit (Evidence or Not)	(Hypothetical) Ecological Context of Transfer (Exact Donor/Recipient Rarely Evident)	Ref.
FUM	16	Fumonisin biosynthesis	?	Soil/phytobiome community interactions	Khaldi and Wolfe (2011) and Proctor et al. (2013)
BIK	6	Bikaverin biosynthesis	Defense against microbes	Soil/phytobiome community interactions	Campbell, Rokas, and Slot (2012) and Schumacher et al. (2013)
VIR	6	Viridin and viridiol biosynthesis	?	Phytobiome community interactions	Gibbons et al. (2012)
CLC	11	Chaetoglobosin-like compound biosynthesis	Resource acquisition/ predation	Endophytism, plant pathogenesis	Dhillon et al. (2015)
ERG	7 core genes	Ergotamine biosynthesis	Herbivore deterrence	Endophytobiome community interactions	Marcet-Houben and Gabaldón (2016)
LOL	6 core genes	Loline biosynthesis	Herbivore deterrence	Endophytobiome community interactions	Marcet-Houben and Gabaldón (2016)
DEP	6	Depudecin biosynthesis	Resource acquisition/ predation	Endophytobiome community interactions	Reynolds et al. (2017)
CTB	10	Cercosporin biosynthesis	Resource acquisition/ predation	Plant pathogen community interactions	de Jonge et al. (2017)

MGC loss and degradation can lead to underestimation of MGC HGT, as is demonstrated by BIK and GAL MGCs (Campbell et al., 2013; Slot & Rokas, 2010). For example, out of 10 *B. cinerea* species descended from an ancestor that received a complete BIK cluster from *Fusarium*, 5 leave no trace of BIK, and 3 would suggest the transfer involved a single gene (the NMR-like regulatory protein, BIK4), if pseudogenes were not thoroughly investigated. Similarly, for multiple fungi with DEP MGCs thought to have been horizontally acquired, there are congeneric species that have retained only two or fewer of the six DEP genes (Reynolds et al., 2017).

Such whole pathway transfers could suggest HGT of bacterial operons to fungi is a mechanism of MGC origin, which was initially hypothesized as the origin of the Penicillin MGC when high isopenicillin N synthetase sequence identity was found between Streptomyces and fungi (Weigel et al., 1988). HGT of bacterial genes is usually found to involve single genes instead (Hall, Brachat, & Dietrich, 2005; Marcet-Houben & Gabaldón, 2010). Furthermore, most fungal MGCs with bacterial homologs, like tyrosine and porphyrin metabolism, are found to be convergently derived (Greene et al., 2014; Zhang et al., 2012).

Analyses that support horizontal transfer of entire structural regions have strengthened support for HGT of the individual genes they contain, and have buoyed the greater case for HGT as a significant contributor to fungal evolution (Gluck-Thaler & Slot, 2015). Multigene segments of chromatin that have consistent signals of incongruence with the species phylogeny not only make it unlikely that such signals are artifacts of sampling or methodology, but they also provide additional evidence of HGT. Synteny is not well conserved through vertical evolution in fungi, so shared synteny across an unexpectedly large phylogenetic divide is additional evidence of HGT. Slot and Rokas (2011) found that the relative conservation of synteny between *A. nidulans* and *Podospora anserina* sterigmatocystin clusters is much greater than expected when compared to a random selection of comparable loci. Further, AflR-binding motifs are shared in 13 of 23 intergenic spaces between *A. nidulans* and *P. anserina*. Because synteny is often conserved following HGT of MGCs, and observed transfers usually comply with the boundaries of MGCs (possibly after trimming of functionally unrelated flanking genes by genetic drift), regions of genomes with shared HGT signatures, and unexpectedly conserved synteny may be used to identify novel MGCs (Fig. 2). Current methods for MGC detection in fungi rely on prior knowledge of gene functions and models of cluster composition, and are optimized for detection of SM biosynthesis MGCs (Blin et al., 2013;

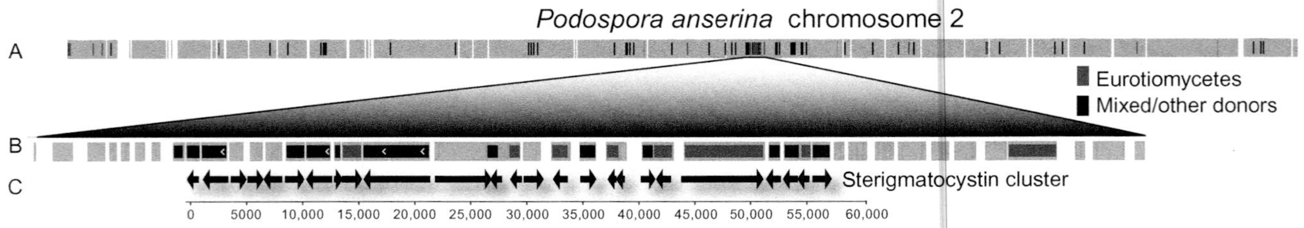

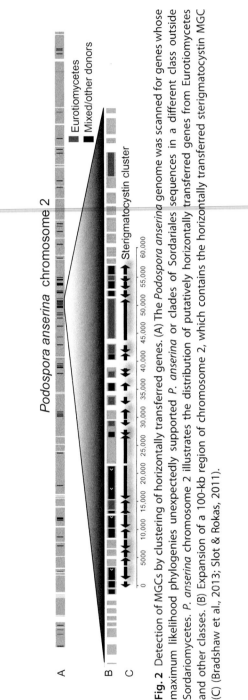

Fig. 2 Detection of MGCs by clustering of horizontally transferred genes. (A) The *Podospora anserina* genome was scanned for genes whose maximum likelihood phylogenies unexpectedly supported *P. anserina* or clades of Sordariales sequences in a different class outside Sordariomycetes. *P. anserina* chromosome 2 illustrates the distribution of putatively horizontally transferred genes from Eurotiomycetes and other classes. (B) Expansion of a 100-kb region of chromosome 2, which contains the horizontally transferred sterigmatocystin MGC (C) (Bradshaw et al., 2013; Slot & Rokas, 2011).

Khaldi et al., 2010). But the genetic mechanisms underlying much of fungal metabolism, particularly catabolic pathways, are not known, so methods that are agnostic with regard to gene homology and function are warranted.

It has alternatively been argued that discontinuous patterns of gene cluster distribution and gene phylogenies that conflict with species relationships can be explained by vertical birth and death processes (Kroken, Glass, Taylor, Yoder, & Turgeon, 2003) or by fungal pan-genomes maintaining MGCs at a very low level. But as fungal genomes have become more broadly and deeply analyzed, these objections have not accumulated sufficient data to reject HGT. One reason birth and death processes were not likely to explain the discontinuous distributions and phylogenetic incongruence of MGCs is that massive ancestral genomes would have to be inferred to account for current paralogs (Szöllősi, Davín, Tannier, Daubin, & Boussau, 2015). Low levels of MGC occurrence in extant populations might have suggested these large numbers could be maintained in the pan-genome; however, in-depth sequencing of fungal species has instead increased both the number of MGCs to account for in ancestors, and the instances of MGC HGT inferred. For example, a recent comparison of 48 SM MGCs across 18 *Aspergillus* spp. found most MGCs have phylogenetically restricted distributions and unstable synteny (De Vries et al., 2017; Lind et al., 2015), suggesting recent origins rather than ancestral presence. On the other hand, Koczyk, Dawidziuk, and Popiel (2015) suggest that while HGT is a major recent mechanism of polyketide MGC dispersal, the core multimodular enzymes were well established before modern lineages diversified. Could these enzymes be the legacy of ancient MGC evolutionary processes?

4. EVOLUTIONARY MECHANISMS CONTRIBUTING TO BIRTH AND DISPERSAL OF MGCs

4.1 Mutation and Selection in the Origin of MGCs

For genes located on the same chromatin scaffold, selection against recombination between coadapted alleles may be able to drive and preserve clustering of functionally related genes (Fisher, 1930; Yeaman, 2013). However, eukaryotic genes are by default distributed among several chromosomes, which represent a barrier to gene movement by simple infidelity in homologous recombination, inversions, and deletions. Further, while stochastic rearrangement of chromatin by translocation and transposition can provide the appropriate types of variation to generate clustering, including the adjacency of functionally linked genes, such vanishingly rare events are unlikely to be fixed by the weak selection acting on eukaryotic genomes (Lynch, 2006). Genome reordering processes alone may not provide sufficient variability to assemble large gene clusters, so other mechanisms may be required. It has been argued, for example, that the localization of chromatin in cellular space during the expression of adaptive phenotypes may increase the rates of the specific types of mutations that lead to clustering of functionally related genes (Hurst, Pal, & Lercher, 2004). For example, transcription factor complexes may correspond to locations where chromosomally distant genes meet to be cotranscribed (transcription factories). Coexpressed genes in transcription factories will have a higher probability of linkage through random strand break and repair (Sutherland & Bickmore, 2009). The combination of selection for coadapted alleles, increased probability of mutations that join functionally related genes, and a sufficient effective population size to fix beneficial mutations may account for the tight clustering genes in fungi.

4.2 Direct Selection on Gene Order

In addition to coselection or coadaptation, evidence is accumulating that the clustered state can directly enhance fitness. For example, coordinated gene regulation facilitated by local chromatin structure and modifications may improve the efficiency of metabolic processes (Gacek & Strauss, 2012; Shwab et al., 2007; Tsochatzidou, Malliarou, Papanikolaou, Roca, & Nikolaou, 2017). Genes in fungal MGCs are known to share bidirectional promoters (Punt et al., 1995), and recent work suggests that fungal operons are more prevalent than initially thought (Yue et al., 2015). It is also possible that rearrangements resulting in shared promoters bypass the need for selection for coregulation of functionally linked genes in the birth and growth of gene clusters. Price, Huang, Arkin, and Alm (2005) suggested coregulation drives operon formation because newly born operons have more conserved regulatory sequence over their evolutionary history. However, as *cis*-regulation is not required for coordination of expression, this explanation is not satisfying as a general evolutionary mechanism for clustering.

In order to account for a sufficient fitness advantage to drive the clustering of coordinately expressed genes, it has been proposed that accumulation of reactive or inhibitory metabolic intermediates leads to toxic, disadvantaged phenotypes. In support of this hypothesis it was found that clusters of gene pairs that are simultaneously nearest metabolic and chromosomal neighbors have producer–consumer relationships with toxic intermediate compounds (McGary, Slot, & Rokas, 2013). This effect was strongest for divergently transcribed gene pairs, which often share promoters. For example, galactokinase (EC:2.7.1.6) produces galactose-1-P, which interferes with glycolysis, and is paired divergently with one or both of the genes simultaneously required for metabolizing galactose-1-P: UDP-glucose 4-epimerase (EC:5.1.3.2) and galactose-1-P uridyl transferase (EC:2.7.7.12). Two yeast species, *Naumovozyma castellii* and *Vanderwaltozyma polymorpha*, have both downstream genes paired divergently with alternate paralogs of galactokinase (Slot & Rokas, 2010). Interestingly, metabolic intermediates that accumulate in human genetic diseases, including galactosemia and tyrosinemia, correspond to gene pairs that are clustered in fungi, and gene clustering in fungi is correlated with coordinated expression of genes in human tissues (Eidem, McGary, & Rokas, 2015). In another case, Elmore et al. (2015) characterized the convergent evolution and proliferation of clusters comprised of divergently transcribed cyanase (EC:4.2.1.104) and carbonic anhydrase (EC:4.2.1.1) which are thought to enable detoxification of the fungicide, cyanate. They argued that these enzymes are mutually dependent due to the requirement of cyanase for bicarbonate supplied by carbonic anhydrase, an enzyme that is inhibited by cyanate. Finally, McGary et al. (2013) suggested clustering of biotin genes may be selected by the surprising toxicity (by competitive inhibition) of racemized (*S*)-8-amino-7-oxononanoate, which occurs rapidly following production of its enantiomerically pure form.

Two genes involved in gentisate-mediated phenolic catabolism, gentisate 1,2 dioxygenase (EC:1.13.11.4) and fumarylpyruvate hydrolase (EC:3.7.1.5), are the most commonly clustered metabolic neighbors, possibly because of plant-generated pressure to degrade voluminous and diverse phenolic compounds. These enzymes have neither been found as the metabolic/chromosomal nearest neighbor gene pairs that are strongly associated with toxic metabolic intermediates (McGary et al., 2013), nor been found clustered with maleylpyruvate isomerase (EC:5.2.1.4), which catalyzes the metabolic step between them. This could suggest that clustering of these genes is the result of codependent functions rather than selection for precise enzyme dosage to mitigate maleylpyruvate accumulation. However, both of these genes are commonly found in clusters with aldehyde dehydrogenase, and stilbene dioxygenase (EC:1.13.11.43), which cleaves stilbenes (e.g., resveratrol) into phenolic aldehydes (Greene et al., 2014). Aldehyde groups are common reactive intermediate states handled by divergently transcribed metabolic network neighbors (McGary et al., 2013).

Lang and Botstein (2011) pioneered the use of experimental evolution of GAL to address the role of coordinated regulation in gene clustering. In this experiment, no fitness advantage was found for genes regulated from a common promoter. In fact, there may even have been a fitness benefit to unclustered genes under the selection regime provided. However, in this experiment, the same promoter was used to express genes in alternate locations. It may be argued that the failure of yeast lineages without clustered genes to persist is not because promoter sharing is inherently advantageous (to the contrary, it may cause competition for the promoter), but because evolutionary divergence in gene expression is more likely when multiple promoters can mutate differently (Lynch, 2006). The ability of a yeast species to coordinately adapt at multiple promoters under selection to increase gene expression will further be limited by available variation at two loci and the size of a recombining population to simultaneously select for multiple variants. Coadaptation of the promoters would be in competition with coadaptation of enzymes. However, over the course of several generations in these experiments, the unclustered GAL loci would not have an opportunity to diverge. Following up these experiments with artificially mutated promoters causing divergent expression between gene pairs may yet show a fitness defect in unclustered GAL MGCs. Including alternate constructs in long-term evolution experiments will improve our understanding of the longer-term effects of clustering on coordination of expression.

4.3 Hooked on a (Unclustered) Pathway

GAL pathways are more likely to be lost when clustered than when unclustered, and are highly persistent in lineages where they are unclustered. However, GAL clustering and MGC loss are also associated with their occurrence in yeasts, which could be a confounding variable. A cursory look at another MGC, HANT, in three filamentous lineages reveals a pattern consistent with the GAL results: In Eurotiomycetes nitrite reductase is only found in 60% of 82 predicted proteomes, but it is clustered with other HANT genes 100% of the time it is present. By contrast, nitrite reductase is present in 94% of 92 predicted Sordariomycetes proteomes, but only clustered in 9%, while Agaricomycetes are intermediate in both factors. The dispersal of genes in the genome may prevent stochastic loss of these pathways by selection against transitional phenotypes due to the production of toxic intermediate metabolites. Consequently, the unclustering of metabolic pathways may represent a form of genetic addiction, contributing to the vertical inheritance of the pathway without regard for its impact on the host genome, as in toxin–toxin antidote addiction modules in bacteria and yeast (Yeo, Abu Bakar, Chan, Espinosa, & Harikrishna, 2016).

4.4 Toxicity and Antidotes

Genes acting as toxin antidotes may also give critical fitness benefits to SM MGCs, which might produce compounds that are at least incidentally antifungal (Keller, 2015). Transporters are common components of MGCs, but for very few are precise functions known. Their role may be to release metabolites into the environment to the detriment of their competitors, but they may also provide critical relief to the fungus producing them, either through secretion or subcellular compartmentalization of intermediate steps (Boenisch et al., 2017). Consequently, the observed acquisition or retention of the transporter regulator modules of SM MGCs may be an effective solution to neutralizing the chemical weaponry of fungal haplotypes in the same community or heterokaryon, under the environmental conditions of their expression (Reynolds et al., 2017). Other toxin antidotes may function by producing a toxin-resistant target, or making a self-protective modification to the toxin prior to its release (Abe et al., 2002; Keller, 2015; Scharf et al., 2010). In order for a complex SM pathway requiring a detox mechanism to have emerged, the detoxification mechanism(s) may have coadapted with biosynthetic genes since the emergence of their cognate toxicity. This could suggest that the origin of many pathways is the pairing of a novel enzyme with the mechanism of its detoxification. From there, MGC/pathway growth must be accompanied by compensatory modifications in the toxin/detoxification systems. Therefore, diversity of SMs may be driven by the inherent "selfishness" of MGCs, and not due to inherent adaptive properties of individual metabolites.

4.5 Pathway Specialization by MGC

Like all genetic information, MGCs must result from the interaction of neutral and selective mechanisms, and the selection for tight gene linkage may subsequently be enhanced by interactions and specialization among the enzymes they encode. There is evidence that agglomeration or channeling of enzymes in a metabolic pathway increases reaction efficiency, which would reduce inference toxic intermediate accumulation (Castellana et al., 2014). Selection for rapid metabolite exchange between enzymes by structural accommodations requires the coordinated evolution of protein secondary structure among enzymes. Recombination between MGCs might disrupt refined multienzyme complexes that encode several types of selective advantages, including cospecialization on molecule classes, and sequestration from pathways with similar intermediates. Gene fusions, such as seen in bifunctional GAL10 in *S. cerevisiae* (Scott & Timson, 2007), the AROM pentafunctional domain for synthesis of aromatic amino acids (Duncan, Edwards, & Coggins, 1987), and modular PKS and NRPS enzymes (Weissman, 2015), may also accomplish tight functional coupling of catalytic domains.

Specialization of enzyme function is associated with the increased and rapidly changing pathway flux that channeling is thought to permit (Nam et al., 2012). SMs with biomolecular activity are expected to stimulate selection to increase metabolic flux in both producers and targets of the metabolite. Fitness of metabolite producers can be enhanced by, for instance, secreting more of a toxin or inhibitor of competing organisms. On the other hand, a fungus adapting to a host-produced toxin is in a race to eliminate host metabolites produced in high quantities to avoid succumbing. It is especially advantageous for pathways that are derived from or parallel to core metabolism to be sequestered in order to avoid upsetting homeostasis with too much or imperfect substrates (Machado et al., 2017). The structural/kinetic optimization required to coordinate enzymes for specialized functions and high metabolic flux implies coadaptation among genes. Clustering may therefore be a consequence of specialization at the pathway scale, which may account for the high rates of loss of enzymes, but not *cis*-regulators. However, an empirical link between channeling and gene clustering has not yet been established, and there is need for a systematic evaluation of specialization of clustered enzymes.

4.6 MGCs as Hotspots of Evolution

Wisecaver et al. (2014) argued that MGCs are hotspots for fungal genome evolution because they are associated with increased rates of both gene duplication and HGT, which has also been demonstrated for prokaryotic gene clusters (Medema, Cimermancic, Sali, Takano, & Fischbach, 2014). The reasons for this association are not entirely clear. It is possible that all of these processes are coassociated with the metabolic functions that are most influenced by ecological selection. However, it has also been argued that emergent properties of MGCs make them special contributors to chromatin modification.

The most important emergent property of MGCs may be modularity, which facilitates their birth, remodeling, and loss. The ability of a fungus' metabolic toolkit to evolve (i.e., its "evolvability") is constrained by the impact any individual change will have on the organism's fitness (Pepper, 2003). MGCs with *cis*-regulatory control and optimization hypothetically require few interactions to be established among the constituent genes and existing networks, but more likely a single interaction between a master regulator and a *cis*-regulator will be sufficient to acquire their encoded pathways. The highly conserved roles of global regulators like LaeA among filamentous Ascomycota suggest a large number of SM MGCs require little to no modification to be integrated into fungal developmental programs following HGT between lineages (Bok & Keller, 2004). Other barriers to integration of metabolic pathways include interference with existing metabolism or other toxicities from metabolic products and intermediates. Among the "self-protection" mechanisms encoded in MGCs, which enable integration of modular pathways, are transporters, potentially involved in export and compartmentalization of intermediate and final pathway products, and resistant copies of the endogenous targets of the bioactive product of an MGC discussed previously. The ability to occupy genomic space without evicting genes that are integrated into core networks is another essential feature of gene clusters recently demonstrated. Pan-genomic chromatin is distributed in multiple ways in different fungal populations. In lineages with a defined sexual cycle, MGCs that approximate the average distance between recombination breakpoints may replace similar sized, dispensible gene clusters in a haploid and give rise to MGC polymorphisms. Plissonneau et al. (2016) demonstrated the high tolerance that highly recombining populations of *Zymoseptoria tritici* have for "orphan loci" that roughly correspond with MGCs in length, and may serve as entry points for novel MGCs. In lineages that rely predominantly on the parasexual cycle to generate genetic diversity, MGCs may tend to be shuttled into genomes through supernumerary chromosomes and replacement of low-complexity subtelomeric chromatin (Vlaardingerbroek et al., 2016). Transposable element activity in these regions may then generate enough variability in MGC location to integrate highly beneficial clusters into more stable genomic neighborhoods (Faino et al., 2016), as long as clusters do not exceed the size constraints of transposition.

Baquero (2004) posited three components required for assembly of genetic modules and gene clusters by natural selection in prokaryotes: operative components, such as enzymes, transporters, and regulators; translocative components, such as transposable elements and other mechanisms of chromatin rearrangement; and dispersive components, such as plasmids and phages that can move genetic material between individuals. In this model, operative components that are simultaneously beneficial under some selective pressure become simultaneously enriched in a metagenome. This increases the probability that dispersive components will deliver the elements to the same individuals. Simultaneously selected genes will be made adjacent by translocative components more frequently under these regimes, potentially interacting to form a complex (or "winning pattern") that provides a greater selective benefit when dispersed to new host genomes. In fungi, operative genes are highly diverse, and fungal genomes are under more or less constant rearrangement by transposable elements and recombination. TEs are particularly active in fast-evolving compartments near telomeres and centromeres.

Dispersion is considered a more difficult process in eukaryotes in general due to larger genes requiring intron splicing, informational barriers like codon preferences, and physical barriers such as the nuclear envelope (Richards, Leonard, Soanes, & Talbot, 2011). Despite considerable evidence for its occurrence, the mechanisms responsible for HGT of MGCs remain an outstanding question in fungal genomics. There are no natural dispersive elements (fungal plasmids or viruses) currently known that are capable of transporting chromatin on the magnitude of gene clusters. However, recent discoveries suggest fungi may be especially adept at acquiring foreign DNA from the community and environment. *Botrytis cinerea* sclerotia, which enable overwintering in soil, appear to be remarkably transformation competent when wounded in the presence of dissolved DNA (Ish Shalom, Gafni, Lichter, & Levy, 2011). Long linear strands of DNA would likely be degraded rapidly in the soil. It has been shown that circular DNA can serve as a stable vector for gene transfer in natural yeast populations (Galeote et al., 2011). TEs may similarly stabilize genetic information through chromatin circularization in Pezizomycotina where they are particularly active (Paun & Kempken, 2015). Consistent with this role of TEs, they are occasionally identified at the flanks of gene clusters suggesting a role in MGC assembly and/or movement (Greene et al., 2014; Proctor et al., 2013). The occurrence of genomic "compartments" with differential rates of remodeling and sequence evolution may provide a physical mechanism for cluster assembly (Fig. 3). Direct fungus-to-fungus transfer through transient hyphal fusions between unrelated species has also been proposed (Soanes & Richards, 2014). Between vegetatively compatible members of a species complex, conditionally dispensable chromosomes

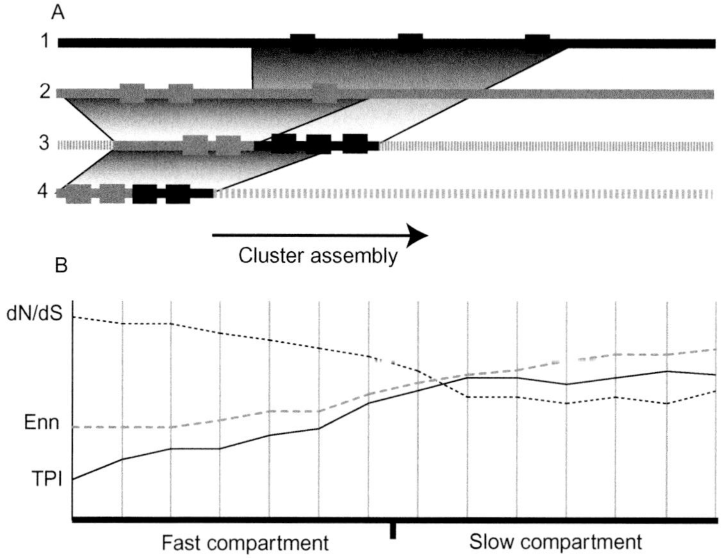

Fig. 3 Model for the birth and growth of MGCs by the interaction between selection and turnover in fungal genomes. MGCs might form gradually in populations, in rapidly evolving compartments of the genome such as subtelomeres, supernumerary chromosomes. Proposed here is a "chromatin conveyor" model in which vertically and horizontally duplicated chromatin is continually accreted to these compartments where genes under simultaneous selection are driven together by deletion of their neighbors that confer no fitness advantages, and gradual movement into less rapidly evolving compartments. However, most detailed cases of MGC birth within a lineage appear to involve horizontal transfers from other fungal lineages. (A) Chromatin from other locations within (1) and outside the genome (2) is continuously inserted into fast-evolving genomic compartments including subtelomeres (3) through elevated transposable element activity. Through the loss of intervening sequence under relaxed selection, coselected genes are driven together and away from the fast-evolving compartments. Repeated transposition and transfer of regions with coselected genes to additional fast-evolving compartments (4) will tighten the linkage and increase selectability of newly born gene clusters. (B) Properties of chromatin compartments that would be consistent with clustering being driven by movement against transposition gradients within a genome. Fast-evolving compartments are expected to exhibit the least purifying selection (dN/dS approaching 1) in the same regions where genes have the shortest average time postinsertion (TPI) due to higher transposable element activity. The birth of clusters may then be modeled by the average coexpression of nearest neighbor genes (Enn).

may be transferred during the parasexual cycle (Keller & Hohn, 1997; Vlaardingerbroek et al., 2016); however, this is unlikely to account for HGT between distantly related fungi. The search for the mechanism(s) of fungal HGT will go on.

5. WHAT'S AHEAD? BEYOND GENE COUNTING

Recent advances make it clear that the structural organization of fungal genomes can provide a more in-depth understanding of genome function than can gene composition alone. What all can be learned from differences in gene clustering and gene cluster content remains to be demonstrated. Not only do clusters enable a more resolved picture of the functional roles of genes from large gene families, but by their very structure and composition they may also indicate the selective pressures facing fungi both from intracellular constraints and environmental conditions. While gene clusters clearly contribute to the evolvability of fungi, it is not clear whether they are part of an evolvability mechanism, as some global regulatory networks would suggest, or merely a reflection of the structure of evolved biochemical networks. Significant challenges remain in understanding the significance of MGCs: is there truly a physiological benefit to the clustered state? Does clustering of genes in chromosomes indicate clustering of transcription and/or enzymes in cellular space? What is the full scope of metabolic pathways that are clustered? Are there higher orders of clustering (i.e., clusters of clusters analogous to genomic "islands" in bacteria) in fungi? Will the profiling of MGC composition bring us closer to modeling fungal guilds and niches? What is the greater impact of MGCs on fungal biodiversity? With the tremendous new opportunities to close fungal genomes on a large scale through single-strand sequencing technologies, the characterization of genome structure and structural evolution will provide a more nuanced and perhaps transformative understanding of the changes in gene composition that have been detailed in fungal gene families, fungal lineages, and fungal nutritional modes.

REFERENCES

Abe, Y., Suzuki, T., Mizuno, T., Ono, C., Iwamoto, K., Hosobuchi, M., et al. (2002). Effect of increased dosage of the ML-236B (compactin) biosynthetic gene cluster on ML-236B production in Penicillium citrinum. *Molecular Genetics and Genomics*, *268*(1), 130–137.

Abrahamian, M., Ah Fong, A. M. V., Davis, C., Andreeva, K., & Judelson, H. S. (2016). Gene expression and silencing studies in *Phytophthora infestans* reveal infection-specific nutrient transporters and a role for the nitrate reductase pathway in plant pathogenesis. *PLoS Pathogen*, *12*, e1006097.

Ahmad, S., Lee, S. Y., Kong, H. G., Jo, E. J., Choi, H. K., Khan, R., et al. (2016). Genetic determinants for pyomelanin production and its protective effect against oxidative stress in Ralstonia solanacearum. *PLoS One*, *11*, e0160845.

Ahn, J. H., & Walton, J. D. (1996). Chromosomal organization of TOX2, a complex locus controlling host-selective toxin biosynthesis in Cochliobolus carbonum. *Plant Cell*, *8*, 887–897.

Arndt, B., Studt, L., Wiemann, P., Osmanov, H., Kleigrewe, K., Köhler, J., et al. (2015). Genetic engineering, high resolution mass spectrometry and nuclear magnetic resonance spectroscopy elucidate the bikaverin biosynthetic pathway in Fusarium fujikuroi. *Fungal Genetics and Biology*, *84*, 26–36.

Arst, H. N., & MacDonald, D. W. (1975). A gene cluster in Aspergillus nidulans with an internally located cis-acting regulatory region. *Nature*, *254*, 26–31.

Baquero, F. (2004). From pieces to patterns: Evolutionary engineering in bacterial pathogens. *Nature Reviews. Microbiology*, *2*, 510–518.

Bills, G. F., Gloer, J. B., & An, Z. (2013). Coprophilous fungi: Antibiotic discovery and functions in an underexplored arena of microbial defensive mutualism. *Current Opinion in Microbiology*, *16*, 549–565.

Blin, K., Medema, M. H., Kazempour, D., Fischbach, M. A., Breitling, R., Takano, E., et al. (2013). antiSMASH 2.0—A versatile platform for genome mining of secondary metabolite producers. *Nucleic Acids Research*, *41*, W204–W212.

Boenisch, M. J., Broz, K. L., Purvine, S. O., Chrisler, W. B., Nicora, C. D., Connolly, L. R., et al. (2017). Structural reorganization of the fungal endoplasmic reticulum upon induction of mycotoxin biosynthesis. *Scientific Reports*, 7, 44296.

Bok, J. W., & Keller, N. P. (2004). LaeA, a regulator of secondary metabolism in Aspergillus spp. *Eukaryotic Cell*, *3*, 527–535.

Bradshaw, R. E., Slot, J. C., Moore, G. G., Chettri, P., de Wit, P. J. G. M., Ehrlich, K. C., et al. (2013). Fragmentation of an aflatoxin-like gene cluster in a forest pathogen. *New Phytologist*, *198*, 525–535.

Brown, D. W., Butchko, R. A. E., Busman, M., & Proctor, R. H. (2012). Identification of gene clusters associated with fusaric acid, fusarin, and perithecial pigment production in Fusarium verticillioides. *Fungal Genetics and Biology*, *49*, 521–532.

Brown, D. W., & Proctor, R. H. (2016). Insights into natural products biosynthesis from analysis of 490 polyketide synthases from Fusarium. *Fungal Genetics and Biology*, *89*, 37–51.

Brown, D. W., Yu, J. H., Kelkar, H. S., Fernandes, M., Nesbitt, T. C., Keller, N. P., et al. (1996). Twenty-five coregulated transcripts define a sterigmatocystin gene cluster in Aspergillus nidulans. *Proceedings of the National Academy of Sciences of the United States of America*, *93*, 1418–1422.

Bushley, K. E., Raja, R., Jaiswal, P., Cumbie, J. S., Nonogaki, M., Boyd, A. E., et al. (2013). The genome of Tolypocladium inflatum: Evolution, organization, and expression of the cyclosporin biosynthetic gene cluster. *PLoS Genetics*, *9*, e1003496.

Campbell, M. A., Rokas, A., & Slot, J. C. (2012). Horizontal transfer and death of a fungal secondary metabolic gene cluster. *Genome Biology and Evolution*, *4*, 289–293.

Campbell, M. A., Staats, M., van Kan, J., & Rokas, A. (2013). Repeated loss of an anciently horizontally transferred gene cluster in Botrytis. *Mycologia*, *105*, 1126–1134.

Castellana, M., Wilson, M. Z., Xu, Y., Joshi, P., Cristea, I. M., Rabinowitz, J. D., et al. (2014). Enzyme clustering accelerates processing of intermediates through metabolic channeling. *Nature Biotechnology*, *32*, 1011–1018.

Cimermancic, P., Medema, M. H., Claesen, J., Kurita, K., Brown, L. C. W., Mavrommatis, K., et al. (2014). Insights into secondary metabolism from a global analysis of prokaryotic biosynthetic gene clusters. *Cell*, *158*, 412–421.

Croll, D., Zala, M., & Mcdonald, B. A. (2013). Breakage-fusion-bridge cycles and large insertions contribute to the rapid evolution of accessory chromosomes in a fungal pathogen. *PLoS Genetics*, *9*, e1003567.

Crutcher, F. K., Parich, A., Schuhmacher, R., Mukherjee, P. K., Zeilinger, S., & Kenerley, C. M. (2013). A putative terpene cyclase, vir4, is responsible for the biosynthesis of volatile terpene compounds in the biocontrol fungus Trichoderma virens. *Fungal Genetics and Biology*, *56*, 67–77.

de Jonge, R., Ebert, M. K., Huitt-Roehl, C. R., Pal, P., Suttle, J. C., Neubauer, J. D., et al. (2017). Ancient duplication and horizontal transfer of a toxin gene cluster reveals novel mechanisms in the cercosporin biosynthesis pathway. *bioRxiv*, 100545, https://doi.org/10.1101/100545.

De Vries, R. P., Riley, R., Wiebenga, A., Aguilar-Osorio, G., Amillis, S., Uchima, C. A., et al. (2017). Comparative genomics reveals high biological diversity and specific adaptations in the industrially and medically important fungal genus Aspergillus. *Genome Biology*, *18*, 28.

Dhillon, B., Feau, N., Aerts, A. L., Beauseigle, S., Bernier, L., Copeland, A., et al. (2015). Horizontal gene transfer and gene dosage drives adaptation to wood colonization in a tree pathogen. *Proceedings of the National Academy of Sciences of the United States of America*, *112*, 3451–3456.

Díez, B., Gutierrez, S., Barredo, J. L., van Solingen, P., van der Voort, L. H., & Martin, J. F. (1990). The cluster of penicillin biosynthetic genes. Identification and characterization of the pcbAB gene encoding the alpha-aminoadipyl-cysteinyl-valine synthetase and linkage to the pcbC and penDE genes. *The Journal of Biological Chemistry*, *265*, 16358–16365.

Douglas, H. C., & Hawthorne, D. C. (1966). Regulation of genes controlling synthesis of the galactose pathway enzymes in yeast. *Genetics*, *54*, 911–916.

Duncan, K., Edwards, R. M., & Coggins, J. R. (1987). The pentafunctional arom enzyme of Saccharomyces cerevisiae is a mosaic of monofunctional domains. *The Biochemical Journal*, *246*, 375–386.

Eidem, H. R., McGary, K. L., & Rokas, A. (2015). Shared selective pressures on fungal and human metabolic pathways lead to divergent yet analogous genetic responses. *Molecular and Biological Evolution*, *32*, 1449–1455.

Elmore, M. H., McGary, K. L., Wisecaver, J. H., Slot, J. C., Geiser, D. M., Sink, S., et al. (2015). Clustering of two genes putatively involved in cyanate detoxification evolved recently and independently in multiple fungal lineages. *Genome Biology and Evolution*, *7*, 789–800.

Faino, L., Seidl, M. F., Shi-Kunne, X., Pauper, M., van den Berg, G. C. M., Wittenberg, A. H. J., et al. (2016). Transposons passively and actively contribute to evolution of the two-speed genome of a fungal pathogen. *Genome Research*, *26*, 1091–1100.

Firn, R. D., & Jones, C. G. (2003). Natural products—A simple model to explain chemical diversity. *Natural Product Reports*, *20*, 382–391.

Fisher, R. A. (1930). *The genetical theory of natural selection*. New York: Oxford University Press.

Floudas, D., Binder, M., Riley, R., Barry, K., Blanchette, R. A., Henrissat, B., et al. (2012). The Paleozoic origin of enzymatic lignin decomposition reconstructed from 31 fungal genomes. *Science*, *336*, 1715–1719.

Gacek, A., & Strauss, J. (2012). The chromatin code of fungal secondary metabolite gene clusters. *Applied Microbiology and Biotechnology*, *95*, 1389–1404.

Galeote, V., Bigey, F., Beyne, E., Novo, M., Legras, J.-L., Casaregola, S., et al. (2011). Amplification of a Zygosaccharomyces bailii DNA segment in wine yeast genomes by extrachromosomal circular DNA formation. *PLoS One*, *6*, e17872.

Gibbons, J. G., Salichos, L., Slot, J. C., Rinker, D. C., McGary, K. L., King, J. G., et al. (2012). The evolutionary imprint of domestication on genome variation and function of the filamentous fungus Aspergillus oryzae. *Current Biology*, *22*, 1403–1409.

Giles, N. H. (1978). The organization, function, and evolution of gene clusters in eukaryotes. *The American Naturalist*, *112*(986), 641–657.

Giles, N. H., Case, M. E., & Jacobson, J. W. (1973). Genetic regulation of quinate-shikimate catabolism in Neurospora crassa. In B. A. Hamkalo & J. Papaconstantinou (Eds.), *Molecular cytogenetics* (pp. 309–314). New York: Springer.

Glenn, A. E., Davis, C. B., Gao, M., Gold, S. E., Mitchell, T. R., Proctor, R. H., et al. (2016). Two horizontally transferred xenobiotic resistance gene clusters associated with detoxification of benzoxazolinones by Fusarium species. *PLoS One*, *11*, e0147486.

Gluck-Thaler, E., & Slot, J. C. (2015). Dimensions of horizontal gene transfer in eukaryotic microbial pathogens. *PLoS Pathogens*, *11*, e1005156.

Greene, G. H., McGary, K. L., Rokas, A., & Slot, J. C. (2014). Ecology drives the distribution of specialized tyrosine metabolism modules in fungi. *Genome Biology and Evolution*, *6*, 121–132.

Hall, C., Brachat, S., & Dietrich, F. S. (2005). Contribution of horizontal gene transfer to the evolution of Saccharomyces cerevisiae. *Eukaryotic Cell*, *4*, 1102–1115.

Hall, C., & Dietrich, F. S. (2007). The reacquisition of biotin prototrophy in Saccharomyces cerevisiae involved horizontal gene transfer, gene duplication and gene clustering. *Genetics*, *177*, 2293–2307.

Hammerbacher, A., Schmidt, A., Wadke, N., Wright, L. P., Schneider, B., Bohlmann, J., et al. (2013). A common fungal associate of the spruce bark beetle metabolizes the stilbene defenses of Norway spruce. *Plant Physiology*, *162*(3), 1324–1336.

Hane, J. K., Rouxel, T., Howlett, B. J., Kema, G. H. J., Goodwin, S. B., & Oliver, R. P. (2011). A novel mode of chromosomal evolution peculiar to filamentous Ascomycete fungi. *Genome Biology*, *12*, R45.

Hittinger, C. T., Gonçalves, P., Sampaio, J. P., Dover, J., Johnston, M., & Rokas, A. (2010). Remarkably ancient balanced polymorphisms in a multi-locus gene network. *Nature*, *464*, 54–58.

Hoffmeister, D., & Keller, N. P. (2007). Natural products of filamentous fungi: Enzymes, genes, and their regulation. *Natural Product Reports*, *24*, 393–416.

Hurst, L., Pal, C., & Lercher, M. (2004). The evolutionary dynamics of eukaryotic gene order. *Nature Reviews. Genetics*, *5*, 299–310.

Ish Shalom, S., Gafni, A., Lichter, A., & Levy, M. (2011). Transformation of Botrytis cinerea by direct hyphal blasting or by wound-mediated transformation of sclerotia. *BMC Microbiology*, *11*, 266.

Jeffries, T. W., & Van Vleet, J. R. H. (2009). Pichia stipitis genomics, transcriptomics, and gene clusters. *FEMS Yeast Research*, *9*, 793–807.

Johnstone, I. L., McCabe, P. C., Greaves, P., Gurr, S. J., Cole, G. E., Brow, M. A., et al. (1990). Isolation and characterisation of the crnA-niiA-niaD gene cluster for nitrate assimilation in Aspergillus nidulans. *Gene*, *90*, 181–192.

Jönsson, L. J., Alriksson, B., & Nilvebrant, N.-O. (2013). Bioconversion of lignocellulose: Inhibitors and detoxification. *Biotechnology for Biofuels*, *6*, 16.

Kanehisa, M., & Goto, S. (2000). KEGG: Kyoto encyclopedia of genes and genomes. *Nucleic Acids Research*, *28*, 27–30.

Kappel, L., Gaderer, R., Flipphi, M., & Seidl-Seiboth, V. (2015). The N-acetylglucosamine catabolic gene cluster in Trichoderma reesei is controlled by the Ndt80-like transcription factor RON1. *Molecular Microbiology*, *99*, 640–657.

Keller, N. P. (2015). Translating biosynthetic gene clusters into fungal armor and weaponry. *Nature Chemical Biology*, *11*, 671–677.

Keller, N., & Hohn, T. (1997). Metabolic pathway gene clusters in filamentous fungi. *Fungal Genetics and Biology*, *21*, 17–29.

Kemppainen, M. J., Alvarez Crespo, M. C., & Pardo, A. G. (2010). fHANT-AC genes of the ectomycorrhizal fungus Laccaria bicolor are not repressed by l-glutamine allowing simultaneous utilization of nitrate and organic nitrogen sources. *Environmental Microbiology Reports*, *2*(4), 541–553.

Kennedy, J., Auclair, K., Kendrew, S. G., Park, C., Vederas, J. C., & Hutchinson, C. R. (1999). Modulation of polyketide synthase activity by accessory proteins during lovastatin biosynthesis. *Science*, *284*, 1368–1372.

Khaldi, N., Collemare, J. R. M., Lebrun, M.-H., & Wolfe, K. H. (2008). Evidence for horizontal transfer of a secondary metabolite gene cluster between fungi. *Genome Biology*, *9*, R18.

Khaldi, N., Seifuddin, F. T., Turner, G., Haft, D., Nierman, W. C., Wolfe, K. H., et al. (2010). SMURF: Genomic mapping of fungal secondary metabolite clusters. *Fungal Genetics and Biology*, *47*, 736–741.

Khaldi, N., & Wolfe, K. H. (2011). Evolutionary origins of the fumonisin secondary metabolite gene cluster in Fusarium verticillioides and Aspergillus niger. *International Journal of Evolutionary Biology*, *2011*, 423821.

Koczyk, G., Dawidziuk, A., & Popiel, D. (2015). The distant siblings? A phylogenomic roadmap illuminates the origins of extant diversity in fungal aromatic polyketide biosynthesis. *Genome Biology and Evolution*, 7, 3132–3154.

Kohler, A., Kuo, A., Nagy, L. G., Morin, E., Barry, K. W., Buscot, F., et al. (2015). Convergent losses of decay mechanisms and rapid turnover of symbiosis genes in mycorrhizal mutualists. *Nature Genetics*, *47*, 410–415.

Kokko, H., Chaturvedi, A., Croll, D., Fischer, M. C., Guillaume, F., Karrenberg, S., et al. (2017). Can evolution supply what ecology demands? *Trends in Ecology & Evolution*, *32*, 187–197.

Kraemer, S. A., Morgan, A. D., Ness, R. W., Keightley, P. D., & Colegrave, N. (2016). Fitness effects of new mutations in Chlamydomonas reinhardtii across two stress gradients. *Journal of Evolutionary Biology*, *29*(3), 583–593.

Kroken, S., Glass, N., Taylor, J., Yoder, O., & Turgeon, B. (2003). Phylogenomic analysis of type I polyketide synthase genes in pathogenic and saprobic ascomycetes. *Proceedings of the National Academy of Sciences of the United States of America*, *100*, 15670–15675.

Kusari, S., Singh, S., & Jayabaskaran, C. (2014). Rethinking production of Taxol. *Trends in Biotechnology*, *32*, 304–311.

Lang, G. I., & Botstein, D. (2011). A test of the coordinated expression hypothesis for the origin and maintenance of the GAL cluster in yeast. *PLoS One*, *6*, e25290.

Langille, M. G. I., Hsiao, W. W. L., & Brinkman, F. S. L. (2010). Detecting genomic islands using bioinformatics approaches. *Nature Reviews. Microbiology*, *8*, 373–382.

Lattanzio, V., Lattanzio, V. M., & Cardinali, A. (2006). Role of phenolics in the resistance mechanisms of plants against fungal pathogens and insects. *Phytochemistry: Advances in Research*, *661*, 23–67.

Lawrence, J., & Roth, J. (1996). Selfish operons: Horizontal transfer may drive the evolution of gene clusters. *Genetics*, *143*, 1843–1860.

Li, Y. F., Tsai, K. J., Harvey, C. J., Li, J. J., Ary, B. E., Berlew, E. E., et al. (2016). Comprehensive curation and analysis of fungal biosynthetic gene clusters of published natural products. *Fungal Genetics and Biology*, *89*, 18–28.

Lind, A. L., Wisecaver, J. H., Smith, T. D., Feng, X., Calvo, A. M., & Rokas, A. (2015). Examining the evolution of the regulatory circuit controlling secondary metabolism and development in the fungal genus Aspergillus. *PLoS Genetics*, *11*, e1005096.

Lynch, M. (2006). Streamlining and simplification of microbial genome architecture. *Annual Review of Microbiology*, *60*, 327–349.

Machado, C. M., De-Souza, E. A., De-Queiroz, A. L. F. V., Pimentel, F. S. A., Silva, G. F. S., Gomes, F. M., et al. (2017). The galactose-induced decrease in phosphate levels leads to toxicity in yeast models of galactosemia. *Biochimica et Biophysica Acta (BBA)—Molecular Basis of Disease*, *1863*, 1403–1409.

Marcet-Houben, M., & Gabaldón, T. (2010). Acquisition of prokaryotic genes by fungal genomes. *Trends in Genetics*, *26*, 5–8.

Marcet-Houben, M., & Gabaldón, T. (2015). Beyond the whole-genome duplication: Phylogenetic evidence for an ancient interspecies hybridization in the Baker's yeast lineage. *PLoS Biology*, *13*, e1002220.
Marcet-Houben, M., & Gabaldón, T. (2016). Horizontal acquisition of toxic alkaloid synthesis in a clade of plant associated fungi. *Fungal Genetics and Biology*, *86*, 71–80.
Marsit, S., & Dequin, S. (2015). Diversity and adaptive evolution of Saccharomyces wine yeast: A review. *FEMS Yeast Research*, *15*, fov067.
Matasyoh, J. C., Dittrich, B., Schueffler, A., & Laatsch, H. (2010). Larvicidal activity of metabolites from the endophytic Podospora sp. against the malaria vector Anopheles gambiae. *Parasitology Research*, *108*, 561–566.
McGary, K. L., Slot, J. C., & Rokas, A. (2013). Physical linkage of metabolic genes in fungi is an adaptation against the accumulation of toxic intermediate compounds. *Proceedings of the National Academy of Sciences of the United States of America*, *110*, 11481–11486.
Medema, M. H., Cimermancic, P., Sali, A., Takano, E., & Fischbach, M. A. (2014). A systematic computational analysis of biosynthetic gene cluster evolution: Lessons for engineering biosynthesis. *PLoS Computational Biology*, *10*, e1004016.
Nam, H., Lewis, N. E., Lerman, J. A., Lee, D. H., Chang, R. L., Kim, D., et al. (2012). Network context and selection in the evolution to enzyme specificity. *Science*, *337*, 1101–1104.
Novo, M., Bigey, F., Beyne, E., Galeote, V., Gavory, F., Mallet, S., et al. (2009). Eukaryote-to-eukaryote gene transfer events revealed by the genome sequence of the wine yeast Saccharomyces cerevisiae EC1118. *Proceedings of the National Academy of Sciences of the United States of America*, *106*(38), 16333–16338.
Pacheco-Arjona, J. R., & Ramirez-Prado, J. H. (2014). Large-scale phylogenetic classification of fungal chitin synthases and identification of a putative cell-wall metabolism gene cluster in Aspergillus genomes. *PLoS One*, *9*, e104920.
Panaccione, D. G., Cipoletti, J. R., Sedlock, A. B., Blemings, K. P., Schardl, C. L., Machado, C., et al. (2006). Effects of ergot alkaloids on food preference and satiety in rabbits, as assessed with gene-knockout endophytes in perennial ryegrass (Lolium perenne). *Journal of Agricultural and Food Chemistry*, *54*, 4582–4587.
Patron, N. J., Waller, R. F., Cozijnsen, A. J., Straney, D. C., Gardiner, D. M., Nierman, W. C., et al. (2007). Origin and distribution of epipolythiodioxopiperazine (ETP) gene clusters in filamentous ascomycetes. *BMC Evolutionary Biology*, 7, 174.
Paun, L., & Kempken, F. (2015). Fungal transposable elements. In K. Maruthachalam & M. A. van den Berg (Eds.), *Genetic transformation systems in fungi: Vol. 2.* Switzerland: Springer International Publishing.
Pepper, J. W. (2003). The evolution of evolvability in genetic linkage patterns. *Biosystems*, *69*, 115–126.
Plissonneau, C., Stürchler, A., & Croll, D. (2016). The evolution of orphan regions in genomes of a fungal pathogen of wheat. *mBio*, 7, e01231–16.
Pophaly, S. D., & Tellier, A. L. (2015). Population level purifying selection and gene expression shape subgenome evolution in maize. *Molecular Biology and Evolution*, *32*(12), 3226–3235.
Price, M. N., Huang, K. H., Arkin, A. P., & Alm, E. J. (2005). Operon formation is driven by co-regulation and not by horizontal gene transfer. *Genome Research*, *15*, 809–819.
Proctor, R. H., McCormick, S. P., Alexander, N. J., & Desjardins, A. E. (2009). Evidence that a secondary metabolic biosynthetic gene cluster has grown by gene relocation during evolution of the filamentous fungus Fusarium. *Molecular Microbiology*, *74*, 1128–1142.
Proctor, R. H., Van Hove, F., Susca, A., Stea, G., Busman, M., van der Lee, T., et al. (2013). Birth, death and horizontal transfer of the fumonisin biosynthetic gene cluster during the evolutionary diversification of Fusarium. *Molecular Microbiology*, *90*, 290–306.

Punt, P., Strauss, J., Smit, R., Kinghorn, J., van den Hondel, C., & Scazzocchio, C. (1995). The intergenic region between the divergently transcribed niiA and niaD genes of Aspergillus nidulans contains multiple NirA binding sites which act bidirectionally. *Molecular and Cellular Biology*, *15*, 5688–5699.

Purcell, J., Brelsford, A., Wurm, Y., Perrin, N., & Chapuisat, M. (2014). Convergent genetic architecture underlies social organization in ants. *Current Biology*, *24*, 2728–2732.

Quin, M. B., Flynn, C. M., & Schmidt-Dannert, C. (2014). Traversing the fungal terpenome. *Natural Product Reports*, *31*, 1449–1473.

Raguso, R. A., Agrawal, A. A., Douglas, A. E., Jander, G., Kessler, A., Poveda, K., et al. (2015). The raison d'être of chemical ecology. *Ecology*, *96*(3), 617–630.

Reynolds, H., Slot, J. C., Divon, H. H., Lysøe, E., Proctor, R. H., & Brown, D. W. (2017). Differential retention of gene functions in a secondary metabolite cluster. *Molecular Biology and Evolution*, *34*, 2002–2015.

Richards, T. A., Leonard, G., Soanes, D. M., & Talbot, N. J. (2011). Gene transfer into the fungi. *Fungal Biology Reviews*, *25*(2), 98–110.

Richards, T. A., & Talbot, N. J. (2013). Horizontal gene transfer in osmotrophs: Playing with public goods. *Nature Genetics*, *11*, 720–727.

Rosewich, U. L., & Kistler, H. C. (2000). Role of horizontal gene transfer in the evolution of fungi. *Annual Review of Phytopathology*, *38*(1), 325–363.

Samils, N., Oliva, J., & Johannesson, H. (2014). Nuclear interactions in a heterokaryon: Insight from the model Neurospora tetrasperma. *Proceedings of the Royal Society B: Biological Sciences*, *281*, 20140084.

Scharf, D. H., Remme, N., Heinekamp, T., Hortschansky, P., Brakhage, A. A., & Hertweck, C. (2010). Transannular disulfide formation in gliotoxin biosynthesis and its role in self-resistance of the human pathogen Aspergillus fumigatus. *Journal of the American Chemical Society*, *132*, 10136–10141.

Schmaler-Ripcke, J., Sugareva, V., Gebhardt, P., Winkler, R., Kniemeyer, O., Heinekamp, T., et al. (2009). Production of pyomelanin, a second type of melanin, via the tyrosine degradation pathway in Aspergillus fumigatus. *Applied and Environmental Microbiology*, *75*, 493–503. Available from: http://www.ncbi.nlm.nih.gov/pmc/articles/PMC2620705/pdf/2077-08.pdf.

Schumacher, J., Gautier, A., Morgant, G., Studt, L., Ducrot, P.-H., Le Pêcheur, P., et al. (2013). A functional bikaverin biosynthesis gene cluster in rare strains of Botrytis cinerea is positively controlled by VELVET. *PLoS One*, *8*, e53729.

Scott, A., & Timson, D. J. (2007). Characterization of the Saccharomyces cerevisiae galactose mutarotase/UDP-galactose 4-epimerase protein, Gal10p. *FEMS Yeast Research*, *7*, 366–371.

Shwab, E. K., Bok, J. W., Tribus, M., Galehr, J., Graessle, S., & Keller, N. P. (2007). Histone deacetylase activity regulates chemical diversity in Aspergillus. *Eukaryotic Cell*, *6*, 1656–1664.

Siewers, V. (2006). Identification of an abscisic acid gene cluster in the grey mold Botrytis cinerea. *Applied and Environmental Microbiology*, *72*, 4619–4626.

Slot, J. C., Hallstrom, K. N., Matheny, P. B., Hosaka, K., Mueller, G., Robertson, D. L., et al. (2010). Phylogenetic, structural and functional diversification of nitrate transporters in three ecologically diverse clades of mushroom-forming fungi. *Fungal Ecology*, *3*, 160–177.

Slot, J. C., & Hibbett, D. S. (2007). Horizontal transfer of a nitrate assimilation gene cluster and ecological transitions in fungi: A phylogenetic study. *PLoS One*, *2*, e1097.

Slot, J. C., & Rokas, A. (2010). Multiple GAL pathway gene clusters evolved independently and by different mechanisms in fungi. *Proceedings of the National Academy of Sciences of the United States of America*, *107*, 10136–10141.

Slot, J. C., & Rokas, A. (2011). Horizontal transfer of a large and highly toxic secondary metabolic gene cluster between fungi. *Current Biology*, *21*, 134–139.
Soanes, D., & Richards, T. A. (2014). Horizontal gene transfer in eukaryotic plant pathogens. *Annual Review of Phytopathology*, *52*, 583–614.
Spiering, M. J. (2004). Gene clusters for insecticidal loline alkaloids in the grassendophytic fungus Neotyphodium uncinatum. *Genetics*, *169*, 1403–1414.
Strobel, G., & Daisy, B. (2003). Bioprospecting for microbial endophytes and their natural products. *Microbiology and Molecular Biology Reviews*, *67*, 491–502.
Sutherland, H., & Bickmore, W. A. (2009). Transcription factories: Gene expression in unions? *Nature Reviews. Genetics*, *10*, 457–466.
Szöllősi, G. J., Davín, A. A., Tannier, E., Daubin, V., & Boussau, B. (2015). Genome-scale phylogenetic analysis finds extensive gene transfer among fungi. *Philosophical Transactions of the Royal Society of London. Series B, Biological Sciences*, *370*, 20140335.
Teixeira, M. M., Moreno, L. F., Stielow, B. J., Muszewska, A., Hainaut, M., Gonzaga, L., et al. (2017). Exploring the genomic diversity of black yeasts and relatives (Chaetothyriales, Ascomycota). *Studies in Mycology*, *86*, 1–28.
Thompson, M. J., & Jiggins, C. D. (2014). Supergenes and their role in evolution. *Heredity*, *113*, 1–8.
Throckmorton, K., Wiemann, P., & Keller, N. P. (2015). Evolution of chemical diversity in a group of non-reduced polyketide gene clusters: Using phylogenetics to inform the search for novel fungal natural products. *Toxins*, 7(9), 3572–3607.
Tsochatzidou, M., Malliarou, M., Papanikolaou, N., Roca, J., & Nikolaou, C. (2017). Genome urbanization: Clusters of topologically co-regulated genes delineate functional compartments in the genome of Saccharomyces cerevisiae. *Nucleic Acids Research*, *45*, 5818–5828.
Tudzynski, B., & Hölter, K. (1998). Gibberellin biosynthetic pathway in Gibberella fujikuroi: Evidence for a gene cluster. *Fungal Genetics and Biology*, *25*, 157–170.
Van Bael, S., Estrada, C., & Arnold, A. E. (2017). Chapter 6: Foliar endophyte communities and leaf traits in tropical trees. In J. Dighton & J. F. White (Eds.), *The fungal community: Its organization and role in the ecosystem* (pp. 79–94). Boca Raton, FL: CRC Press.
Vlaardingerbroek, I., Beerens, B., Rose, L., Fokkens, L., Cornelissen, B. J. C., & Rep, M. (2016). Exchange of core chromosomes and horizontal transfer of lineage-specific chromosomes in Fusarium oxysporum. *Environmental Microbiology*, *18*, 3702–3713.
Walton, J. D. (2000). Horizontal gene transfer and the evolution of secondary metabolite gene clusters in fungi: An hypothesis. *Fungal Genetics and Biology*, *30*, 167–171.
Wawrzyn, G. T., Quin, M. B., Choudhary, S., López-Gallego, F., & Schmidt-Dannert, C. (2012). Draft genome of Omphalotus olearius provides a predictive framework for sesquiterpenoid natural product biosynthesis in Basidiomycota. *Chemistry & Biology*, *19*, 772–783.
Weigel, B. J., Burgett, S. G., Chen, V. J., Skatrud, P. L., Frolik, C. A., Queener, S. W., et al. (1988). Cloning and expression in Escherichia coli of isopenicillin N synthetase genes from Streptomyces lipmanii and Aspergillus nidulans. *Journal of Bacteriology*, *170*(9), 3817–3826.
Weissman, K. J. (2015). The structural biology of biosynthetic megaenzymes. *Nature Chemical Biology*, *11*, 660–670.
Wiemann, P., Guo, C. J., Palmer, J. M., Sekonyela, R., Wang, C. C., & Keller, N. P. (2013). Prototype of an intertwined secondary-metabolite supercluster. *Proceedings of the National Academy of Sciences of the United States of America*, *110*(42), 17065–17070.
Wiemann, P., Willmann, A., Straeten, M., Kleigrewe, K., Beyer, M., Humpf, H. U., et al. (2009). Biosynthesis of the red pigment bikaverin in Fusarium fujikuroi: Genes, their function and regulation. *Molecular Microbiology*, *72*(4), 931–946.

Wight, W. D., Labuda, R., & Walton, J. D. (2013). Conservation of the genes for HC-toxin biosynthesis in Alternaria jesenskae. *BMC Microbiology*, *13*, 165.

Wightman, R. (2001). *The origin and significance of the THI5 gene family in* Saccharomyces cerevisiae. University of Leicester. Doctoral thesis.

Wisecaver, J. H., Borowsky, A. T., Tzin, V., Jander, G., Kliebenstein, D. J., & Rokas, A. (2017). A global coexpression network approach for connecting genes to specialized metabolic pathways in plants. *The Plant Cell*, *29*, 944–959.

Wisecaver, J. H., Slot, J. C., & Rokas, A. (2014). The evolution of fungal metabolic pathways. *PLoS Genetics*, *10*, e1004816.

Wolfe, K. H., Armisén, D., Proux-Wéra, E., ÓhÉigeartaigh, S. N. S., Azam, H., Gordon, J. L., et al. (2015). Clade- and species-specific features of genome evolution in the Saccharomycetaceae. *FEMS Yeast Research*, *15*, fov035.

Wong, S., & Wolfe, K. (2005). Birth of a metabolic gene cluster in yeast by adaptive gene relocation. *Nature Genetics*, *37*, 777–782.

Yeaman, S. (2013). Genomic rearrangements and the evolution of clusters of locally adaptive loci. *Proceedings of the National Academy of Sciences of the United States of America*, *110*, E1743–E1751.

Yeo, C., Abu Bakar, F., Chan, W., Espinosa, M., & Harikrishna, J. (2016). Heterologous expression of toxins from bacterial toxin-antitoxin systems in eukaryotic cells: Strategies and applications. *Toxins (Basel)*, *8*, 49.

Yu, J., Chang, P., Bhatnagar, D., & Cleveland, T. (2000). Cloning of a sugar utilization gene cluster in Aspergillus parasiticus. *Biochimica et Biophysica Acta (BBA)—Gene Structure and Expression*, *1493*, 211–214.

Yue, Q., Chen, L., Li, Y., Bills, G. F., Zhang, X., Xiang, M., et al. (2015). Functional operons in secondary metabolic gene clusters in Glarea lozoyensis (Fungi, Ascomycota, Leotiomycetes). *mBio*, *6*, e00703.

Zhang, H., Rokas, A., & Slot, J. C. (2012). Two different secondary metabolism gene clusters occupied the same ancestral locus in fungal dermatophytes of the Arthrodermataceae. *PLoS One*, *7*, e41903.

CHAPTER FIVE

Deciphering Pathogenicity of *Fusarium oxysporum* From a Phylogenomics Perspective

Yong Zhang*, Li-Jun Ma*[,1]
*University of Massachusetts Amherst, Amherst, MA, United States
[1]Corresponding author: e-mail address: lijun@biochem.umass.edu

Contents

Abstract

Fusarium oxysporum is a large species complex of both plant and human pathogens that attack a diverse array of species in a host-specific manner. Comparative genomic studies have revealed that the host-specific pathogenicity of the *F. oxysporum* species complex (FOSC) was determined by distinct sets of supernumerary (SP) chromosomes. In contrast to common vertical transfer, where genetic materials are transmitted via cell division, SP chromosomes can be transmitted horizontally between phylogenetic lineages, explaining the polyphyletic nature of the host-specific pathogenicity of the FOSC. The existence of a diverse array of SP chromosomes determines the broad host range of this species complex, while the conserved core genome maintains essential housekeeping functions. Recognition of these SP chromosomes enables the functional and structural compartmentalization of *F. oxysporum* genomes. In this review, we examine the impact of this group of cross-kingdom pathogens on agricultural productivity and human health. Focusing on the pathogenicity of *F. oxysporum* in the phylogenomic

Advances in Genetics, Volume 100
ISSN 0065-2660
https://doi.org/10.1016/bs.adgen.2017.09.010

framework of the genus *Fusarium*, we elucidate the evolution of pathogenicity within the FOSC. We conclude that a population genomics approach within a clearly defined phylogenomic framework is essential not only for understanding the evolution of the pathogenicity mechanism but also for identifying informative candidates associated with pathogenicity that can be developed as targets in disease management programs.

ABBREVIATIONS

ADON DON-acetylated derivative
AVR avirulence gene
cAMP–PKA cyclic adenosine monophosphate–protein kinase A
CWDE cell wall-degrading enzyme
DON deoxynivalenol
FB1 fumonisin B1
FHB *Fusarium* head blight
Foc *F. oxysporum* f. sp. *cubense*
Fol *F. oxysporum* f. sp. *lycopersici*
Fom *F. oxysporum* f. sp. *meloni*
Fon *F. oxysporum* f. sp. *niveum*
FOSC *Fusarium oxysporum* species complex
f. sp. *forma specialis*
FUM fumonisin biosynthetic gene
GWAS genome-wide association study
HGT horizontal gene transfer
HUT homoserine utilization
NIV nivalenol
NRPS nonribosomal peptide synthetase
PKS polyketide synthase
QTN quantitative trait nucleotide
SMB secondary metabolite biosynthesis
SP supernumerary
STR4 subtropical race4
TR4 tropical race4
TRI trichothecene biosynthetic gene
TS terpene synthase
Sge SIX gene expression
SIX secreted in xylem
TE transposable element
TOM tomatinase

1. CROSS-KINGDOM VIRULENCE OF THE SPECIES COMPLEX *FUSARIUM OXYSPORUM*

F. oxysporum is an economically important filamentous fungus in the phylum Ascomycota, class Sordariomycetes, order Hypocreales, family

Nectriaceae, and genus *Fusarium*. *F. oxysporum* constitutes a large species complex (hereafter FOSC) that is widely distributed in diverse environments, including soil, indoors, and aquatic habitats (Bell & Khabbaz, 2013; Brandt & Park, 2013; Palmero et al., 2009). Members within the FOSC not only cause destructive and intractable vascular diseases in diverse plant species but also cause infectious diseases in patients with suppressed immunity and eye infections even in individuals with competent immune systems, posing serious threats to food safety and public health.

As a species complex, *F. oxysporum* has a broad host range, causing *Fusarium* vascular wilt diseases in over 100 plant species (Ma, Geiser, et al., 2013), ranging from gymnosperms to angiosperms and monocots to dicots (Michielse & Rep, 2009). Listed as one of the top 10 most economically damaging fungal pathogens, *F. oxysporum* challenges the production of numerous economically important crops, including banana (*Musa acuminata*), cotton (*Gossypium hirsutum*), canola (*Brassica napus*), and tomato (*Solanum lycopersicum*) (Ma, Geiser, et al., 2013). One notorious example of the devastating effects of *F. oxysporum* infection is Panama disease of banana (O'Donnell, Kistler, Cigelnik, & Ploetz, 1998; Ploetz, Kema, & Ma, 2015). The impact of this disease was demonstrated in its first epidemics in the 1950s, which almost destroyed banana production based on the "Gros Michel" cultivar. That crisis was resolved by replacing "Gros Michel" with the race 1-resistant cultivar "Cavendish" (Ploetz, 2015). Unfortunately, the newly emerged tropical race4 (TR4) is highly virulent to Cavendish banana, and a recent outbreak of Panama disease spreads from South-East Asia to Mozambique and Jordan, greatly dampening the agricultural production of banana (García-Bastidas et al., 2013). Since locally consumed bananas and plantains are significant staple foods and represent a primary dietary source of carbohydrates in Africa, South-East Asia, and tropical America, the spread of the disease could potentially create localized food shortages, intensifying world hunger, and exacerbating poverty in developing nations (Ploetz et al., 2015).

The pathogenicity of *F. oxysporum* is highly host specific; a single pathogenic form usually infects a single plant species. Based on its host specificity, an artificial naming system—*forma specialis* (f. sp.)—was used to group isolates that are capable of causing disease in a specific plant host (Baayen et al., 2000; O'Donnell et al., 1998). For instance, isolates that are pathogenic to banana plants are classified as *F. oxysporum* f. sp. *cubense* (*Foc*), and all tomato pathogens are classified as *F. oxysporum* f. sp. *lycopersici* (*Fol*). Interestingly, *F. oxysporum forma specialis* consists of multiple, independent lineages that evolved polyphyletically through convergent evolution (Baayen et al., 2000; O'Donnell et al., 1998).

In addition to its agricultural impact, *F. oxysporum* is a causal agent of fusariosis (Anaissie et al., 2001; O'Donnell et al., 2004), the top emerging opportunistic mycosis and the second most common opportunistic mold infection after aspergillosis (Guarro, 2013). *Fusarium* infection is also the major cause of fungal keratitis, which causes over a million new cases of blindness annually (Kredics et al., 2015). Pathogenic *F. oxysporum* isolates have an exceptional ability to penetrate host tissues. Positive fungal cultures were recovered from the blood samples of more than 50% of infected patients, even though most infections began with either skin lesions or inhalation of airborne organisms (Nucci et al., 2013). Highly invasive infections caused by *Fusarium* spp. usually result in hepatitis, meningitis, and other disseminated diseases (Boutati & Anaissie, 1997).

Advanced medical treatments have increased the complexity of patient populations with immunodeficiency disorders. For instance, chemotherapy, which is toxic to innate immune system cells, increases the survival rate and life expectancy of cancer patients (Mukherjee et al., 2010), and successful management of immunosuppression prolongs the life expectancy of organ transplant recipients (Holgersson, Rydberg, & Breimer, 2014). Consequently, opportunistic fungi have emerged as important causes of morbidity and mortality in immunocompromised patients (Brandt & Park, 2013; Low & Rotstein, 2011). Each year, fungi infect over 1 billion people and fungal infections claim around 1.5 million lives worldwide (Brown, Denning, et al., 2012; Fisher et al., 2012). Given the expansion of the vulnerable patient population compounded with limited treatment options, opportunistic fungal infections, including fusariosis, pose an increasing threat to public health. Currently, three major commercial drug classes of antifungal agents (i.e., azoles, echinocandins, and polyenes) are available (http://www.f2g.com/antifungal-market) and most of these drugs perturb cell membrane integrity. Since 2006, no new class of antifungal agent has been approved (Denning & Bromley, 2015). This stagnation in antifungal drug discovery is partly due to the challenge of identifying drug targets unique to eukaryotic pathogens and the lack of innovation in designing screening platforms to identify such agents.

The scarcity of antifungal agents poses even a greater problem for combatting these cross-kingdom pathogens, as fungicides constantly used in the field to control plant diseases have the same targets as those prescribed in clinics. For instance, the azoles (propiconazole, bromuconazole, epoxiconazole, difenoconazole, and tebuconazole) are widely used to protect plants in the field because of their broad-spectrum antifungal activity and long stability. However, resistance has developed rapidly among *F. oxysporum* isolates (Price, Parker, Warrilow, Kelly, & Kelly, 2015). As antifungals used in the clinic to treat patients share the same mode of action as fungicides used in the field to control plant diseases, *F. oxysporum* clinical isolates showed broad resistance to most antifungals available on the market (Al-Hatmi, Hagen, Menken, Meis, & de Hoog, 2016; Short, O'Donnell, Zhang, Juba, & Geiser, 2011), even without previous exposure to the antifungal. Consequently, the mortality from fungal infections exceeds 50% (Hahn-Ast et al., 2010). The US Centers for Disease Control and Prevention recently deemed antifungal resistance "one of our most serious health threats" (http://www.cdc.gov/fungal/antifungal-resistance.html). There is an urgent need to decipher the mechanisms underlying fungal pathogenesis and to develop effective and diverse antifungal therapies to control this group of eukaryotic pathogens.

In this review, we present an overview of *F. oxysporum* pathogenicity in the phylogenomic framework of the genus *Fusarium*. Sequences that determine host-specific pathogenicity are often encoded in the accessory component and share high sequence similarity between strains within the same *forma specialis*. The unique compartmentalization of the *Fusarium* genome provides an excellent perspective from which to dissect the genetic mechanisms underlying pathogenesis. We describe how the phylogenomic framework based on a clear understanding of genomic structure elucidates the evolution of pathogenicity within the FOSC. Building on this framework, the combination of comparative, population and functional genomics will not only shed light on the mechanisms underlying the evolution of the pathogenicity but also enable the identification of informative pathogenicity-associated candidates that can be used as targets to develop pathotype-specific diagnostic tools and novel therapeutic and disease management programs.

2. PHYLOGENOMICS FRAMEWORK OF THE GENUS *FUSARIUM*

2.1 Phylogenetic Framework of the Genus *Fusarium*

Fusarium is a large genus that includes the most important phytopathogenic and toxigenic fungi (Booth, 1971). Research pioneered by Dr. O'Donnell, using a number of single-copy and phylogenetically informative loci, established a well-defined phylogenetic framework (Balmas et al., 2010; Chang et al., 2006; Kistler et al., 1998; Ma, Geiser, et al., 2013; O'Donnell et al., 2009). This phylogenetic framework proposed the existence of a monophyletic group that encompasses all economically important *Fusarium* species (O'Donnell et al., 2013), including the FOSC and other well-supported monophyletic species complexes (Fig. 1) (Ma, Geiser, et al., 2013), providing a valuable blueprint for dissecting the evolution and diversity of the genus. The diversification of the genus *Fusarium* was calculated to have arisen about 91.3 Mya (Fig. 1), coinciding with the emergence of flowering plants (Soltis, Bell, Kim, & Soltis, 2008). Based on this calculation, the FOSC is thought to have diverged from its close relative during the Miocene Epoch, about 11 Mya (Ma, Geiser, et al., 2013).

In addition to vascular wilts, *Fusarium* also causes blights, rots, and cankers of many horticultural, field, ornamental, and forest plants in both agricultural and natural ecosystems. For instance, *Fusarium* head blight (FHB), an economically important disease affecting cereals, is caused by *Fusarium graminearum* (Gale, Chen, Hernick, Takamura, & Kistler, 2002; Goswami & Kistler, 2005; Guo & Ma, 2014). First reported in England in 1884, there have been countless outbreaks of FHB in Europe, Asia, Canada, and the United States. In the United States alone, the direct and indirect economic losses resulting from FHB epidemics of 1991–1997 (Johnson, Flaskerud, Taylor, & Satyanarayana, 1998) and 1998–2000 (Nganje, Bangsund, Leistritz, Wilson, & Tiapo, 2004) were estimated to be around 1.3 and 2.7 billion USD, respectively. In China, FHB has devastated wheat production in over 7 million hectares of farmland and led to yield losses of 1 million tons during severe outbreaks (Bai & Shaner, 2004). Ear rot and stalk rot of maize caused by *F. graminearum* and *F. verticillioides* seriously decrease the yield and affect the quality of corn production all around the world (White, 1999).

In addition, numerous *Fusarium* spp. also infect humans and animals, causing a wide spectrum of diseases known as fusariosis (O'Donnell et al., 2004; Short et al., 2011; Zhang et al., 2006). Unlike infections in the healthy host that are mostly limited to skin and other superficial lesions, fusariosis in the immunocompromised patient is typically invasive and could be fatal. Due to the increase in susceptible and immunocompromised patients, reported cases of disseminated fusariosis are increasing worldwide (Cilo et al., 2015; Low & Rotstein, 2011). Ten *Fusarium* species complexes were reported to cause opportunistic infections in humans, including the *F. solani* species complex, *F. oxysporum* species complex, *F. fujikuroi* species complex,

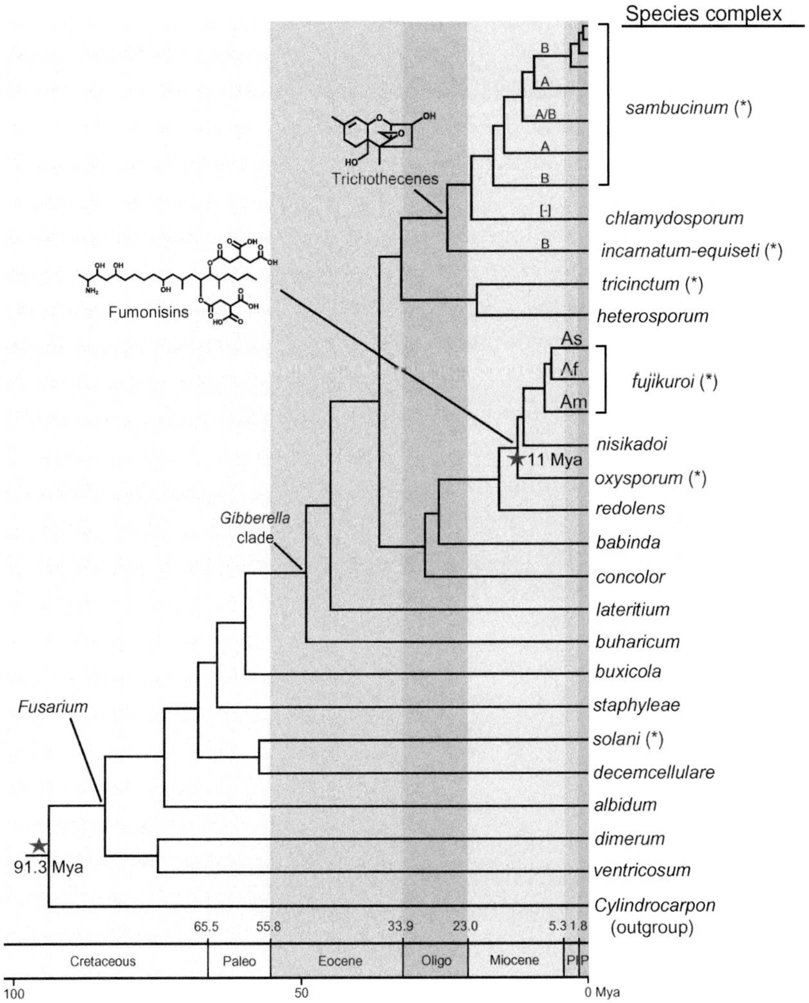

Fig. 1 Time series estimating the origins and divergence time of the 20 *Fusarium* species complexes. Ten species complexes highlighted with *blue* color indicate that the capability of human infection has been reported in at least one species within the species complex. The *stars* by the species complex are used to show that at least one whole-genome sequence has been revealed within the species complex. The numbers within the clade are showing the estimation time of the species that diversifies from its relative close species. The two important *Fusarium* mycotoxins are inferred on the phylogeny with the estimating evolutionary origins. A, B, and [−] within the clade of trichothecene toxin-producing fusaria indicate the production of type A trichothecenes, type B trichothecenes, and no trichothecenes, respectively. Three biogeographically structured clades within the *fujikuroi* complex are identified as follows: *Af*, African; *Am*, American; *As*, Asian. Abbreviations: *Oligo*, Oligocene; *P*, Pleistocene; *Paleo*, Paleocene; *Pl*, Pliocene. *Modified with permission from Ma, L. J., Geiser, D. M., Proctor, R. H., Rooney, A. P., O'Donnell, K., Trail, F., et al. (2013). Fusarium pathogenomics. Annual Review of Microbiology, 67, 399–416.*

F. incarnatum-equiseti species complex, *F. dimerum* species complex, *F. chlamydosporum* species complex, *F. sambucinum* species complex, *F. concolor* species complex, *F. lateritium* species complex, and *F. tricinctum* species complex. Of these, the *F. solani* and *F. oxysporum* species complexes are the most frequently isolated from infected humans (Al-Hatmi, Meis, & de Hoog, 2016).

The genetic and ecological diversity of the *Fusarium* genus is also reflected in their reproductive strategies. Species within the *Fusarium* genus are capable of producing three types of asexual spores: microconidia, macroconidia, and chlamydospores. Both micro- and macroconidia are produced in the aerial mycelium; however, small, single-celled, fusiform or ovoid microconidia are produced from a single spore-forming structure (conidiophore) and large, multiseptated, boat-shaped macroconidia are produced in the sporodochium, a superficial, cushion-shaped asexual fruiting body consisting of a cluster of conidiophores. Under unfavorable conditions, some fusaria species also produce thick-walled dormant chlamydospores. However, not all *Fusarium* species can produce all known asexual spore types (Ma, Geiser, et al., 2013).

The sexual cycle has been characterized in ~20% of *Fusarium* spp. Among all carefully inspected species, all are heterothallic, except for *F. graminearum*, which is homothallic. Determined by the mating-type locus (idiomorphs, *MAT1-1* and *MAT1-2*), heterothallic individuals possess either *MAT1-1* or *MAT1-2* genes at the *MAT1* locus, and mating isolates must be of different idiomorphs. The flanking sequences of idiomorphs are conserved in both the *MAT1-1* and *MAT1-2* mating types. *F. graminearum* possesses both *MAT1-1* and *MAT1-2* idiomorphs at the mating-type locus (Leslie & Summerell, 2006). The homothallic lifestyle likely evolved in the *F. graminearum* lineage from a self-sterile ancestor, potentially providing the advantage of establishing infection without the dependency of searching for a mating partner.

2.2 Phylogenomic Framework of the Genus *Fusarium*

While *Fusarium* phylogenetics based on the analysis of house-keeping genes provides a valuable framework for studying the diversity of different species complexes, there are unresolved issues, most distinctively the polyphyletic origin of host specificity. Some *Fusarium* species have a narrow host range, while others can infect a diverse array of hosts. For example, *F. graminearum* infects mainly cereal species and *F. verticillioides* primarily infects maize and sorghum, whereas members of the FOSC infect over 100 plant species, ranging from gymnosperms to angiosperms and monocots to dicots. For species complexes, like the FOSC, many *formae speciales* are polyphyletic and host-specific pathogenicity for each pathotype is the result of convergent evolution. Therefore, phylogenomics that considers both divergent and convergent evolutionary processes within the genomic content will be critical to convey information that shapes the genetic diversity of this group of pathogens.

The FOSC has a highly dynamic genome organization. Notable features of the FOSC and several species complexes within the *Fusarium* genus are chromosomal polymorphism (Boehm, Ploetz, & Kistler, 1994) and the presence of abundant mobile transposable elements (Daboussi & Capy, 2003). Large-scale chromosomal polymorphism within the *F. oxysporum* and *F. solani* (teleomorph *Nectria haematococca*) species complexes was documented in the early 1990s (Boehm et al., 1994; Miao, Matthews, & VanEtten, 1991; Migheli, Berio, Gullino, & Garibaldi, 1995; Rosewich, Pettway, Katan, & Kistler, 1999; Taga, Murata, & VanEtten, 1999; Temporini & VanEtten, 2002). A subsequent study in *F. solani* confirmed that only a few chromosomes contributed to the chromosomal polymorphism (Han, Liu, Benny, Kistler, & VanEtten, 2001; Miao, Covert, & VanEtten, 1991). Removing these chromosomes from the genome did not affect fungal fitness; therefore, the terms dispensable chromosomes or supernumerary (SP) chromosomes were used to describe these chromosomes (Han et al., 2001; Hatta et al., 2002; Miao, Covert, et al., 1991; Rodriguez-Carres, White, Tsuchiya, Taga, & Vanetten, 2008; VanEtten, Jorgensen, Enkerli, & Covert, 1998). Even though SP chromosomes are dispensable for normal growth, they are indispensable in determining host-specific virulence (Han et al., 2001; Miao, Covert, et al., 1991) or rhizosphere niche competitiveness (Rodriguez-Carres et al., 2008).

Comparative genomic studies confirmed that SP chromosomes exist in the *F. oxysporum* (Ma et al., 2010) and *F. solani* (Coleman et al., 2009) genomes. Comparative genomics revealed that the sequenced *Fusarium* genomes vary considerably in genome size and chromosome number. For instance, *Fol* strain 4287 has the largest genome (61 Mb), mainly due to the presence of four SP chromosomes, as well as lineage-specific extensions to the two largest chromosomes (Ma et al., 2010). The supernumerary nature of three chromosomes, 14, 15, and 17, in the genome of *F. solani* strain 77-13-4 was confirmed (Coleman et al., 2009).

In agreement with their dispensable nature, SP chromosomes in both *F. solani* and *F. oxysporum* genomes are deficient in genes associated with house-keeping functions (Coleman et al., 2009; Ma et al., 2010), but enriched for genes that determine host-specific virulence and influence niche adaptation, resulting in both structural and functional compartmentalization of a genome. Despite the variation in genome size and chromosome number, the subdivision of a genome into two components provides insight into the different evolutionary origins of the genome and the different patterns of evolution of genes that reside in the genome. The core component shared by all *Fusarium* species encodes functions for basic growth and survival, and is characterized by high levels of synteny with related species, few repetitive sequences, and a high density of house-keeping genes involved in primary metabolism. In contrast to the core components, the accessory components are typically arranged on smaller chromosomes and harbor numerous transposable elements and a high density of pathogenic factor-encoding genes that are under diversifying selection, with biased codon usage.

Within the same *forma specialis*, host-specific virulence factors are largely shared among strains. This pattern of distribution has been repeatedly observed, such as among *F. oxysporum* f. sp. *lycopersici* (*Fol*) tomato pathogens, *F. oxysporum* f. sp. *niveum* (*Fon*) watermelon pathogens (van Dam et al., 2016), and *F. oxysporum* f. sp. *cubense* (*Foc*) banana pathogens (our unpublished results). The nature of the polyphyletic host-specific pathogenicity underscores the importance of establishing the phylogenomics framework of the genus *Fusarium*. For instance, strains from different *formae speciales* of *F. oxysporum* could be phylogenetically closer than strains belonging to the same *forma specialis*, as illustrated in Fig. 2. Therefore, it is not reliable to determine the host specificity of a *F. oxysporum* strain based on conserved gene sequences.

3. DETERMINANTS OF PATHOGENICITY

There are several known determinants of pathogenicity in the genus *Fusarium*, including mycotoxins, effectors, and other regulatory elements (Ma, Kistler, & Rep, 2012; van der Does & Rep, 2007). The phylogenomic framework of the genus *Fusarium* enables the identification of SP chromosomes, providing a de facto division of the genome into two components: a core component that encodes functions necessary for growth and survival and an accessory component that encodes pathogenicity (or virulence)

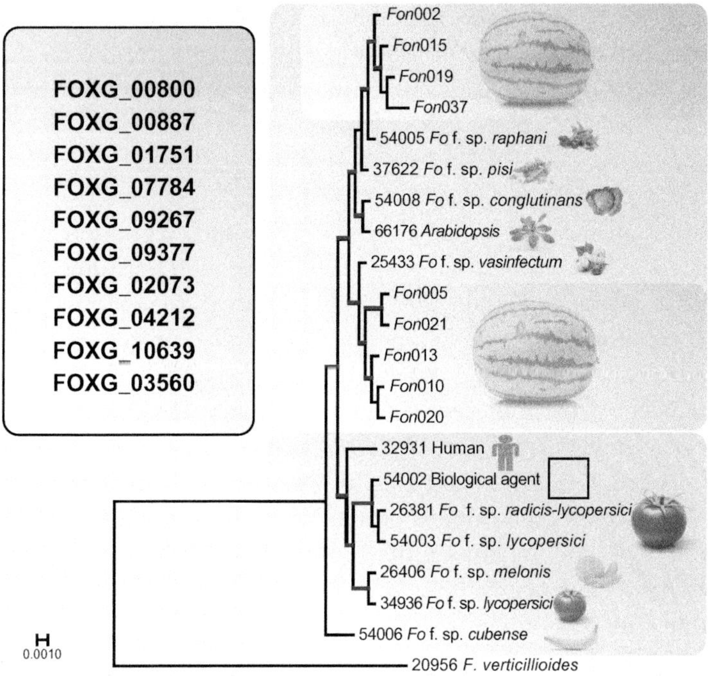

Fig. 2 Phylogenetic framework of FOSC is showing the polyphyletic origins of *F. oxysporum* strains within the same *forma specialis*. The phylogeny is inferred from the maximum likelihood analysis of 10 conserved single-copy orthologous genes within the *Fusarium* genus, rooted on the sequence of *F. verticillioides*. Branches with >95 bootstrap value (based on 100 bootstrap iterations) are indicated in *thickened red bars*. Two *light green boxes* highlight the two distinct clonal lineages of *F. oxysporum* f. sp. *niveum* (*Fon*). The five-digit strain number represents the accession number in the Agricultural Research Service (ARS) Culture Collection (NRRL). The 10 gene IDs in the *left corner* represent the 10 conserved orthologous genes in NRRL34936.

factors. This feature essentially institutes the concept of genomic compartmentalization at both the structural and functional levels, narrowing the search space for characterizing host-specific pathogenicity (Ma et al., 2010; Michielse & Rep, 2009).

3.1 Mycotoxins as the Virulence Factor in Fusarial Pathogenicity

Many species within the *Fusarium* genus produce mycotoxins derived from diverse biochemical pathways, including the polyketide, terpenoid, and amino acid metabolic pathways. Many known mycotoxins are virulence factors. The mycotoxins trichothecenes and fumonisins, produced by *F. graminearum* and *F. verticillioides*, respectively, are notorious for their devastating effects on crop yields and for their toxicity that threatens human and animal health.

3.1.1 Trichothecenes

Trichothecenes are sesquiterpenoids and the most common mycotoxins produced by *F. graminearum* and related species (Desjardins & Proctor, 2007; Goswami & Kistler, 2005). Due to their toxicity and economic impact, trichothecenes are among the best-studied mycotoxins. Over 150 trichothecene derivatives, grouped into two categories, have been identified to date. Type A trichothecenes include T-2 toxin, HT-2 toxin, NX-2, and neosolaniol, whereas Type B trichothecenes include nivalenol (NIV), deoxynivalenol (DON), and the DON-acetylated derivatives ADONs. Type B trichothecenes inhibit protein translation (Cundliffe, Cannon, & Davies, 1974) and are among the most toxic mycotoxin compounds and best-studied virulence factors. Humans and animals that have consumed trichothecene mycotoxins exhibit various symptoms, including vomiting, dizziness, diarrhea, and spontaneous abortion (Rocha, Ansari, & Doohan, 2005). In addition to their toxic effects, trichothecenes serve as virulent factors in plant–fungal interactions (Goswami & Kistler, 2005) and elicit plant defense responses. Even though trichothecene production is not a prerequisite for *Fusarium* to develop on and penetrate the host, trichothecenes are necessary compounds for the spread of the pathogen after initial colonization (Boenisch & Schäfer, 2011; Ilgen, Hadeler, Maier, & Schafer, 2009). Mutants defective in trichothecene production (e.g., the DON and NIV strains) have reduced virulence on wheat.

3.1.2 Fumonisins

The fumonisins are a group of mycotoxins produced by *F. verticillioides* and related *Fusarium* species (Desjardins & Proctor, 2007). Over 15 different fumonisins, divided into four categories (A, B, C, and P), have been identified (Ahangarkani, Rouhi, & Azizi, 2014). The most notable form is fumonisin B1 (FB1) due to its toxicity on humans and animals. FB1 is also the most abundant and important contaminant of maize and maize-related products. A structural analog to sphingolipids, FB1, disturbs sphingolipid metabolism by inhibiting the enzyme ceramide synthase, leading to degeneration of the sphingolipid-rich tissues and changes in cell membrane composition (Domijan, 2012). FB1 production in association with disease manifestation varies markedly depending on the infected tissue type and host conditions (Stockmann-Juvala & Savolainen, 2008). Although fumonisin produced by *F. verticillioides* has a limited effect on maize ear rot, it does have a significant impact on maize seedling blight. The successful transformation of the fumonisin-producing genes into an endophytic, nonfumonisin producing *F. verticillioides* strain has converted this endophyte into a pathogen that causes seedling blight disease in maize (Glenn et al., 2008), strongly supporting the notion that fumonisin is a pathogenicity factor in maize seedling infection.

3.1.3 Secondary Metabolite Biosynthetic Gene Clusters

As most mycotoxins are nonessential for their producer, they are also referred to as secondary metabolites. In fungi, genes responsible for secondary metabolite biosynthesis (SMB) are often clustered and can be identified through the presence of three classes of enzymes that are typically responsible for a fundamental step in the biosynthesis of the metabolite, such as terpene synthase (TS), polyketide synthase (PKS), and nonribosomal peptide synthetase (NRPS) (Cole & Schweikert, 2003; Keller, Turner, & Bennett, 2005). These enzymes catalyze the condensation or rearrangement of simple molecules to form more complex structures. Clusters also typically include the core genes responsible for structural modifications of the initial metabolite, transporters for metabolite transport, and transcription factors for coordinated transcriptional regulation of genes in the cluster. These initial chemical products typically undergo multiple enzymatic modifications to form biologically active secondary metabolites and are transported to their site of activity. Characterization of SMB clusters provides insight into the conservation of secondary metabolic pathways among species, into enzyme and cluster evolution, and into the functions of secondary metabolites in the ecology of the producing fungal species.

The gene cluster responsible for trichothecene biosynthesis is the trichothecene biosynthetic gene (*TRI*) cluster. In *F. graminearum*, a core *TRI* gene cluster includes 12 genes (*TRI3* to *TRI14*) on chromosome 2 (Alexander, Proctor, & McCormick, 2009; Ward, Bielawski, Kistler, Sullivan, & O'Donnell, 2002). Among them, *TRI6* and *TRI10* are major transcriptional regulators of *TRI* expression (Seong et al., 2009). Genetic polymorphisms within the trichothecene biosynthesis core gene cluster could contribute to some structural variations of the Type B trichothecenes. For instance, polymorphisms of *TRI8* resulted in the chemotype of ADON (Alexander, McCormick, Waalwijk, van der Lee, & Proctor, 2011; Goswami, Xu, Trail, Hilburn, & Kistler, 2006).

Similarly, fumonisin production is regulated by the fumonisin biosynthetic gene (*FUM*) cluster, including 16 genes encoding biosynthetic enzymes, a transcription factor, and an ABC transporter (Alexander et al., 2009; Proctor et al., 2013). *FUM1* encodes a polyketide synthase responsible for the synthesis of a linear polyketide that forms the backbone structure of fumonisins. *FUM8* catalyzes the condensation of the linear polyketide with alanine to produce fumonisins. *FUM21* encodes a Zn(II)2Cys6 DNA-binding transcription factor that positively regulates *FUM* expression (Brown, Butchko, Busman, & Proctor, 2007). The cluster also encodes an ABC transporter (*FUM19*) that imparts self-protection by exporting the toxin and reducing its cellular concentration.

Based on the genomic characterization of SMB clusters, *Fusarium* spp. have the potential to produce diverse secondary metabolites (Ma et al., 2012, 2010). Compared to the number of the identified SMB gene clusters, characterized secondary metabolites only represent a small proportion of compounds that a species may potentially produce. Few SMB gene clusters have been functionally characterized in *Fusarium*. For instance, the genome of *F. graminearum*—the best-studied mycotoxin-producing fungus—encodes a total of 43 SMB gene clusters, including 16 PKSs, 19 NRPSs, and 8 TSs. However, only eight secondary metabolites, including aurofusarin, butenolide, chlamydosporol, culmorin, cyclonerodiol, fusarins, trichothecenes, and zearalenones, have been characterized (Cole & Schweikert, 2003). Interestingly, microarray analysis confirmed coregulation of 14 out of 16 *F. graminearum* PKS-encoding SMB gene clusters, even though the function of these clusters is unknown. These novel SMB clusters can be used to study undiscovered secondary metabolites and their impacts on organism pathogenicity. The coregulated expression of these clusters, especially the novel clusters, suggests their potential functionality and will illuminate the importance of some novel secondary metabolites.

Despite its increased genome size and large protein-encoding gene set compared to *F. graminearum* and *F. verticillioides*, the *Fol* genome has the fewest SMB gene clusters, accounting for a total of 24 SMB gene clusters, including 11 PKSs, 12 NRPSs, and 1 TS based on the genome annotation (Ma et al., 2012, 2010). Compared to *F. graminearum* and *F. verticillioides*, we know little about the mycotoxins produced by FOSC strains and their functional importance, suggesting that mycotoxins may be less crucial for the pathogenicity of FOSC strains.

3.2 Effector Biology in the Genus *Fusarium*

Effectors are small, secreted, cysteine-rich proteins known to play important roles in fungal virulence, especially during fungal–plant interactions. By contrast to mycotoxins, effectors were first recognized to confer gene-for-gene resistance in *F. oxysporum* (Michielse, van Wijk, Reijnen, Cornelissen, & Rep, 2009; Rep et al., 2003), suggesting their fundamental importance.

3.2.1 SIX Proteins

Initially isolated and characterized from the xylem sap of infected tomato plants, *F. oxysporum* effector proteins are named *SIX* (secreted in xylem) proteins. The genome of the tomato-infecting pathogen *Fol* encodes 14 *SIX* proteins, named according to the order in which they were characterized (Rep, 2005; Rep et al., 2003). *SIX1*, *SIX3*, *SIX5*, and *SIX6* are required for the full virulence of *Fol*. *SIX1* and *SIX3* proteins were detected during root colonization and hyphal growth in the xylem, respectively (van der Does et al., 2008). Deletion of *SIX1*, *SIX2*, *SIX3*, and *SIX6* reduces, but does not eliminate, virulence toward a generally susceptible tomato line (Houterman et al., 2009; Rep et al., 2004).

Pathogens secrete effector proteins to promote host colonization. At the same time, plants recognize these effectors and induce a host defense response. Some *SIX* genes are recognized as pathogen-specific avirulence genes (*AVR*) that interact with specific host resistance genes, resulting in resistance of the host to the pathogen (Houterman et al., 2009). For instance, *SIX4*, also known as *AVR1*, is required for R gene *I-1*-mediated resistance (Houterman, Cornelissen, & Rep, 2008). *SIX1*, also known as *AVR3*, is recognized by the tomato receptor-like kinase protein I-3 and induces *I-3*-mediated resistance of tomato against *Fusarium* wilt disease (Rep, Meijer, Houterman, van der Does, & Cornelissen, 2005). *SIX3*, referred to as *AVR2*, is recognized by the tomato I-2 resistance protein in the nucleus, resulting in the induction of host defenses (Ma, Cornelissen, & Takken, 2013).

Playing crucial roles at the frontier of the plant–fungal interactions, SIX proteins were also used to document the evolutionary arms race between *Fol* and its tomato host (Houterman et al., 2008; Lievens et al., 2009; van Dam et al., 2016). For instance, *Fol AVR1* is only present in race 1 strains, but is absent in both race 2 and 3 *Fol* strains, suggesting that the deletion or functional disruption of *Fol AVR1* enabled the pathogen to escape host recognition (Houterman et al., 2009). In support of this notion, the *AVR1* deletion mutant in a race 1 background gained virulence toward the I-1 tomato line (Houterman et al., 2008). Interestingly, a single point mutation in *AVR2* allowed the pathogen to evade host recognition (van Dam et al., 2016).

Some *SIX* genes identified in *Fol*, including *SIX1*, *SIX4*, *SIX6*, *SIX7*, and *SIX8*, have also been detected in other *F. oxysporum formae speciales*

(Chakrabarti et al., 2011; Lievens et al., 2009). The genome of *Fusarium oxysporum* f. sp. *medicaginis* contains homologs of *SIX1*, *SIX8*, *SIX9*, and *SIX13*, all of which were significantly upregulated in planta (Thatcher, Williams, Garg, Buck, & Singh, 2016). Homologs of *SIX1*, *SIX4*, *SIX8*, and *SIX9* were identified in the genome of the Arabidopsis-infecting isolate Fo5176. The *SIX4* homolog (termed Fo5176-SIX4) was induced during infection of Arabidopsis. Transgenic Arabidopsis plants constitutively expressing Fo5176-SIX4 had increased disease symptoms with Fo5176. Conversely, Fo5176-SIX4 gene knockout mutants (Δ*six4*) had significantly reduced virulence on Arabidopsis (Thatcher, Gardiner, Kazan, & Manners, 2012). *SIX6* is present in *Fo* species infecting melon, watermelon, passion fruit, cucumber, and cotton. Homologs have also been found in two *Colletotrichum* species (Chakrabarti et al., 2011; Lievens et al., 2009). Among the banana pathogen *Foc*, three *SIX1* homologs and multiple *SIX8* were identified in *Foc* TR4, the most recent pathogenic form that broke down resistance of Cavendish banana. All three *SIX1* genes are highly expressed during fungal infection (our unpublished data).

3.2.2 Genome Compartmentalization and Effector Biology

In a compartmentalized *Fol* genome, all crucial *SIX* genes involved in the evolutionary arms race between the tomato pathogen and its tomato plant are encoded by a single SP chromosome (chromosome 14). The exclusive presence of *Fol*-effector genes underscores the involvement of SP chromosomes in host-specific virulence. Furthermore, the transfer of this particular chromosome between strains of *F. oxysporum* converts a nonpathogenic strain into a pathogen (Ma et al., 2010).

Transferring SP chromosomes that carry a host-specific virulence factor into phylogenetically distinct strains offers a perfect illustration of the polyphyletic nature of *F. oxysporum formae speciales* within the species complex. Under such circumstances, phylogeny based on house-keeping genes does not provide information on host specificity among the species complex. By contrast, sequences that determine host-specific pathogenicity are often encoded in the SP chromosomes and share high levels of sequence similarity between strains within the same *forma specialis*. For instance, *Fol* effectors are largely unique to this *forma specialis*, although homologs of *SIX1*, *SIX4*, *SIX6*, *SIX7*, and *SIX8* have been found in a few other *formae speciales* (Chakrabarti et al., 2011; Lievens et al., 2009).

Besides *Fol*, the core and accessory genome compartments have been defined for other *F. oxysporum formae speciales*, such as *cucumerinum*, *niveum*, *melonis*, and *radicis-cucumerinum*. As observed in the *Fol* strains, unique sequences associated with virulence and host-specific pathogenicity were identified. Based on the phylogeny and the presence/absence patterns of candidate effectors, members of the same *forma specialis* were grouped together, incongruent with the conventional phylogenetic results (van Dam et al., 2016), further emphasizing the importance of establishing the phylogenomics framework that traces down the evolutionary origin of each genomic compartment and identifies candidate virulence factors.

3.2.3 Genome Compartmentalization and Human Infection

Distinct infection strategies are employed by plant vs animal pathogens. As hemibiotrophic phytopathogens, *F. oxysporum* strains employ effectors encoded in the SP chromosomes to manipulate plant host immunity to promote infection. For pathogens infecting humans, the fungus could enter immunocompromised hosts through inhalation of contaminated air and cause disseminated infection (Moretti et al., 2014). To cause disease in a human host, the pathogen has to survive the specific physiological conditions, such as elevated body temperature, alkaline pH, and limited access to iron and other micronutrients. The analysis of the genome of a *F. oxysporum* clinical isolate collected from the blood of an infected patient revealed the presence of unique SP chromosomes.

Interestingly, the SP chromosomes from this human-infecting *F. oxysporum* clinical isolate are distinct from any SP chromosomes found in plant pathogens. For instance, all plant *F. oxysporum* pathogens have *SIX* or *SIX*-like effectors. In addition, a miniature impala (mimp), which is a Miniature Inverted-repeat TE (MITE) of the nonautonomous DNA transposon, is present in the upstream region of these effectors. Strikingly, neither SIX proteins nor mimp transposons are encoded in the genome of this human pathogenic isolate (our unpublished data). Rather, functional annotation revealed that genes encoded in these SP chromosomes may contribute to pathogen adaptation to human body conditions and result in infection (Ma, unpublished results).

3.2.4 Effectors in Other Fusarium Genomes

The function of SP chromosomes was first documented in *F. solani* f. sp. *pisi* (Han et al., 2001; Liu, Inlow, & VanEtten, 2003; Miao, Covert, et al., 1991; Rodriguez-Carres et al., 2008; VanEtten, Funnell-Baerg, Wasmann, & McCluskey, 1994; VanEtten et al., 1998), a pea pathogen. These SP chromosomes contribute to karyotype polymorphism in the *F. solani* species complex. Three SP chromosomes, 14, 15, and 17, were identified in the sequenced reference genome, enabling a clear definition of a compartmentalized genomic structure (Coleman et al., 2009). Of those SP chromosomes, the 1.6-Mb chromosome 14 harbors the pea pathogenicity (*PEP*) gene cluster, which encodes several genes that influence lesion size on the epicotyls of the host *Pisum sativum* (Han et al., 2001) and confers resistance to the pea phytoalexin pisatin; therefore, these genes encode genetic determinants of host-specific virulence. This chromosome also carries genes involved in the utilization of homoserine, an amino acid particularly enriched in the pea rhizosphere. A strain of *F. solani* containing the homoserine utilization (HUT) locus on chromosome 14 was more competitive in the pea rhizosphere than was a HUT^{-} mutant (Rodriguez-Carres et al., 2008). Similarly, a chickpea pathogenic *F. solani* genome has a SP chromosome that harbors genes that detoxify the chickpea phytoalexin maackiain, which is shown to be important for the virulence on chickpea.

Effector biology is a rapidly evolving research field. In addition to the effector biology research reported in the FOSC, understanding the importance of effectors among other *Fusarium* genomes is under way. For example, the *F. graminearum* genome lacks SP chromosomes and has few repeat sequences. However, it also possesses accessory components in the middle of the chromosomes, where chromosomal fusions occurred (Cuomo et al., 2007; Kistler, Rep, & Ma, 2013). These regions are highly diversified between isolates and have high recombination rates during the sexual stage. In addition, the secreted protein-encoding genes and many secondary metabolite gene clusters that play important roles in pathogen–host interactions are enriched in these accessory regions, indicating their potential role in specific pathogenicity on the host. Comparative proteomic analyses identified proteins that are secreted during *F. graminearum* infection (Paper, Scott-Craig, Adhikari, Cuomo, & Walton, 2007; Yang et al., 2012), and a secretome study identified more potential novel effectors, many of which are specific to *F. graminearum* (Brown, Antoniw, & Hammond-Kosack, 2012).

3.3 Other Expanded Protein Families Contributing to Fungal Virulence

In addition to the small, cysteine-rich effectors, fusaria also possess a variety of enzymes that target host physical and chemical barriers to facilitate infection and colonization. Many of these enzymes have evolved into multicopy gene families through gene family expansion to enhance virulence.

A notable group of enzymes that overcome the plant's physical barrier are the cell wall-degrading enzymes (CWDEs); these are typically enriched in fungal plant pathogens (Kubicek, Starr, & Glass, 2014; Ma et al., 2010; Roncero et al., 2003; Zhao, Liu, Wang, & Xu, 2013). The expression of these genes is usually induced during host colonization. For example, the *Fol4287* genome encodes eight copies of a CWDE enzyme polygalacturonase, which is secreted during early infection to weaken the plant cell wall pectin network (Di Pietro & Roncero, 1996) and assists in fungal invasion (Bravo Ruiz, Di Pietro, & Roncero, 2016).

A pathogenic fungus also has to adapt to overcome host chemical barriers, as illustrated by *Fol* and its host tomato. For self-defense, tomato plants constitutively produce the antifungal alpha-tomatine that disrupts the integrity of the fungal cell membrane. To overcome this chemical barrier, the tomato pathogen *Fol* encodes a tomatinase (TOM) that cleaves the alpha-tomatine and converts it into the much less toxic tomatidine and lycotetraose, thereby suppressing the plant's defense system (Pareja-Jaime, Roncero, & Ruiz-Roldán, 2008). In the *Fol* genome, five putative tomatinase genes have been identified. Overexpression of one of the *TOM* genes enhances the virulence of *Fol* toward its tomato host, whereas deletion of this gene decreases tomatinase activity and delays disease symptoms in plants (Kamper et al., 2006).

In addition, mechanisms have evolved that modify host enzyme activity to overcome host defense. As part of the defense response, tomato plants secrete chitinases that hydrolyze fungal chitin and disrupt fungal cell wall integrity. To circumvent this defense, pathogenic *F. oxysporum* isolates secrete proteases, such as the serine protease *SEP1* and metalloprotease *MEP1*, which cleave extracellular tomato chitinases and thereby reduce chitinase activity. Double deletion mutations of these two secreted protease genes significantly reduced disease symptoms in tomato (Jashni et al., 2015).

4. GENOMIC PLASTICITY SHEDS LIGHT ON THE EVOLUTION OF PATHOGENICITY

The genus *Fusarium* encompasses great genetic diversity. In several cases, we have observed independent lineages within this genus that evolved polyphyletically through convergent evolution. Phylogenomics enables a systematic biology approach to exploring the evolutionary processes that define the genomic variation within the genus.

4.1 Evolution of Host Pathogenicity Through Horizontal Gene Transfer

Multiple lines of evidence suggest the horizontal transfer of SP chromosomes in both *F. oxysporum* and *F. solani* species complexes (Coleman et al., 2009; Ma et al., 2010). The most direct evidence of this comes from the experimental confirmation of the transfer of two SP chromosomes between two *F. oxysporum* strains. The acquisition of SP chromosomes enables the fungus to overcome host defense mechanisms and adapt to the host environment, while the core genome performs the essential functions. The transfer event successfully converted a nonpathogenic strain into a pathogen (Ma et al., 2010; van der Does & Rep, 2007). Transfer of SP chromosomes between genetically isolated strains explains the polyphyletic origin of host specificity and the emergence of new pathogenic lineages in *F. oxysporum*. However, a recent study revealed that the transfer of the pathogenic chromosome from *Fol* to *Fo47* is accompanied by the exchange of core chromosomes, indicating that the evolution of the core component through horizontal gene transfer (HGT) should not be overlooked (Vlaardingerbroek et al., 2016).

Dispensable sequences from different *F. oxysporum formae speciales* are considerably different even among phylogenetically related strains (our unpublished data). In comparison with the other chromosomes, SP chromosomes contain more repeat sequences, are enriched in duplicated genes, and have a different G+C content and codon bias (Coleman et al., 2009; Ma et al., 2010). These observations provide clear evidence that accessory components were not vertically transmitted from the last common ancestor of *Fusarium*. Instead, they were likely acquired through horizontal transfer.

Comparisons of *F. oxysporum* strains that are pathogenic to other plant hosts revealed the existence of a unique set of SP chromosomes associated with different hosts, which likely contribute to virulence and host specificity, as observed in the *Fol* strain (Ma, unpublished). Although the origin(s) of the extra genes and the SP chromosomes is unknown, the gene expansion and large genome size are consistent with this species' diverse range of habitats. Such genetic diversity captured by SP chromosomes within a single species complex is striking, considering that over a hundred *formae speciales* have been identified (Michielse & Rep, 2009). This remarkable richness of extra genetic material combined with identical effector genes within a *forma specialis* appears to be the result of a combination of long-term evolution of extra chromosomes and horizontal chromosome transfer events.

Potential HGT events may also contribute to the rich diversity of SMB gene clusters. Only a small fraction of *Fusarium* SMB gene clusters identified to date are shared; many are species-specific and exhibit great genetic diversity (Ma et al., 2012, 2010). For instance, among the 16 *F. graminearum* PKS SMB clusters, only 2 are conserved in all species examined (Brown, Butchko, Baker, & Proctor, 2012). By contrast, most *Fusarium* SMB clusters exhibit a discontinuous distribution that does not necessarily correlate with phylogenetic relationships among the fungal species that harbor them. For example, the narrowly distributed fumonisin SMB cluster is present in some but not all species in the *F. fujikuroi* species complex, suggesting it arose through horizontal transfer (Proctor et al., 2013).

4.2 Population Genomics in Understanding the Phylogenomic Evolution

While HGT of SP chromosomes introduced a novel capacity for a fungus to invade a specific host, natural selection also encompasses at nucleotide levels. Population genomics, in which the evolutionary processes of divergence, differentiation, and recombination are studied using a large number of genotyping loci throughout the genome (Croll, Lendenmann, Stewart, & McDonald, 2015; Stukenbrock & Bataillon, 2012; Talas & McDonald, 2015), is better suited to address evolutionary changes of genes and SP chromosomes after integration. This method has been widely used to decipher evolutionary processes involved in pathogenicity, host specialization, fungicide resistance, and adaptation to different environments (Stukenbrock & Bataillon, 2012; Talas, Kalih, Miedaner, & McDonald, 2016). Common approaches used in population genomics include genome-wide association studies (GWASs) (Dalman et al., 2013), genome scans (Cooke et al., 2012; Ellison et al., 2011), and comparative genomics (Menardo et al., 2016; Stukenbrock et al., 2011).

In *Fusarium* species that undergo sexual reproduction, GWASs have been used to identify quantitative trait loci or genes associated with particular phenotypes. One example is the GWAS of 220 *F. graminearum* isolates. Based on approximately 29,000 SNPs across the whole genome, a total of 50, 29, and 74 quantitative trait nucleotides (QTNs) were associated with pathogen aggressiveness, DON production, and sensitivity to the fungicide propiconazole, respectively. Most of the detected QTNs were synonymous substitutions or located in genes with predicted regulatory functions. One highly significant QTN associated with aggressiveness was found in a gene encoding the RAS-GTPase activating protein. Three QTNs associated with propiconazole sensitivity were located in genes not previously associated with azole sensitivity. These genes are strong candidates for studying mechanisms of pathogenicity and fungicide resistance in *F. graminearum* populations (Talas et al., 2016).

For an asexual fungus like *F. oxysporum*, genome scans of individuals within a population have the potential to detect genetic variation and identify genes involved in pathogenicity. This method was used to identify *AVRFOM2*, a gene in *F. oxysporum* f. sp. *meloni* (*Fom*) that encodes a small secreted protein that is recognized by melon plant resistance gene *Fom-2* (Schmidt et al., 2016). Losing *AVRFOM2*, the pathogen avoids host R gene recognition and escapes R gene-induced resistance.

Comparative population genomics was used to study the origins and the underlying processes driving the emergence of novel clones or lineages among *Foc*. Genomic sequences for 33 *Foc* isolates were generated (our unpublished results). Three distinct populations have been identified and the recent evolving race 4 strains that broke down Cavendish banana's resistance against *Fusarium* wilt pathogen formed a distinct monophyletic clade distinguished from other *Foc* strains. Interestingly, genes, especially pathogenesis-related genes, are under diversifying selection. This type of selection drives evolutionary processes among TR4 strains that increase diversity. Comparative genomic studies among three legume-infecting *formae speciales*, *F. oxysporum* f. sp. *ciceris*, f. sp. *pisi*, and f. sp. *medicaginis*, have also revealed limited conservation among shared dispensable sequences. Even though these isolates can all infect the same host, limited conservation could be attributed to either completely different origins or large-scale rearrangements of dispensable sequences (Williams et al., 2016).

4.3 Systemic Regulation of Fungal Pathogenicity

In a compartmentalized genome, virulence factors may have two origins, i.e., conserved genes with pathogenic function or unique genes located on the SP chromosomes. However, to function as an integrated organism, a genome must coordinately regulate these two distinct genomic components, directly or indirectly.

Many basic cell signaling transduction pathways that are highly conserved and directly impact organism fitness, such as MAPK signaling pathways, G-protein signaling pathways, and cAMP pathways, are also involved in fungal pathogenicity. A study of the cyclic adenosine monophosphate–protein kinase A (cAMP–PKA) pathway linked this conserved pathway to evolving functions related to fungal pathogenesis (Guo, Breakspear, et al., 2016; Guo, Zhao, et al., 2016). Comparative transcriptomics confirmed highly correlated expression patterns for most orthologues (80%) between two related *Fusarium* species, *F. graminearum* and *F. verticillioides*, suggesting the overall functional conservation of this signaling pathway. Furthermore, the study showed that 6% of the orthologous genes had distinct expression patterns, indicating functional divergence. Interestingly, these functionally diverged orthologues are enriched for genes regulating the production and detoxification of secondary metabolites unique to each species. As mentioned above, *F. graminearum* and *F. verticillioides* have distinct mycotoxin profiles, many of which are species-specific. Through integrating the production and detoxification of species-specific toxins with the conserved cAMP–PKA signaling pathway, both species are able to control toxin production. This study suggests that the divergent evolution of conserved signaling pathways contributes to fungal divergence and niche adaptation.

Many virulence factors, such as effectors, CWDEs, and mycotoxins, are often transcriptionally coregulated during host colonization. Such synchronized regulation determines the outcome of pathogen–host interactions. Genes on SP chromosomes can be regulated by transcription factors from either the core or the accessory components of the genome. The *Fol* SP chromosomes contain 13 predicted transcription factors. The predicted DNA-binding sites for these transcription factors are enriched among SP chromosomes, suggesting at least partially transcriptionally autonomous of the SP chromosomes (van der Does et al., 2016). Upregulation of effectors upon infection is regulated by Sge1 (SIX gene expression) (Michielse et al., 2009), a transcription factor conserved among ascomycete fungi. Apart from being required for expression of effector genes, Sge1 homologs mediate the production of certain secondary metabolites in other genomes (Jonkers et al., 2014) and regulate the dimorphic switch in human pathogenic fungi (Michielse et al., 2009). Clearly, there are extensive transcriptional connections between core and accessory chromosomes (van der Does et al., 2016). With the greatly increased sequencing capacity and available transcriptomics data, regulatory networks have been successfully reconstructed to study functional pathogenesis in species such as *F. graminearum* (Guo, Zhao, et al., 2016). With accumulated expression data, the gene regulatory network will be a powerful tool to study functional compartmentalization among FOSC genomes.

5. CLOSING REMARKS

As a cross-kingdom pathogen, the FOSC is ecologically diverse and economically important, as revealed by the phylogenetic framework of the genus *Fusarium*. Although many studies have examined the nature of polyphyletic host-specific pathogenicity and the function of virulence factors, the phylogeny of house-keeping genes does not provide information on host specificity among *F. oxysporum* strains. Comparative genomic studies have revealed the existence of *F. oxysporum* SP chromosomes and their roles as genetic determinants of host-specific pathogenicity. Horizontal gene transfer between *F. oxysporum* strains offers a logical explanation of the polyphyletic origins of host-specific pathogenicity within the FOSC. Genome compartmentalization provides new insight into how different strains adapted to different hosts. Genomic subdivision through population genomics enables the structural compartmentalization of a fusarial genome. The sequences that determine host-specific pathogenicity are often encoded in the SP chromosomes and share high sequence similarity between strains within the same *forma specialis*. Applying population genomic techniques is not only essential for deciphering the evolution of pathogenicity mechanisms but also for identifying informative candidates associated with pathogenicity that can be developed as targets in disease management programs.

REFERENCES

Ahangarkani, F., Rouhi, S., & Azizi, I. G. (2014). A review on incidence and toxicity of fumonisins. *Toxin Reviews*, *33*, 95–100.

Alexander, N. J., McCormick, S. P., Waalwijk, C., van der Lee, T., & Proctor, R. H. (2011). The genetic basis for 3-ADON and 15-ADON trichothecene chemotypes in *Fusarium*. *Fungal Genetics and Biology*, *48*, 485–495.

Alexander, N. J., Proctor, R. H., & McCormick, S. P. (2009). Genes, gene clusters, and biosynthesis of trichothecenes and fumonisins in *Fusarium*. *Toxin Reviews*, *28*, 198–215.

Al-Hatmi, A. M. S., Hagen, F., Menken, S. B. J., Meis, J. F., & de Hoog, G. S. (2016). Global molecular epidemiology and genetic diversity of *Fusarium*, a significant emerging group of human opportunists from 1958 to 2015. *Emerging Microbes and Infections*, *5*, e124.

Al-Hatmi, A. M., Meis, J. F., & de Hoog, G. S. (2016). *Fusarium*: Molecular diversity and intrinsic drug resistance. *PLoS Pathogens*, *12*, e1005464.

Anaissie, E. J., Kuchar, R. T., Rex, J. H., Francesconi, A., Kasai, M., Muller, F. M., et al. (2001). Fusariosis associated with pathogenic *Fusarium* species colonization of a hospital water system: A new paradigm for the epidemiology of opportunistic mold infections. *Clinical Infectious Diseases*, *33*, 1871–1878.

Baayen, R. P., O'Donnell, K., Bonants, P. J., Cigelnik, E., Kroon, L. P., Roebroeck, E. J., et al. (2000). Gene genealogies and AFLP analyses in the *Fusarium oxysporum* complex identify monophyletic and nonmonophyletic formae speciales causing wilt and rot disease. *Phytopathology*, *90*, 891–900.

Bai, G., & Shaner, G. (2004). Management and resistance in wheat and barley to *Fusarium* head blight. *Annual Review of Phytopathology, 42*, 135–161.

Balmas, V., Migheli, Q., Scherm, B., Garau, P., O'Donnell, K., Ceccherelli, G., et al. (2010). Multilocus phylogenetics show high levels of endemic fusaria inhabiting Sardinian soils (Tyrrhenian Islands). *Mycologia, 102*, 803–812.

Bell, B. P., & Khabbaz, R. F. (2013). Responding to the outbreak of invasive fungal infections: The value of public health to Americans. *JAMA, 309*, 883–884.

Boehm, E., Ploetz, R. C., & Kistler, H. C. (1994). Statistical analysis of electrophoretic karyotype variation among vegetative compatibility groups of *Fusarium oxysporum* f. sp. *cubense*. *Molecular Plant Microbe Interactions*, 7, 196–207.

Boenisch, M. J., & Schäfer, W. (2011). *Fusarium graminearum* forms mycotoxin producing infection structures on wheat. *BMC Plant Biology, 11*, 110.

Booth, C. (1971). *The genus* Fusarium. Surrey, England: Commonwealth Mycological Institute, Key.

Boutati, E. I., & Anaissie, E. J. (1997). *Fusarium*, a significant emerging pathogen in patients with hematologic malignancy: Ten years' experience at a cancer center and implications for management. *Blood, 90*, 999–1008.

Brandt, M. E., & Park, B. J. (2013). Think fungus-prevention and control of fungal infections. *Emerging Infectious Diseases, 19*, 1688–1689.

Bravo Ruiz, G., Di Pietro, A., & Roncero, M. I. G. (2016). Combined action of the major secreted exo-and endopolygalacturonases is required for full virulence of *Fusarium oxysporum*. *Molecular Plant Pathology, 17*, 339–359.

Brown, N. A., Antoniw, J., & Hammond-Kosack, K. E. (2012). The predicted secretome of the plant pathogenic fungus *Fusarium graminearum*: A refined comparative analysis. *PLoS One*, 7, e33731.

Brown, D. W., Butchko, R. A. E., Baker, S. E., & Proctor, R. H. (2012). Phylogenomic and functional domain analysis of polyketide synthases in *Fusarium*. *Fungal Biology, 116*, 318–331.

Brown, D. W., Butchko, R. A., Busman, M., & Proctor, R. H. (2007). The *Fusarium verticillioides* FUM gene cluster encodes a Zn(II)2Cys6 protein that affects FUM gene expression and fumonisin production. *Eukaryotic Cell, 6*, 1210–1218.

Brown, G. D., Denning, D. W., Gow, N. A., Levitz, S. M., Netea, M. G., & White, T. C. (2012). Hidden killers: Human fungal infections. *Science Translational Medicine, 4*, 165rv13.

Chakrabarti, A., Rep, M., Wang, B., Ashton, A., Dodds, P., & Ellis, J. (2011). Variation in potential effector genes distinguishing Australian and non-Australian isolates of the cotton wilt pathogen *Fusarium oxysporum* f. sp. *vasinfectum*. *Plant Pathology, 60*, 232–243.

Chang, D. C., Grant, G. B., O'Donnell, K., Wannemuehler, K. A., Noble-Wang, J., Rao, C. Y., et al. (2006). Multistate outbreak of *Fusarium keratitis* associated with use of a contact lens solution. *JAMA, 296*, 953–963.

Cilo, B. D., Al-Hatmi, A., Seyedmousavi, S., Rijs, A., Verweij, P., Ener, B., et al. (2015). Emergence of fusarioses in a university hospital in Turkey during a 20-year period. *European Journal of Clinical Microbiology & Infectious Diseases, 34*, 1683–1691.

Cole, R. J., and Schweikert, M. A. (2003). "Handbook of secondary fungal metabolites, Vol. I," Academic Press, San Diego.

Coleman, J. J., Rounsley, S. D., Rodriguez-Carres, M., Kuo, A., Wasmann, C. C., Grimwood, J., et al. (2009). The genome of *Nectria haematococca*: Contribution of supernumerary chromosomes to gene expansion. *PLoS Genetics, 5*, e1000618.

Cooke, D. E., Cano, L. M., Raffaele, S., Bain, R. A., Cooke, L. R., Etherington, G. J., et al. (2012). Genome analyses of an aggressive and invasive lineage of the Irish potato famine pathogen. *PLoS Pathogens, 8*, e1002940.

Croll, D., Lendenmann, M. H., Stewart, E., & McDonald, B. A. (2015). The impact of recombination hotspots on genome evolution of a fungal plant pathogen. *Genetics*, *201*, 1213–1228.

Cundliffe, E., Cannon, M., & Davies, J. (1974). Mechanism of inhibition of eukaryotic protein synthesis by trichothecene fungal toxins. *Proceedings of the National Academy of Sciences of the United States of America*, *71*, 30–34.

Cuomo, C. A., Guldener, U., Xu, J. R., Trail, F., Turgeon, B. G., Di Pietro, A., et al. (2007). The *Fusarium graminearum* genome reveals a link between localized polymorphism and pathogen specialization. *Science*, *317*, 1400–1402.

Daboussi, M.-J., & Capy, P. (2003). Transposable elements in filamentous fungi. *Annual Reviews in Microbiology*, *57*, 275–299.

Dalman, K., Himmelstrand, K., Olson, A., Lind, M., Brandstrom-Durling, M., & Stenlid, J. (2013). A genome-wide association study identifies genomic regions for virulence in the non-model organism *Heterobasidion annosum* s.s. *PLoS One*, *8*, e53525.

Denning, D. W., & Bromley, M. J. (2015). Infectious disease. How to bolster the antifungal pipeline. *Science*, *347*, 1414–1416.

Desjardins, A. E., & Proctor, R. H. (2007). Molecular biology of *Fusarium* mycotoxins. *International Journal of Food Microbiology*, *119*, 47–50.

Di Pietro, A., & Roncero, M. I. (1996). Purification and characterization of an exo-polygalacturonase from the tomato vascular wilt pathogen *Fusarium oxysporum* f.sp. *lycopersici*. *FEMS Microbiology Letters*, *145*, 295–299.

Domijan, A. M. (2012). Fumonisin B(1): A neurotoxic mycotoxin. *Arhiv za Higijenu Rada i Toksikologiju*, *63*, 531–544.

Ellison, C. E., Hall, C., Kowbel, D., Welch, J., Brem, R. B., Glass, N. L., et al. (2011). Population genomics and local adaptation in wild isolates of a model microbial eukaryote. *Proceedings of the National Academy of Sciences of the United States of America*, *108*, 2831–2836.

Fisher, M. C., Henk, D. A., Briggs, C. J., Brownstein, J. S., Madoff, L. C., McCraw, S. L., et al. (2012). Emerging fungal threats to animal, plant and ecosystem health. *Nature*, *484*, 186–194.

Gale, L. R., Chen, L. F., Hernick, C. A., Takamura, K., & Kistler, H. C. (2002). Population analysis of *Fusarium graminearum* from wheat fields in Eastern China. *Phytopathology*, *92*, 1315–1322.

García-Bastidas, F., Ordóñez, N., Konkol, J., Al-Qasim, M., Naser, Z., Abdelwali, M., et al. (2013). First report of *Fusarium oxysporum* f. sp. *cubense* tropical race 4 associated with panama disease of banana outside Southeast Asia. *Plant Disease*, *98*, 694.

Glenn, A. E., Zitomer, N. C., Zimeri, A. M., Williams, L. D., Riley, R. T., & Proctor, R. H. (2008). Transformation-mediated complementation of a FUM gene cluster deletion in *Fusarium verticillioides* restores both fumonisin production and pathogenicity on maize seedlings. *Molecular Plant-Microbe Interactions*, *21*, 87–97.

Goswami, R. S., & Kistler, H. C. (2005). Pathogenicity and in planta mycotoxin accumulation among members of the *Fusarium graminearum* species complex on wheat and rice. *Phytopathology*, *95*, 1397–1404.

Goswami, R. S., Xu, J. R., Trail, F., Hilburn, K., & Kistler, H. C. (2006). Genomic analysis of host-pathogen interaction between *Fusarium graminearum* and wheat during early stages of disease development. *Microbiology*, *152*, 1877–1890.

Guarro, J. (2013). Fusariosis, a complex infection caused by a high diversity of fungal species refractory to treatment. *European Journal of Clinical Microbiology & Infectious Diseases*, *32*, 1491–1500.

Guo, L., Breakspear, A., Zhao, G., Gao, L., Kistler, H. C., Xu, J. R., et al. (2016). Conservation and divergence of the cyclic adenosine monophosphate-protein kinase A (cAMP-PKA) pathway in two plant-pathogenic fungi: *Fusarium graminearum* and *F. verticillioides*. *Molecular Plant Pathology*, *17*, 196–209.

Guo, L., & Ma, L. J. (2014). Genomics of wheat pathogen *Fusarium graminearum*. In R. Dean & A. Lichens-Park (Eds.), *Genomics of plant-associated fungi and oomycetes* (pp. 102–122). Massachusetts, USA: Springer.

Guo, L., Zhao, G., Xu, J. R., Kistler, H. C., Gao, L., & Ma, L. J. (2016). Compartmentalized gene regulatory network of the pathogenic fungus *Fusarium graminearum*. *The New Phytologist, 211*, 527–541.

Hahn-Ast, C., Glasmacher, A., Muckter, S., Schmitz, A., Kraemer, A., Marklein, G., et al. (2010). Overall survival and fungal infection-related mortality in patients with invasive fungal infection and neutropenia after myelosuppressive chemotherapy in a tertiary care centre from 1995 to 2006. *The Journal of Antimicrobial Chemotherapy, 65*, 761–768.

Han, Y., Liu, X., Benny, U., Kistler, H. C., & VanEtten, H. D. (2001). Genes determining pathogenicity to pea are clustered on a supernumerary chromosome in the fungal plant pathogen *Nectria haematococca*. *The Plant Journal, 25*, 305–314.

Hatta, R., Ito, K., Hosaki, Y., Tanaka, T., Tanaka, A., Yamamoto, M., et al. (2002). A conditionally dispensable chromosome controls host-specific pathogenicity in the fungal plant pathogen *Alternaria alternata*. *Genetics, 161*, 59–70.

Holgersson, J., Rydberg, L., & Breimer, M. E. (2014). Molecular deciphering of the ABO system as a basis for novel diagnostics and therapeutics in ABO incompatible transplantation. *International Reviews of Immunology, 33*, 174–194.

Houterman, P. M., Cornelissen, B. J., & Rep, M. (2008). Suppression of plant resistance gene-based immunity by a fungal effector. *PLoS Pathogens, 4*, e1000061.

Houterman, P. M., Ma, L., van Ooijen, G., de Vroomen, M. J., Cornelissen, B. J., Takken, F. L., et al. (2009). The effector protein Avr2 of the xylem-colonizing fungus *Fusarium oxysporum* activates the tomato resistance protein I-2 intracellularly. *The Plant Journal, 58*, 970–978.

Ilgen, P., Hadeler, B., Maier, F. J., & Schafer, W. (2009). Developing kernel and rachis node induce the trichothecene pathway of *Fusarium graminearum* during wheat head infection. *Molecular Plant-Microbe Interactions, 22*, 899–908.

Jashni, M. K., Dols, I. H., Iida, Y., Boeren, S., Beenen, H. G., Mehrabi, R., et al. (2015). Synergistic action of a metalloprotease and a serine protease from *Fusarium oxysporum* f. sp. *lycopersici* cleaves chitin-binding tomato chitinases, reduces their antifungal activity, and enhances fungal virulence. *Molecular Plant-Microbe Interactions, 28*, 996–1008.

Johnson, D. D., Flaskerud, G. K., Taylor, R. D., & Satyanarayana, V. (1998). Economic impacts of *Fusarium* head blight in wheat. In *Agricultural economics report 396*.

Jonkers, W., Xayamongkhon, H., Haas, M., Olivain, C., van der Does, H. C., Broz, K., et al. (2014). EBR1 genomic expansion and its role in virulence of *Fusarium* species. *Environmental Microbiology, 16*, 1982–2003.

Kamper, J., Kahmann, R., Bolker, M., Ma, L. J., Brefort, T., Saville, B. J., et al. (2006). Insights from the genome of the biotrophic fungal plant pathogen Ustilago maydis. *Nature, 444*, 97–101.

Keller, N. P., Turner, G., & Bennett, J. W. (2005). Fungal secondary metabolism—From biochemistry to genomics. *Nature Reviews. Microbiology, 3*, 937–947.

Kistler, H. C., Alabouvette, C., Baayen, R. P., Bentley, S., Brayford, D., Coddington, A., et al. (1998). Systematic numbering of vegetative compatibility groups in the plant pathogenic fungus *Fusarium oxysporum*. *Phytopathology, 88*, 30–32.

Kistler, H. C., Rep, M., & Ma, L.-J. (2013). Structural dynamics of *Fusarium* genomes. In D. W. Brown & R. H. Proctor (Eds.), *Fusarium: Genomics, molecular and cellular biology*. Horizon Scientific Press: Norwich, UK.

Kredics, L., Narendran, V., Shobana, C. S., Vagvolgyi, C., Manikandan, P., & Indo-Hungarian Fungal Keratitis Working Group. (2015). Filamentous fungal infections of the cornea: A global overview of epidemiology and drug sensitivity. *Mycoses, 58*, 243–260.

Kubicek, C. P., Starr, T. L., & Glass, N. L. (2014). Plant cell wall-degrading enzymes and their secretion in plant-pathogenic fungi. *Annual Review of Phytopathology*, *52*, 427–451.

Leslie, J. F., & Summerell, B. A. (2006). *Fusarium* laboratory workshops—A recent history. *Mycotoxin Research*, *22*, 73–74.

Lievens, B., van Baarlen, P., Verreth, C., van Kerckhove, S., Rep, M., & Thomma, B. P. (2009). Evolutionary relationships between *Fusarium oxysporum* f. sp. *lycopersici* and *F. oxysporum* f. sp. *radicis-lycopersici* isolates inferred from mating type, elongation factor-1alpha and exopolygalacturonase sequences. *Mycological Research*, *113*, 1181–1191.

Liu, X., Inlow, M., & VanEtten, H. D. (2003). Expression profiles of pea pathogenicity (PEP) genes in vivo and in vitro, characterization of the flanking regions of the PEP cluster and evidence that the PEP cluster region resulted from horizontal gene transfer in the fungal pathogen *Nectria haematococca*. *Current Genetics*, *44*, 95–103.

Low, C. Y., & Rotstein, C. (2011). Emerging fungal infections in immunocompromised patients. *F1000 Medicine Reports*, *3*, 14.

Ma, L., Cornelissen, B. J., & Takken, F. L. (2013). A nuclear localization for Avr2 from *Fusarium oxysporum* is required to activate the tomato resistance protein I-2. *Frontiers in Plant Science*, *4*, 94.

Ma, L. J., Geiser, D. M., Proctor, R. H., Rooney, A. P., O'Donnell, K., Trail, F., et al. (2013). *Fusarium* pathogenomics. *Annual Review of Microbiology*, *67*, 399–416.

Ma, L.-J., Kistler, H. C., & Rep, M. (2012). Evolution of plant pathogenicity in *Fusarium* species. In L. D. Sibley, B. J. Howlett, & J. Heitman (Eds.), *Evolution of virulence in eukaryotic microbes* (pp. 486–498). Hoboken, NJ: Wiley & Sons.

Ma, L. J., van der Does, H. C., Borkovich, K. A., Coleman, J. J., Daboussi, M. J., Di Pietro, A., et al. (2010). Comparative genomics reveals mobile pathogenicity chromosomes in *Fusarium*. *Nature*, *464*, 367–373.

Menardo, F., Praz, C. R., Wyder, S., Ben-David, R., Bourras, S., Matsumae, H., et al. (2016). Hybridization of powdery mildew strains gives rise to pathogens on novel agricultural crop species. *Nature Genetics*, *48*, 201–205.

Miao, V. P., Covert, S. F., & VanEtten, H. D. (1991). A fungal gene for antibiotic resistance on a dispensable ("B") chromosome. *Science*, *254*, 1773–1776.

Miao, V. P., Matthews, D. E., & VanEtten, H. D. (1991). Identification and chromosomal locations of a family of cytochrome P-450 genes for pisatin detoxification in the fungus *Nectrla haematococca*. *Molecular and General Genetics*, *226*, 214–223.

Michielse, C. B., & Rep, M. (2009). Pathogen profile update: *Fusarium oxysporum*. *Molecular Plant Pathology*, *10*, 311–324.

Michielse, C. B., van Wijk, R., Reijnen, L., Cornelissen, B. J., & Rep, M. (2009). Insight into the molecular requirements for pathogenicity of *Fusarium oxysporum* f. sp. *lycopersici* through large-scale insertional mutagenesis. *Genome Biology*, *10*, R4.

Migheli, Q., Berio, T., Gullino, M. L., & Garibaldi, A. (1995). Electrophoretic karyotype variation among pathotypes of *Fusarium oxysporum* f. sp. *dianthi*. *Plant Pathology*, *44*, 308–315.

Moretti, M. L., Busso-Lopes, A., Moraes, R., Muraosa, Y., Mikami, Y., Trabasso, P., et al. (2014). Environment as a potential source of *Fusarium* spp. invasive infections in immunocompromised patients. In *Open forum infectious diseases*, Vol. 1 (p. S38). Oxford, England, UK: Oxford University Press.

Mukherjee, S. D., Swystun, L. L., Mackman, N., Wang, J. G., Pond, G., Levine, M. N., et al. (2010). Impact of chemotherapy on thrombin generation and on the protein C pathway in breast cancer patients. *Pathophysiology of Haemostasis and Thrombosis*, *37*, 88–97.

Nganje, W. E., Bangsund, D. A., Leistritz, F. L., Wilson, W. W., & Tiapo, N. M. (2004). Regional economic impacts of *Fusarium* head blight in wheat and barley. *Review of Agricultural Economics*, 332–347.

Nucci, M., Varon, A. G., Garnica, M., Akiti, T., Barreiros, G., Trope, B. M., et al. (2013). Increased incidence of invasive fusariosis with cutaneous portal of entry, Brazil. *Emerging Infectious Diseases*, *19*, 1567–1572.

O'Donnell, K., Gueidan, C., Sink, S., Johnston, P. R., Crous, P. W., Glenn, A., et al. (2009). A two-locus DNA sequence database for typing plant and human pathogens within the *Fusarium oxysporum* species complex. *Fungal Genetics and Biology*, *46*, 936–948.

O'Donnell, K., Kistler, H. C., Cigelnik, E., & Ploetz, R. C. (1998). Multiple evolutionary origins of the fungus causing Panama disease of banana: Concordant evidence from nuclear and mitochondrial gene genealogies. *Proceedings of the National Academy of Sciences of the United States of America*, *95*, 2044–2049.

O'Donnell, K., Rooney, A. P., Proctor, R. H., Brown, D. W., McCormick, S. P., Ward, T. J., et al. (2013). Phylogenetic analyses of RPB1 and RPB2 support a middle Cretaceous origin for a clade comprising all agriculturally and medically important fusaria. *Fungal Genetics and Biology*, *52*, 20–31.

O'Donnell, K., Sutton, D. A., Rinaldi, M. G., Magnon, K. C., Cox, P. A., Revankar, S. G., et al. (2004). Genetic diversity of human pathogenic members of the *Fusarium oxysporum* complex inferred from multilocus DNA sequence data and amplified fragment length polymorphism analyses: Evidence for the recent dispersion of a geographically widespread clonal lineage and nosocomial origin. *Journal of Clinical Microbiology*, *42*, 5109–5120.

Palmero, D., Iglesias, C., de Cara, M., Lomas, T., Santos, M., & Tello, J. C. (2009). Species of *Fusarium* isolated from river and sea water of southeastern Spain and pathogenicity on four plant species. *Plant Disease*, *93*, 377–385.

Paper, J. M., Scott-Craig, J. S., Adhikari, N. D., Cuomo, C. A., & Walton, J. D. (2007). Comparative proteomics of extracellular proteins in vitro and in planta from the pathogenic fungus *Fusarium graminearum*. *Proteomics*, 7, 3171–3183.

Pareja-Jaime, Y., Roncero, M. I. G., & Ruiz-Roldán, M. C. (2008). Tomatinase from *Fusarium oxysporum* f. sp. *lycopersici* is required for full virulence on tomato plants. *Molecular Plant-Microbe Interactions*, *21*, 728–736.

Ploetz, R. C. (2015). *Fusarium* wilt of banana. *Phytopathology*, *105*, 1512–1521.

Ploetz, R. C., Kema, G. H., & Ma, L. J. (2015). Impact of diseases on export and smallholder production of banana. *Annual Review of Phytopathology*, *53*, 269–288.

Price, C. L., Parker, J. E., Warrilow, A. G. S., Kelly, D. E., & Kelly, S. L. (2015). Azole fungicides—Understanding resistance mechanisms in agricultural fungal pathogens. *Pest Management Science*, *71*, 1054–1058.

Proctor, R. H., Van Hove, F., Susca, A., Stea, G., Busman, M., van der Lee, T., et al. (2013). Birth, death and horizontal transfer of the fumonisin biosynthetic gene cluster during the evolutionary diversification of *Fusarium*. *Molecular Microbiology*, *90*, 290–306.

Rep, M. (2005). Small proteins of plant-pathogenic fungi secreted during host colonization. *FEMS Microbiology Letters*, *253*, 19–27.

Rep, M., Dekker, H. L., Vossen, J. H., de Boer, A. D., Houterman, P. M., de Koster, C. G., et al. (2003). A tomato xylem sap protein represents a new family of small cysteine-rich proteins with structural similarity to lipid transfer proteins. *FEBS Letters*, *534*, 82–86.

Rep, M., Meijer, M., Houterman, P. M., van der Does, H. C., & Cornelissen, B. J. (2005). *Fusarium oxysporum* evades I-3-mediated resistance without altering the matching avirulence gene. *Molecular Plant-Microbe Interactions*, *18*, 15–23.

Rep, M., van der Does, H. C., Meijer, M., van Wijk, R., Houterman, P. M., Dekker, H. L., et al. (2004). A small, cysteine-rich protein secreted by *Fusarium oxysporum* during colonization of xylem vessels is required for I-3-mediated resistance in tomato. *Molecular Microbiology*, *53*, 1373–1383.

Rocha, O., Ansari, K., & Doohan, F. M. (2005). Effects of trichothecene mycotoxins on eukaryotic cells: A review. *Food Additives and Contaminants*, *22*, 369–378.

Rodriguez-Carres, M., White, G., Tsuchiya, D., Taga, M., & Vanetten, H. D. (2008). The supernumerary chromosome of *Nectria haematococca* that carries pea-pathogenicity genes also carries a trait for rhizosphere competitiveness on pea. *Applied and Environmental Microbiology*, *74*, 3849–3856.

Roncero, M. I. G., Hera, C., Ruiz-Rubio, M., Garcı´, F. I., Madrid, M. P., Caracuel, Z., et al. (2003). *Fusarium* as a model for studying virulence in soilborne plant pathogens. *Physiological and Molecular Plant Pathology*, *62*, 87–98.

Rosewich, U. L., Pettway, R. E., Katan, T., & Kistler, H. C. (1999). Population genetic analysis corroborates dispersal of *Fusarium oxysporum* f. sp. *radicis-lycopersici* from Florida to Europe. *Phytopathology*, *89*, 623–630.

Schmidt, S. M., Lukasiewicz, J., Farrer, R., van Dam, P., Bertoldo, C., & Rep, M. (2016). Comparative genomics of *Fusarium oxysporum* f. sp. *melonis* reveals the secreted protein recognized by the Fom-2 resistance gene in melon. *The New Phytologist*, *209*, 307–318.

Seong, K. Y., Pasquali, M., Zhou, X., Song, J., Hilburn, K., McCormick, S., et al. (2009). Global gene regulation by *Fusarium* transcription factors Tri6 and Tri10 reveals adaptations for toxin biosynthesis. *Molecular Microbiology*, *72*, 354–367.

Short, D. P., O'Donnell, K., Zhang, N., Juba, J. H., & Geiser, D. M. (2011). Widespread occurrence of diverse human pathogenic types of the fungus *Fusarium* detected in plumbing drains. *Journal of Clinical Microbiology*, *49*, 4264–4272.

Soltis, D. E., Bell, C. D., Kim, S., & Soltis, P. S. (2008). Origin and early evolution of angiosperms. *Annals of the New York Academy of Sciences*, *1133*, 3–25.

Stockmann-Juvala, H., & Savolainen, K. (2008). A review of the toxic effects and mechanisms of action of fumonisin B1. *Human & Experimental Toxicology*, *27*, 799–809.

Stukenbrock, E. H., & Bataillon, T. (2012). A population genomics perspective on the emergence and adaptation of new plant pathogens in agro-ecosystems. *PLoS Pathogens*, *8*, e1002893.

Stukenbrock, E. H., Bataillon, T., Dutheil, J. Y., Hansen, T. T., Li, R., Zala, M., et al. (2011). The making of a new pathogen: Insights from comparative population genomics of the domesticated wheat pathogen *Mycosphaerella graminicola* and its wild sister species. *Genome Research*, *21*, 2157–2166.

Taga, M., Murata, M., & VanEtten, H. D. (1999). Visualization of a conditionally dispensable chromosome in the filamentous ascomycete *Nectria haematococca* by fluorescence in situ hybridization. *Fungal Genetics and Biology*, *26*, 169–177.

Talas, F., Kalih, R., Miedaner, T., & McDonald, B. A. (2016). Genome-wide association study identifies novel candidate genes for aggressiveness, deoxynivalenol production, and azole sensitivity in natural field populations of *Fusarium graminearum*. *Molecular Plant-Microbe Interactions*, *29*, 417–430.

Talas, F., & McDonald, B. A. (2015). Genome-wide analysis of *Fusarium graminearum* field populations reveals hotspots of recombination. *BMC Genomics*, *16*, 996.

Temporini, E. D., & VanEtten, H. D. (2002). Distribution of the pea pathogenicity (PEP) genes in the fungus *Nectria haematococca* mating population VI. *Current Genetics*, *41*, 107–114.

Thatcher, L. F., Gardiner, D. M., Kazan, K., & Manners, J. M. (2012). A highly conserved effector in *Fusarium oxysporum* is required for full virulence on *Arabidopsis*. *Molecular Plant-Microbe Interactions*, *25*, 180–190.

Thatcher, L. F., Williams, A. H., Garg, G., Buck, S. G., & Singh, K. B. (2016). Transcriptome analysis of the fungal pathogen *Fusarium oxysporum* f. sp. *medicaginis* during colonisation of resistant and susceptible *Medicago truncatula* hosts identifies differential pathogenicity profiles and novel candidate effectors. *BMC Genomics*, *17*, 860.

van Dam, P., Fokkens, L., Schmidt, S. M., Linmans, J. H., Kistler, H. C., Ma, L. J., et al. (2016). Effector profiles distinguish formae speciales of *Fusarium oxysporum*. *Environmental Microbiology*, *18*, 4087–4102.

van der Does, H. C., Duyvesteijn, R. G., Goltstein, P. M., van Schie, C. C., Manders, E. M., Cornelissen, B. J., et al. (2008). Expression of effector gene SIX1 of *Fusarium oxysporum* requires living plant cells. *Fungal Genetics and Biology*, *45*, 1257–1264.
van der Does, H. C., Fokkens, L., Yang, A., Schmidt, S. M., Langereis, L., Lukasiewicz, J. M., et al. (2016). Transcription factors encoded on core and accessory chromosomes of *Fusarium oxysporum* induce expression of effector genes. *PLoS Genetics*, *12*, e1006401.
van der Does, H. C., & Rep, M. (2007). Virulence genes and the evolution of host specificity in plant-pathogenic fungi. *Molecular Plant-Microbe Interactions*, *20*, 1175–1182.
VanEtten, H., Funnell-Baerg, D., Wasmann, C., & McCluskey, K. (1994). Location of pathogenicity genes on dispensable chromosomes in *Nectria haematococca* MPVI. *Antonie Van Leeuwenhoek*, *65*, 263–267.
VanEtten, H., Jorgensen, S., Enkerli, J., & Covert, S. F. (1998). Inducing the loss of conditionally dispensable chromosomes in *Nectria haematococca* during vegetative growth. *Current Genetics*, *33*, 299–303.
Vlaardingerbroek, I., Beerens, B., Rose, L., Fokkens, L., Cornelissen, B. J., & Rep, M. (2016). Exchange of core chromosomes and horizontal transfer of lineage-specific chromosomes in *Fusarium oxysporum*. *Environmental Microbiology*, *18*, 3702–3713.
Ward, T. J., Bielawski, J. P., Kistler, H. C., Sullivan, E., & O'Donnell, K. (2002). Ancestral polymorphism and adaptive evolution in the trichothecene mycotoxin gene cluster of phytopathogenic *Fusarium*. *Proceedings of the National Academy of Sciences of the United States of America*, *99*, 9278–9283.
White, D. G. (1999). *Compendium of corn diseases*. St. Paul, MN: APS Press.
Williams, A. H., Sharma, M., Thatcher, L. F., Azam, S., Hane, J. K., Sperschneider, J., et al. (2016). Comparative genomics and prediction of conditionally dispensable sequences in legume-infecting *Fusarium oxysporum* formae speciales facilitates identification of candidate effectors. *BMC Genomics*, *17*, 191.
Yang, F., Jensen, J. D., Svensson, B., Jorgensen, H. J., Collinge, D. B., & Finnie, C. (2012). Secretomics identifies *Fusarium graminearum* proteins involved in the interaction with barley and wheat. *Molecular Plant Pathology*, *13*, 445–453.
Zhang, N., O'Donnell, K., Sutton, D. A., Nalim, F. A., Summerbell, R. C., Padhye, A. A., et al. (2006). Members of the *Fusarium solani* species complex that cause infections in both humans and plants are common in the environment. *Journal of Clinical Microbiology*, *44*, 2186–2190.
Zhao, Z., Liu, H., Wang, C., & Xu, J. R. (2013). Comparative analysis of fungal genomes reveals different plant cell wall degrading capacity in fungi. *BMC Genomics*, *14*, 274.

CHAPTER SIX

Multiple Approaches to Phylogenomic Reconstruction of the Fungal Kingdom

Charley G.P. McCarthy, David A. Fitzpatrick[1]
Maynooth University, Maynooth, County Kildare, Ireland
[1]Corresponding author: e-mail address: david.fitzpatrick@nuim.ie

Contents

Abstract

Fungi are possibly the most diverse eukaryotic kingdom, with over a million member species and an evolutionary history dating back a billion years. Fungi have been at the forefront of eukaryotic genomics, and owing to initiatives like the 1000 Fungal Genomes Project the amount of fungal genomic data has increased considerably over the last 5 years, enabling large-scale comparative genomics of species across the kingdom. In this chapter, we first review fungal evolution and the history of fungal genomics.

Advances in Genetics, Volume 100
ISSN 0065-2660
https://doi.org/10.1016/bs.adgen.2017.09.006

We then review in detail seven phylogenomic methods and reconstruct the phylogeny of 84 fungal species from 8 phyla using each method. Six methods have seen extensive use in previous fungal studies, while a Bayesian supertree method is novel to fungal phylogenomics. We find that both established and novel phylogenomic methods can accurately reconstruct the fungal kingdom. Finally, we discuss the accuracy and suitability of each phylogenomic method utilized.

1. INTRODUCTION

1.1 The Phylogeny of the Fungal Kingdom

The fungi are one of the six kingdoms of life sensu Cavalier-Smith, sister to the animal kingdom, and are thought to span approximately 1.5 million species found across a broad range of ecosystems (Baldauf & Palmer, 1993; Berbee & Taylor, 1992; Cavalier-Smith, 1998; Hawksworth, 2001; Nikoh, Hayase, Iwabe, Kuma, & Miyata, 1994). While the overall fossil record of the fungi is poor due to their simple morphology, fungal fossils have been identified dating back to the Ordovician period approximately 400 million years ago (Redecker, 2000) and molecular clock analyses suggest that the fungi originated in the Precambrian eon approximately 0.76–1.06 billion years ago (Berbee & Taylor, 2010). Classic studies into fungal evolution were based on the comparison of morphological or biochemical characteristics; however, the broad range of diversity within the fungal kingdom had limited the efficacy of some of these studies (Berbee & Taylor, 1992; Heath, 1980; Léjohn, 1974; Taylor, 1978). Since the development of phylogenetic approaches within systematics and the incorporation of molecular data into phylogenetic analyses, our understanding of the evolution of fungi has improved substantially (Guarro, Gené, & Stchigel, 1999).

Initial phylogenetic analyses of fungal species had revealed that there were four distinct phyla within the fungal kingdom: the early-diverging Chytridiomycota and Zygomycota, and the Ascomycota and Basidiomycota. The Chytridiomycota grouping was later subject to revision (James et al., 2006), and in their comprehensive classification of the fungal kingdom in 2007 Hibbet et al. formally abandoned the phylum Zygomycota (Hibbett et al., 2007). Instead, Hibbet et al. treated zygomycete species as four *incertae sedis* subphyla (Entomophthoromycotina, Kickellomycotina, Mucoromycotina, and Zoopagomycotina) and subsequently described one subkingdom (the Dikarya) and seven phyla namely Chytridiomycota, Neocallimastigomycota, Blastocladiomycota, Microsporidia, Glomeromycota, Ascomycota, and

Basidiomycota (Hibbett et al., 2007). More recent phylogenetic classification of the zygomycetes has led to the circumscription of the Mucoromycota and Zoopagomycota phyla (Spatafora et al., 2016). Furthermore, recent phylogenetic analyses have shown that *Rozella* species occupy a deep branching position in the fungal kingdom (James et al., 2006; Jones, Forn, et al., 2011), the clade containing these species are now termed the Cryptomycota phylum (Jones, Forn, et al., 2011; Jones, Richards, Hawksworth, & Bass, 2011).

1.2 *Saccharomyces cerevisiae* and the Origin of Modern Fungal Genomics

In terms of genomic data, fungi are by far the highest sampled eukaryotic kingdom, with assembly data available for over 1000 fungal species on the NCBI's GenBank facility as of May 2017. Many of these species also have multiple strains sequenced (the most extreme example being *S. cerevisiae*, which has over 400 strain assemblies available on GenBank). This reflects both the ubiquity of fungi in many areas of biological and medical study and the relative simplicity of sequencing fungal genomes with modern sequencing technology. Fungi have been the exemplar group in eukaryote genetics and genomics, from the first determination of a nucleic acid sequence taken from *S. cerevisiae* by Holley and company in the late 1960s to the sequencing of the first eukaryotic genome in the mid-1990s (Goffeau et al., 1996; Holley et al., 1965). The genome of *S. cerevisiae* was sequenced through a massive international collaboration that grew to involve approximately 600 scientists in 94 laboratories and sequencing centers from across 19 countries between 1989 and 1996 (Engel et al., 2014; Goffeau et al., 1996; Goffeau & Vassarotti, 1991). Throughout the early 1990s, each of the standard 16 nuclear chromosomes of *S. cerevisiae*, sourced from the common laboratory strain 288C and its isogenic derivative strains AB972 and FY1679, was individually sequenced and published by participating researchers (Engel et al., 2014 briefly summarize each of these sequencing projects) with the initial publication of chromosome III involving 35 European laboratories on its own (Oliver et al., 1992). The complete genome sequence of *S. cerevisiae* 288C was finally published in 1996, with 5885 putative protein-coding genes and 275 transfer RNA genes identified across the genome's ~12 million base pairs (Goffeau et al., 1996).

In the intervening years the *S. cerevisiae* 288C reference genome has been constantly updated and refined as individual genes or chromosomes have been reanalyzed or even resequenced, and all of these revisions have been recorded and maintained by the Saccharomyces Genome Database (Fisk et al., 2006). It is worth noting, however, that such was the attention paid to the original sequencing project by its contributors that the most recent major update of the *S. cerevisiae* 288C reference genome, a full resequencing of the derivative AB972 strain using far less labor-intensive modern sequencing and annotation techniques, made only minor alterations to the original genome annotation overall (Engel et al., 2014). Much of our understanding regarding the processes of genome evolution in eukaryotes since 1996 has also been derived from the study of the *S. cerevisiae* 288C genome, including the confirmation that the *S. cerevisiae* genome had undergone a whole-genome duplication (WGD) event (Kellis, Birren, & Lander, 2004; Wolfe & Shields, 1997), the effect of interspecific hybridization on genome complexity (De Barros Lopes, Bellon, Shirley, & Ganter, 2002), evidence that interdomain horizontal gene transfer (HGT) from prokaryotes into eukaryotes has occurred (Hall & Dietrich, 2007), to the ongoing development of an entirely synthetic genome through the Sc2.0 project (Annaluru et al., 2014).

1.3 Fungal Genomics and Phylogenomics Beyond the Yeast Genome

As more model organisms from other eukaryotic kingdoms had their genomes sequenced, *S. cerevisiae* 288C provided a useful comparison as the reference fungal genome, even for more complex eukaryotes like *Drosophila melanogaster*. However, the later sequencing of other model fungal species *Schizosaccharomyces pombe* and *Neurospora crassa* showed the limits of relying solely on *S. cerevisiae* as a reference for the entire fungal kingdom, particularly the latter; *N. crassa* was found to have a far larger genome than either *S. cerevisiae* or *S. pombe* and over 57% of genes predicted in *N. crassa* had no homolog in either of the other two sequenced fungal genomes (Galagan et al., 2003; Galagan, Henn, Ma, Cuomo, & Birren, 2005; Wood et al., 2002). Borne out of a lull in fungal genomic advances and the increasing sophistication of sequencing technology, the Fungal Genome Initiative (FGI) was set up by a number of research organizations in the early 2000s, under the aegis of the Broad Institute (Cuomo & Birren, 2010). Collaborators within the FGI were tasked with the sequencing and annotating the genomes of over 40 species from across the fungal kingdom, with a broad scope of species selected for analysis, medically significant human fungal pathogens like *Candida albicans* and *Aspergillus fumigatus*, commercially important species such as *Penicillium chrysogenum* and *Sclerotinia sclerotiorum*, as well as basal fungal species such as *Phycomyces blakesleeanus* (Cuomo & Birren, 2010). Between 2004 and 2012, in approximately the same amount of time it had taken to sequence each individual chromosome of *S. cerevisiae* 288C in the 1990s, over 100 fungal genomes were sequenced and made publicly available on facilities like GenBank and the Joint Genome Institute (JGI)'s Genome Portal website (Benson et al., 2013; Grigoriev, Nordberg, et al., 2011).

The steady increase in genomic data available for fungi from the first decade of this century on, while still sampled mainly from the Ascomycota and Basidiomycota phyla, allowed for a greater range of fungal genomic analyses to be conducted. This included phylogenomic analyses of the fungal kingdom using a variety of different methods (which we will discuss in detail in the following section) and comparative investigations such as analysis of the evolution of pathogenicity in genera like *Candida* or *Aspergillus* (Butler et al., 2009; Galagan, Calvo, Cuomo, et al., 2005; Jackson et al., 2009), the extent of inter-/intrakingdom HGT both to and from fungal genomes (Fitzpatrick, Logue, & Butler, 2008; Marcet-Houben & Gabaldón, 2010; Richards et al., 2011; Szöllősi, Davín, Tannier, Daubin, & Boussau, 2015), identification of clusters of secondary metabolites (Keller, Turner, & Bennett, 2005; Khaldi et al., 2010), and syntenic relationships across *Saccharomyces* and *Candida* (Byrne & Wolfe, 2005; Fitzpatrick, O'Gaora, Byrne, & Butler, 2010). The wealth of genomic data available for some fungal orders or classes has allowed for easier automation of the sequencing and annotation of novel-related species, through the development of reference transcriptomic or proteomic data for gene prediction software such as AUGUSTUS or quality assessment software for genome assembly such as BUSCO (Simão, Waterhouse, Ioannidis, Kriventseva, & Zdobnov, 2015; Stanke, Steinkamp, Waack, & Morgenstern, 2004).

1.4 The 1000 Fungal Genomes Project

The recent deluge of genomic data available for the fungal kingdom comes as a result of the 1000 Fungal Genomes Project, an initiative headed by the JGI. The project (which can be found at http://genome.jgi.doe.gov/pages/fungi-1000-projects.jsf) aims to provide genomic sequence data from at least one species from every circumscribed fungal family, either from projects headed by the JGI, projects which have been incorporated into the MycoCosm database or through community-led nomination and provision of sequencing material. The project has an inbuilt preference for sequencing projects arising from families with no sequenced species to date, or only one other reference genome at the time of nomination. Assembly and annotation data are then hosted at the JGI's MycoCosm facility as well as other publically available databases (Grigoriev et al., 2014). This community-wide effort has led to a staggering increase in the number of fungal genomes available within the last 5 years; Grigoriev et al. (2014) quoted the number of genomes present in MycoCosm at over 250 at the end of 2013; as of May 2017 there are 772 fungal genomes available to download from the facility, with another 500 species nominated for sequencing. The project has seen a large increase particularly in the amount of data available from fungal phyla outside of the Dikarya, with 58 genomes currently available from the zygomycetes, the Chytridiomycota, Neocallimastigomycota, and Blastocladiomycota. There are many other fungal families with species yet to be nominated for sequencing, including many families from the Pezizomycotina subphylum within Ascomycota and the Chytridiomycota phylum. It is hoped that the wealth of fungal genomic data arising from the 1000 Fungal Genomes Project will help, among countless other scenarios, to fuel the search for novel biosynthetic products and to better understand the ecological effects of different families within the fungal kingdom (Grigoriev, Cullen, et al., 2011). The initiative will also enable the large-scale comparative analysis of hundreds of fungal species from across the fungal kingdom, including kingdom-level phylogenomic reconstructions.

2. PHYLOGENOMIC RECONSTRUCTIONS OF THE FUNGAL KINGDOM

Phylogenetic inference arising from molecular data has, in the past, predominately relied on single genes or small numbers of highly conserved genes or nuclear markers. While usually these markers make for robust individual phylogenies, potential conflicts can occur between individual phylogenies depending on the marker(s) used. The selection of such markers may also overlook other gene families which may be phylogenetically informative, such as gene duplication events or HGT events (Bininda-Emonds, 2004). With the advent of genome sequencing and the increasing sophistication of bioinformatics software and techniques, it has become common practice to reconstruct the evolutionary relationships of species by utilizing large amounts of phylogenetically informative genomic data. Such data can include ubiquitous or conserved genes, individual orthologous and paralogous gene phylogenies, shared genomic content, or compositional signatures of genomes (Fig. 1). Methods of phylogenomic analysis, in other words phylogenetic reconstruction of species using genome-scale data, have

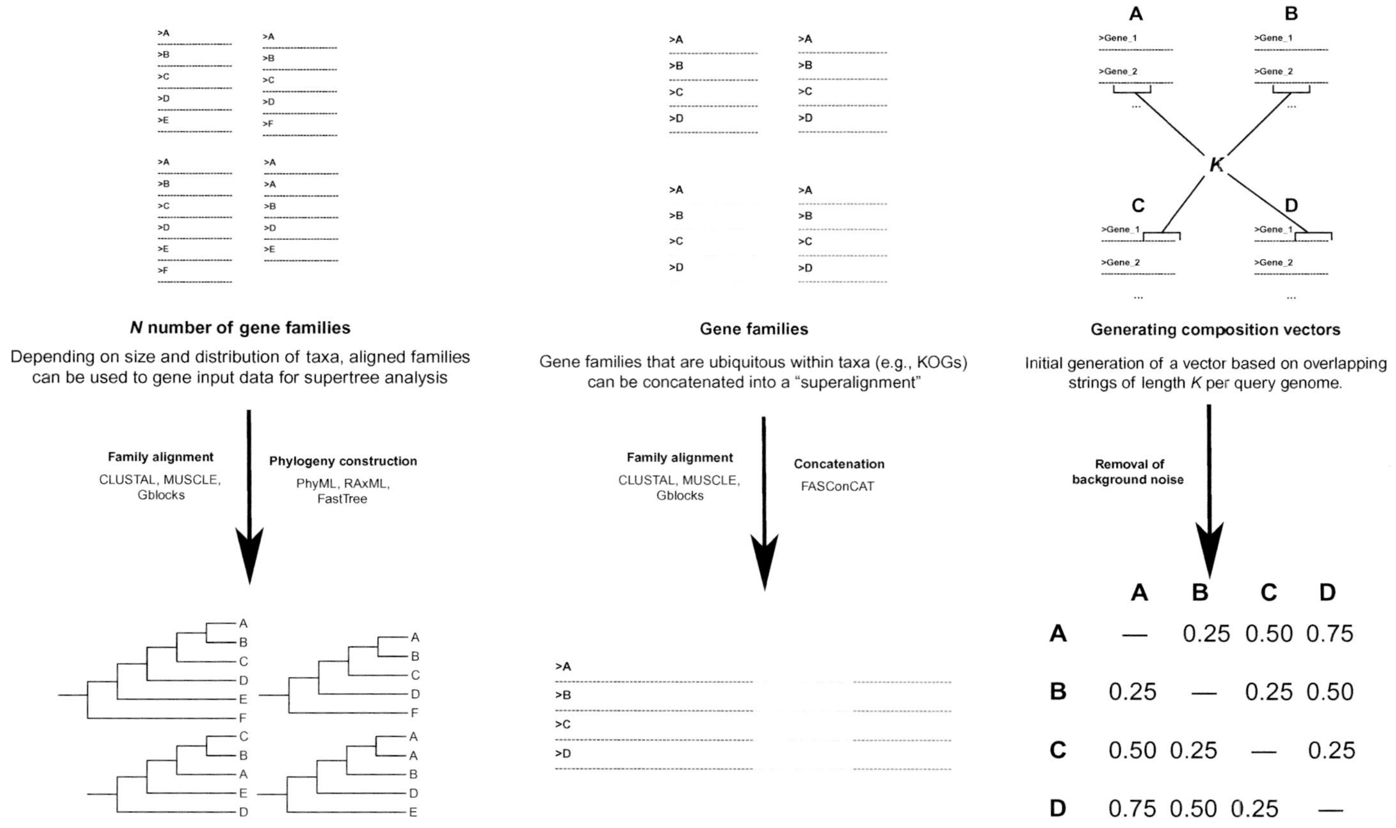

Fig. 1 See figure legend on next page

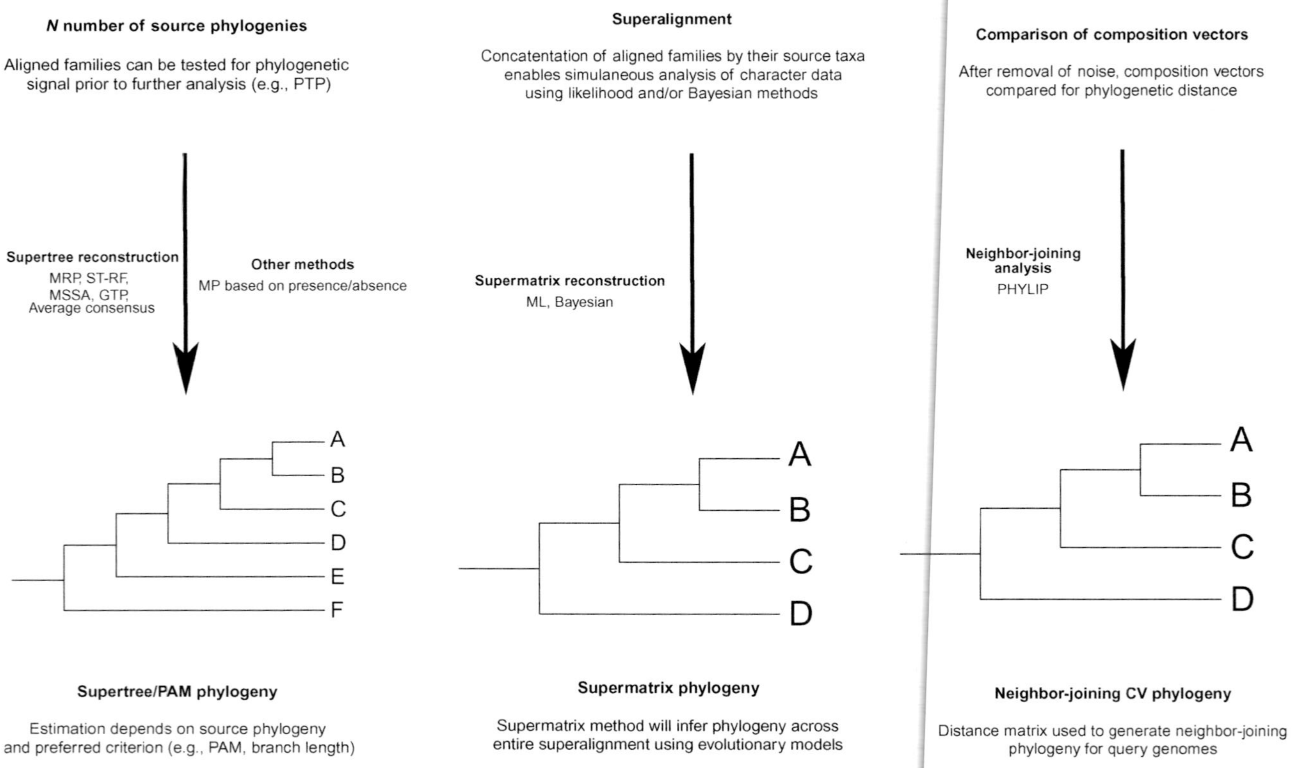

Fig. 1 Illustrative comparison of common phylogenomic methods. *Left*: supertree and presence–absence methods, *middle*: supermatrix methods, and *right*: composition vector methods.

all been developed for each of these types of potential phylogenetic marker and each comes with their advantages and disadvantages. Many phylogenomic analyses of the fungal kingdom have been carried out using these methods.

In this section, we review in turn each established approach to phylogenomic reconstruction from molecular data present in the literature and review each approach's application in previous fungal phylogenomic analyses. To demonstrate both the application and accuracy of all of these approaches to reconstructing phylogeny from genome-scale data, we have conducted our own phylogenomic analyses of the fungal kingdom using each method (Fig. 2). We have carried out such analyses to take advantage of both the greater coverage of the fungal kingdom arising from the 1000 Fungal Genomes Project and the advances in phylogenetic methodologies in the years following many of the analyses that we review below. In total, 84 fungal genomes from across 8 phyla (Table 1) were selected for our

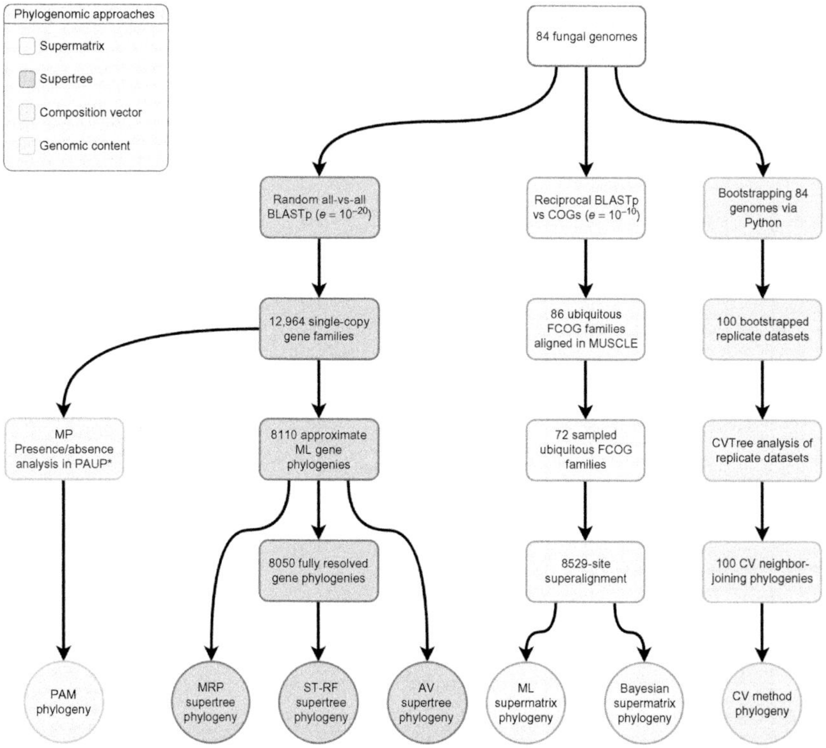

Fig. 2 Summary of the methodology of all 7 phylogenomic analyses of 84 fungal species carried out in this review.

Table 1 List of Species Used in Phylogenomic Analysis

Species	Phylum	Subphylum	Class	MycoCosm ID
Bipolaris maydis	Ascomycota	Pezizomycotina	Dothideomycetes	CocheC4_1
Cenococcum geophilum	Ascomycota	Pezizomycotina	Dothideomycetes	Cenge3
Hysterium pulicare	Ascomycota	Pezizomycotina	Dothideomycetes	Hyspu1_1
Zymoseptoria tritici	Ascomycota	Pezizomycotina	Dothideomycetes	Mycgr3
Aspergillus niger	Ascomycota	Pezizomycotina	Eurotiomycetes	Aspni7
Coccidioides immitis	Ascomycota	Pezizomycotina	Eurotiomycetes	Cocim1
Endocarpon pusillum	Ascomycota	Pezizomycotina	Eurotiomycetes	EndpusZ1
Exophiala dermatitidis	Ascomycota	Pezizomycotina	Eurotiomycetes	Exode1
Phaeomoniella chlamydospora	Ascomycota	Pezizomycotina	Eurotiomycetes	Phach1
Blumeria graminis	Ascomycota	Pezizomycotina	Leotiomycetes	Blugr1
Botrytis cinerea	Ascomycota	Pezizomycotina	Leotiomycetes	Botci1
Arthrobotrys oligospora	Ascomycota	Pezizomycotina	Orbiliomycetes	Artol1
Dactylellina haptotyla	Ascomycota	Pezizomycotina	Orbiliomycetes	Monha1
Pyronema omphalodes	Ascomycota	Pezizomycotina	Pezizomycetes	Pyrco1
Tuber melanosporum	Ascomycota	Pezizomycotina	Pezizomycetes	Tubme1
Coniochaeta ligniaria	Ascomycota	Pezizomycotina	Sordariomycetes	Conli1
Hypoxylon sp. *EC38*	Ascomycota	Pezizomycotina	Sordariomycetes	HypEC38_3

Magnaporthe grisea	Ascomycota	Pezizomycotina	Sordariomycetes	Maggr1
Neurospora crassa	Ascomycota	Pezizomycotina	Sordariomycetes	Neucr_trp3_1
Ophiostoma piceae	Ascomycota	Pezizomycotina	Sordariomycetes	Ophpic1
Phaeoacremonium minimum	Ascomycota	Pezizomycotina	Sordariomycetes	Phaal1
Xylona heveae	Ascomycota	Pezizomycotina	Xylonomycetes	Xylhe1
Candida albicans	Ascomycota	Saccharomycotina	Saccharomycetes	Canalb1
Lipomyces starkeyi	Ascomycota	Saccharomycotina	Saccharomycetes	Lipst1_1
Ogataea polymorpha	Ascomycota	Saccharomycotina	Saccharomycetes	Hanpo2
Saccharomyces cerevisiae	Ascomycota	Saccharomycotina	Saccharomycetes	SacceM3707_1
Saitoella complicata	Ascomycota	Taphrinomycotina	N/A	Saico1
Pneumocystis jirovecii	Ascomycota	Taphrinomycotina	Pneumocystidomycetes	Pneji1
Schizosaccharomyces cryophilus	Ascomycota	Taphrinomycotina	Schizosaccharomycetes	Schcy1
Schizosaccharomyces japonicus	Ascomycota	Taphrinomycotina	Schizosaccharomycetes	Schja1
Schizosaccharomyces octosporus	Ascomycota	Taphrinomycotina	Schizosaccharomycetes	Schoc1
Schizosaccharomyces pombe	Ascomycota	Taphrinomycotina	Schizosaccharomycetes	Schpo1
Protomyces lactucaedebilis	Ascomycota	Taphrinomycotina	Taphrinomycetes	Prola1
Taphrina deformans	Ascomycota	Taphrinomycotina	Taphrinomycetes	Tapde1_1
Agaricus bisporus	Basidiomycota	Agaricomycotina	Agaricomycetes	Agabi_varbur_1

Continued

Table 1 List of Species Used in Phylogenomic Analysis—cont'd

Species	Phylum	Subphylum	Class	MycoCosm ID
Auricularia subglabra	Basidiomycota	Agaricomycotina	Agaricomycetes	Aurde3_1
Botryobasidium botryosum	Basidiomycota	Agaricomycotina	Agaricomycetes	Botbo1
Fibulorhizoctonia	Basidiomycota	Agaricomycotina	Agaricomycetes	Fibsp1
Gloeophyllum trabeum	Basidiomycota	Agaricomycotina	Agaricomycetes	Glotr1_1
Heterobasidion annosum	Basidiomycota	Agaricomycotina	Agaricomycetes	Hetan2
Jaapia argillacea	Basidiomycota	Agaricomycotina	Agaricomycetes	Jaaar1
Punctularia strigosozonata	Basidiomycota	Agaricomycotina	Agaricomycetes	Punst1
Serendipita indica	Basidiomycota	Agaricomycotina	Agaricomycetes	Pirin1
Serpula lacrymans	Basidiomycota	Agaricomycotina	Agaricomycetes	SerlaS7_3_2
Sistotremastrum suecicum	Basidiomycota	Agaricomycotina	Agaricomycetes	Sissu1
Sphaerobolus stellatus	Basidiomycota	Agaricomycotina	Agaricomycetes	Sphst1
Wolfiporia cocos	Basidiomycota	Agaricomycotina	Agaricomycetes	Wolco1
Calocera cornea	Basidiomycota	Agaricomycotina	Dacrymycetes	Calco1
Dacryopinax primogenitus	Basidiomycota	Agaricomycotina	Dacrymycetes	Dacsp1
Basidioascus undulatus	Basidiomycota	Agaricomycotina	Geminibasidiomycetes	Basun1
Cryptococcus neoformans	Basidiomycota	Agaricomycotina	Tremellomycetes	Cryne_JEC21_1
Cutaneotrichosporon oleaginosus	Basidiomycota	Agaricomycotina	Tremellomycetes	Triol1

Wallemia sebi	Basidiomycota	Agaricomycotina	Wallemiomycetes	Walse1
Leucosporidium creatinivorum	Basidiomycota	Pucciniomycotina	Microbotryomycetes	Leucr1
Microbotryum lychnidis-dioicae	Basidiomycota	Pucciniomycotina	Microbotryomycetes	Micld1
Rhodotorula graminis	Basidiomycota	Pucciniomycotina	Microbotryomycetes	Rhoba1_1
Mixia osmundae	Basidiomycota	Pucciniomycotina	Mixiomycetes	Mixos1
Puccinia graminis	Basidiomycota	Pucciniomycotina	Pucciniomycetes	Pucgr2
Tilletiaria anomala	Basidiomycota	Ustilaginomycotina	Exobasidiomycetes	Tilan2
Malassezia sympodialis	Basidiomycota	Ustilaginomycotina	Malasseziomycetes	Malsy1_1
Sporisorium reilianum	Basidiomycota	Ustilaginomycotina	Ustilaginomycetes	Spore1
Ustilago maydis	Basidiomycota	Ustilaginomycotina	Ustilaginomycetes	Ustma1
Allomyces macrogynus	Blastocladiomycota	N/A	Blastocladiomycetes	GCA_000151295.1
Catenaria anguillulae	Blastocladiomycota	N/A	Blastocladiomycetes	Catan2
Batrachochytrium dendrobatidis	Chytridiomycota	N/A	Chytridiomycetes	GCA_000149865.1
Rhizoclosmatium globosum	Chytridiomycota	N/A	Chytridiomycetes	Rhihy1
Spizellomyces punctatus	Chytridiomycota	N/A	Chytridiomycetes	Spipu1
Gonapodya prolifera	Chytridiomycota	N/A	Monoblepharidomycetes	Ganpr1
Rozella allomycis	Cryptomycota	N/A	N/A	Rozal1_1
Rhizophagus irregularis	Mucoromycota	Glomeromycotina	Glomeromycetes	Gloin1

Continued

Table 1 List of Species Used in Phylogenomic Analysis—cont'd

Species	Phylum	Subphylum	Class	MycoCosm ID
Mortierella elongate	Mucoromycota	Mortierellomycotina	N/A	Morel2
Phycomyces blakesleeanus	Mucoromycota	Mucoromycotina	N/A	Phybl2
Rhizopus oryzae	Mucoromycota	Mucoromycotina	N/A	Rhior3
Umbelopsis ramanniana	Mucoromycota	Mucoromycotina	N/A	Umbra1
Anaeromyces robustus	Neocallimastigomycota	N/A	Neocallimastigomycetes	Anasp1
Neocallimastix californiae	Neocallimastigomycota	N/A	Neocallimastigomycetes	Neosp1
Orpinomyces sp. *C1A*	Neocallimastigomycota	N/A	Neocallimastigomycetes	Orpsp1_1
Piromyces finnis	Neocallimastigomycota	N/A	Neocallimastigomycetes	Pirfi3
Basidiobolus meristosporus	Zoopagomycota	Entomophthoromycotina	Basidiobolomycetes	Basme2finSC
Conidiobolus thromboides	Zoopagomycota	Entomophthoromycotina	Entomophthoromycetes	Conth1
Coemansia reversa	Zoopagomycota	Kickxellomycotina	N/A	Coere1
Linderina pennispora	Zoopagomycota	Kickxellomycotina	N/A	Linpe1
Martensiomyces pterosporus	Zoopagomycota	Kickxellomycotina	N/A	Marpt1
Ramicandelaber brevisporus	Zoopagomycota	Kickxellomycotina	N/A	Rambr1

Genome data from MycoCosm (http://genome.jgi.doe.gov/programs/fungi/index.jsf) has previously been published and MycoCosm ID is given in final column. GENBANK accessions given for *Allomyces macrogynus* and *Batrachochytrium dendrobatidis*.

large-scale phylogenomic reconstructions of the fungal kingdom. Where possible, we included at least one published representative genome from each order covered by the 1000 Fungal Genomes Project in our dataset. All genomic data were ultimately obtained from the JGI's MycoCosm facility (Grigoriev et al., 2014). Our analyses include the first phylogenomic reconstruction of fungi carried out using a Bayesian supertree approach, and the first large-scale gene content approach to fungal phylogenomics that has been conducted in at least a decade. We discuss, in brief, the methodology and the results of each reconstruction and their accuracy (or otherwise) in reconstructing the phylogeny of both basal fungal lineages and the Dikarya. In Section 3, we discuss the overall phylogeny of the fungal kingdom arising from our analyses and compare with previous literature.

2.1 Supermatrix Phylogenomic Analysis of Fungi

The two best-established alignment-based approaches to reconstructing phylogeny on a genomic scale are the "supertree" method, in which a consensus phylogeny is derived from many individual gene phylogenies (discussed in Section 2.2), and the "supermatrix" method which we discuss here. Supermatrix method phylogeny is the simultaneous analysis of a phylogenetic matrix, also referred to as a "superalignment," constructed from all available character data from a given set of taxa. Generally supermatrices are constructed from concatenating highly conserved markers (e.g., rRNA genes, mitochondrial markers) for small-scale multigene phylogenies, and from homologs of conserved orthologous genes (known as COGs, or sometimes KOGs in eukaryotes) for genome-scale phylogenies (de Queiroz & Gatesy, 2007; Koonin et al., 2004). Supermatrix approaches can also incorporate statistically powerful maximum-likelihood and Bayesian methods of phylogenomic analysis. Described in simple terms, given an alignment of sequences and a suitable evolutionary model, maximum-likelihood phylogenetic analysis examines all possible trees by their possible parameters (e.g., topology, site support, branch length) and returns the most likely phylogenetic tree for the alignment (Page & Holmes, 1998). Similarly, Bayesian analysis incorporates phylogenetic likelihoods to calculate the posterior probability of a phylogeny, which is the probability of that phylogeny given the alignment data (Huelsenbeck, Ronquist, Nielsen, & Bollback, 2001).

One advantage of a supermatrix approach to phylogenomic analysis over a supertree approach is the retention of character evidence in analysis in the former approach; most supertree methods can be considered estimations using individual trees based on summarized character data, at least two steps removed from any actual sequence data, whereas a supermatrix approach entails direct analysis of combined character data (Creevey & McInerney, 2009; de Queiroz & Gatesy, 2007). Supermatrix methods also have the potential to resolve deep branches and reveal so-called hidden supports within phylogenies that supertree methods may overlook (de Queiroz & Gatesy, 2007). However, supermatrix analysis requires ubiquitous sequences from all taxa being investigated, which restricts the available pool of character data and may overlook miss important phylogenetic information from phylogenies with gene deletion, gene duplication, or horizontal gene transfer events that supertree methods can utilize (Creevey & McInerney, 2009). Compositional biases may also have an effect on supermatrix methods, though phylogenetic models have been developed which can ameliorate errors that these biases may induce during analysis (Lartillot, Brinkmann, & Philippe, 2007; Lartillot & Philippe, 2004). In practice, many phylogenomic analyses utilize both supertree and supermatrix methods in tandem to reconstruct phylogeny in a "total evidence" approach (Kluge, 1989) and will often comment on the taxonomic congruence (or otherwise) of the resulting phylogenies.

2.1.1 Fungal Phylogenomics Using the Supermatrix Approach

Supermatrix analysis has been widely used in fungal phylogenomics. One of the initial comparisons of individual gene phylogenies with genome-scale species phylogenies used a maximum-parsimony analysis among other methods to reconstruct the phylogeny of seven *Saccharomyces* species and *C. albicans*; the authors showed that incongruence among individual gene phylogenies could be resolved with high support using a concatenated alignment (Rokas, Williams, King, & Carroll, 2003). Initial genome-based phylogenies of Ascomycota using 17 genomes and both supertree and supermatrix methods resolved both Pezizomycotina and Saccharomycotina, as well as placing *S. pombe* as an early-diverging branch within Ascomycota (Robbertse, Reeves, Schoch, & Spatafora, 2006). Robbertse et al. (2006) generated a superalignment of 195,664 amino acid characters in length derived from 781 gene families, which produced identical topologies under both neighbor-joining and maximum-likelihood criteria. The first large-scale phylogenomic analysis of fungi used a 67,101-character superalignment derived from 531 eukaryotic COGs found in 21 fungal genomes, all of which were sampled from Ascomycota and Basidiomycota (Kuramae, Robert, Snel, Weiß, & Boekhout, 2006). A more extensive phylogenomic analysis from the same year produced 2 highly congruent genome phylogenies from 42 fungal genomes using 2 methods: a matrix representation with parsimony (MRP) supertree derived from 4805 single-copy gene families (which we discuss in greater detail in Section 2.2.1), and a 38,000-character superalignment derived from 153 ubiquitous gene families (Fitzpatrick, Logue, Stajich, & Butler, 2006).

Most of the relationships resolved in Fitzpatrick et al. (2006) were further supported by a 31,123-character superalignment from 69 proteins conserved in up to 60 fungal genomes generated by Marcet-Houben, Marceddu, and Gabaldón (2009), although they found a large degree of topological conflict within a 21-species Saccharomycotina clade (Marcet-Houben & Gabaldón, 2009; Marcet-Houben et al., 2009). A later follow-up analysis to Fitzpatrick et al. (2006) by Medina, Jones, and Fitzpatrick (2011) reconstructed the phylogeny of 103 fungal species by performing Bayesian analysis on a 12,267-site superalignment derived from 87 gene families with a phyletic range of over half of their dataset, in addition to supertree analysis (Medina et al., 2011). A recent phylogenomic analysis of 46 fungal genomes, including 25 zygomycetes species, reconstructed the phylogeny of the early-diverging fungal lineages using a 60,383-character superalignment (Spatafora et al., 2016). Another recent phylogenomic analysis used a 28,807-site superalignment derived from 136 gene families from 40 eukaryotic genomes to investigate the evolution of sourcing carbon from algal and plant pectin in early-diverging fungi (Chang et al., 2015). Finally, a comparison of the dynamics of genome evolution between 28 Dikarya species and cyanobacteria used a supermatrix phylogeny of 24,514 amino acid characters from 529 fungal gene families with large phyletic range as a scaffold to infer rates of intrakingdom HGT within Dikarya that were near similar to those within cyanobacteria (Szöllősi et al., 2015).

To extend the analyses described above, we carried out supermatrix analysis using maximum-likelihood and Bayesian methods on a superalignment constructed from orthologous genes conserved throughout 84 species from 8 phyla within the fungal kingdom.

2.1.2 Phylogenomic Reconstruction of 84 Fungal Species From 72 Ubiquitous Gene Families Using Maximum-Likelihood and Bayesian Supermatrix Analysis

A reciprocal BLASTp search was carried out between all protein sequences from our 84-genome dataset and 458 core orthologous genes (COGs) from *S. cerevisiae* obtained from the CEGMA dataset, with an *e*-value cutoff of 10^{-10} (Camacho et al., 2009; Parra, Bradnam, & Korf, 2007), from which

456 COG families were retrieved (2 *S. cerevisiae* COGs did not return any homologs). From these, 86 ubiquitous fungal COG families, i.e., families containing a homolog from all 84 input species, were identified. Each ubiquitous fungal COG family was aligned in MUSCLE, and conserved regions of each alignment were sampled in Gblocks using the default parameters (Castresana, 2000; Edgar, 2004). Fourteen alignments did not retain any character data after Gblocks filtering and were removed from further analysis. The remaining 72 sampled alignments were concatenated into a superalignment of 8529 aligned positions using the Perl program FASconCat (Kück & Meusemann, 2010). This superalignment was bootstrapped 100 times using Seqboot (Felsenstein, 1989), and maximum-likelihood phylogenetic trees were generated for each individual replicate using PhyML with an LG + I + G amino acid substitution model as selected by ProtTest (Darriba, Taboada, Doallo, & Posada, 2011; Guindon et al., 2010). A consensus phylogeny was generated from all 100 individual replicate phylogenies using CLANN (Creevey & McInerney, 2005). Markov Chain Monte Carlo (MCMC) Bayesian phylogenetic inference was carried out on the same superalignment using PhyloBayes MPI with the default CAT + GTR amino acid substitution model, running 2 chains for 1000,000 iterations and sampling every 100 iterations (Lartillot & Philippe, 2004; Lartillot, Rodrigue, Stubbs, & Richer, 2013). Both chains were judged to have converged after 100,000 iterations and a consensus Bayesian phylogeny was generated with a burn-in of 1000 trees. Both supermatrix phylogenies were visualized using the Interactive Tree of Life (iTOL) website and annotated according to the NCBI's taxonomy database (Federhen, 2012; Letunic & Bork, 2016). Both supermatrix phylogenies were rooted at *Rozella allomycis*, which is the most basal species in evolutionary terms in our dataset (Jones, Forn, et al., 2011) and is the root for all the phylogenies we present hereafter (Figs. 3 and 4).

2.1.3 Supermatrix Analyses of 84 Fungal Species Accurately Reconstructs the Fungal Kingdom

We reconstructed the phylogeny of the fungal kingdom by generating a superalignment of 72 concatenated ubiquitous gene families and performing ML analysis using PhyML and Bayesian analysis using a parallelized version of PhyloBayes. Both ML and Bayesian analysis reconstruct the phylogeny of our fungal dataset with a high degree of accuracy relative to other kingdom phylogenies in the literature and in most cases recover the eight fungal phyla in our dataset (Figs. 3 and 4). Here, we discuss the results of both our analyses with regard to the basal fungal lineages, and the two Dikarya

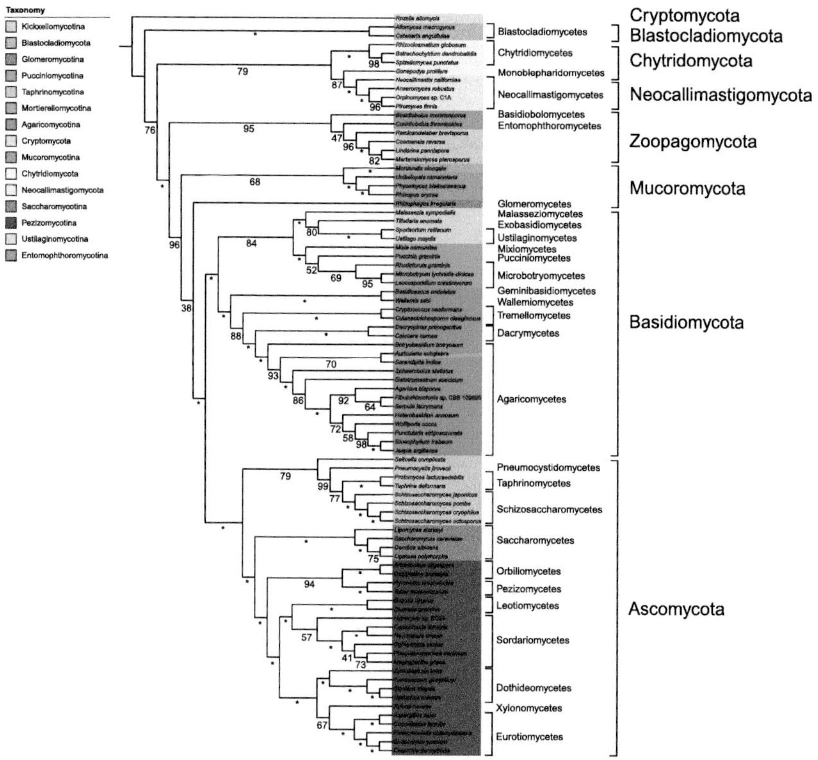

Fig. 3 ML phylogeny of 84 fungal species from a 8529-character superalignment derived from 72 ubiquitous fungal COG families sampled in Gblocks using PhyML with a LG+I+G model. Bootstrap supports shown on branches. Maximum bootstrap support designated with an *asterisk* (*).

phyla. Further in this chapter, we use these supermatrix analyses as the point of comparison for our other phylogenomic methods.

2.1.3.1 Basal Fungi

In our ML supermatrix phylogeny, Blastocladiomycota emerge as the earliest-diverging fungi with maximum bootstrap support (henceforth abbreviated to BP) after rooting at *R. allomycis* (Fig. 3). Chytridiomycota and Neocallimastigomycota are placed as sister clades with 79% BP, surprisingly the Chytridiomycota species *Gonapodya prolifera* branches as sister to Neocallimastigomycota (87% BP). The Chytridiomycetes class is monophyletic with maximum bootstrap support, as is the Neocallimastigomycetes class (Fig. 3). The former zygomycetes phylum Zoopagomycota is strongly supported as a monophyletic clade with 95% BP (Fig. 3). The other former

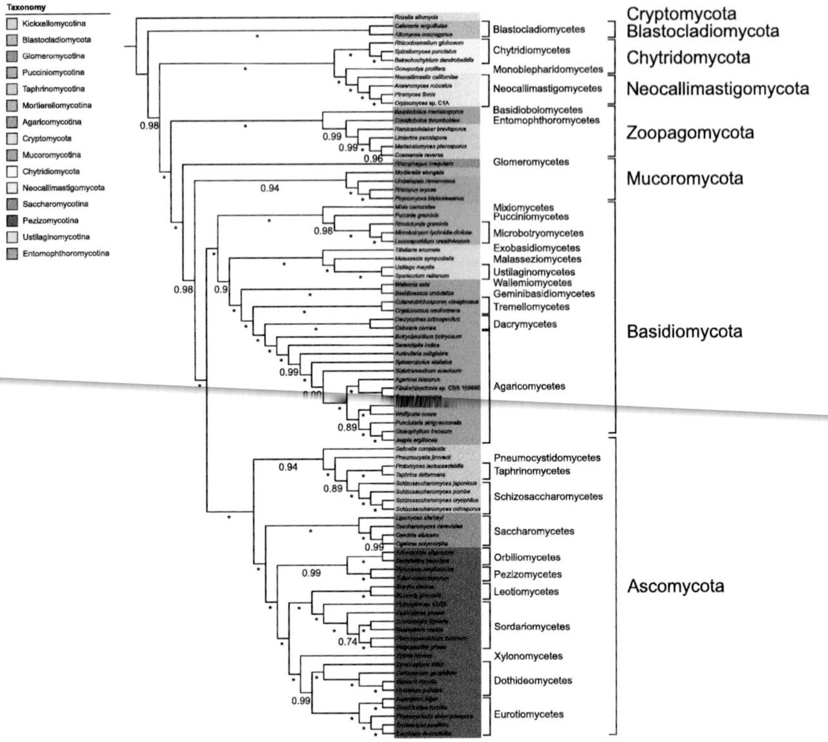

Fig. 4 Bayesian phylogeny of 84 fungal species from an 8529-character superalignment derived from 72 ubiquitous fungal COG families sampled in Gblocks using PhyloBayes MPI with a CAT+GTR model. Posterior probabilities shown on branches with a burn-in of 1000 trees. Maximum posterior probability support designated with an *asterisk* (*).

zygomycetes phylum Mucoromycota is paraphyletic and split between a clade containing the Mucoromycotina and Mortierellomycotina species *Mortierella elongata* that has 68% BP, and the Glomeromycotina species *Rhizophagus irregularis* branching basal to Dikarya with lower support (38% BP). The placement of Mucoromycota as the closest phyla to Dikarya has near-maximum support (96% BP) which matches other analysis (Spatafora et al., 2016).

The Bayesian supermatrix phylogeny is in near-total agreement with the ML phylogeny in resolving the relationships of the basal fungi in our dataset (Fig. 4). The relationship between Chytridiomycota and Neocallimastigomycota in the Bayesian phylogeny mirrors that seen in the ML phylogeny, with all branches receiving maximum support as monophyletic with a Bayesian posterior probability (henceforth abbreviated to PP) equal to 1 (Fig. 4). The

Zoopagomycota are monophyletic with full support, with a topology matching the ML phylogeny with strong branch support throughout (Fig. 4). There is also a close association between the three Mucoromycota subphyla: Glomeromycota branches earlier in the Bayesian phylogeny than in the ML phylogeny, which receives maximum support in the Bayesian phylogeny, and the sister relationship between Mucoromycotina and *M. elongata* receives strong support (0.94 PP) in the Bayesian phylogeny (Fig. 4). Both the ML and Bayesian place the Mucoromycota as the basal phylum that is most closely related to Dikarya (Fig. 4).

2.1.3.2 Basidiomycota

In the ML phylogeny, the three subphyla within Basidiomycota are fully resolved with maximum BP, with 84% BP for the placement of Ustilagomycotina and Pucciniomycotina as sister clades (Fig. 3). *Basidioascus undulatus* and *Wallemia sebi* branch at the base of Agaricomycotina with maximum BP, while the other classes with the subphyla are all fully supported. There is also high support (88% BP) for the placement of Tremellomycetes as sister to Dacrymycetes and Agaricomycetes (Fig. 3). The Tremellomycetes, including *Cryptococcus neoformans*, are monophyletic. The Dacrymycetes are also monophyletic with maximum BP. The forest saprophyte *Botryobasidium botryosum* is placed at the base of the Agaricomycetes, which has some strong intraclade resolution with weaker branch supports toward the tips of the clade (Fig. 3). *Malassezia sympodialis*, a commensal fungi of humans and animals, is placed at the base of the Ustilagomycotina. The Exobasidiomycetes species *Tilletiaria anomala* branches between *M. sympodialis* and the Ustilagomycetes. The Pucciniomycotina are monophyletic with full support (Fig. 3). The most highly represented Pucciniomycotina class, the Microbotryomycetes, are monophyletic with 69% BP (Fig. 3).

The Bayesian phylogeny reflects the ML phylogeny in its resolution of the Basidiomycota as monophyletic with full support (Fig. 4). The phylogeny places Pucciniomycotina at the base of the phylum with maximum support. Resolution of branches within Pucciniomycotina is substantially improved under Bayesian phylogeny (Fig. 4). There is high support (0.9 PP) for a sister relationship between Ustilagomycotina and Agaricomycotina (Fig. 4). The Exobasidiomycetes species *T. anomala* now branches at the base of the Ustilagomycotina, which is resolved with maximum PP. There is maximum support for the placement of *M. sympodialis* as sister to the Ustilagomycetes, which are monophyletic (Fig. 4). As in the ML phylogeny, *B. undulatus* and *W. sebi* branch at the base of Agaricomycotina with maximum support, while the other classes with the subphyla all have maximum support and have similar topology under Bayesian analysis. There is a large improvement in the support of branches in the Agaricomycotina in the Bayesian phylogeny relative to the ML phylogeny (Fig. 4).

2.1.3.3 Ascomycota

Both the ML and Bayesian supermatrix phylogenies display near-identical topologies for the Ascomycota, and Bayesian analysis shows stronger support for some branches toward the tips of the phylogeny than the ML phylogeny does (Figs. 3 and 4). The three subphyla within Ascomycota are fully resolved, with maximum BP support for Saccharomycotina and Pezizomycotina and 79% BP for the monophyly of Taphrinomycotina in the ML phylogeny (contrast with 0.94 PP for the monophyly of Taphrinomycotina in the Bayesian phylogeny; Figs. 3 and 4). The placement of Taphrinomycotina as an ancestral clade within Ascomycota is fully supported, and within Taphrinomycotina, there is high support (77% BP/0.89 PP) for a sister relationship between Schizosaccharomycetes and Taphrinomycetes. Six of the seven classes within Pezizomycotina in our dataset with two or more representatives (i.e., all bar Xylonomycetes) are monophyletic, most of which receive maximum BP and/or PP support. Many of the relationships between classes are also well supported in both phylogenies, with lower support (67% BP) for a sister relationship between the Xylonomycetes species *Xylona heveae* and the Eurotiomycetes class in the ML phylogeny; in the Bayesian phylogeny *X. heveae* branches sister to a clade containing Dothideomycetes and Eurotiomycetes with maximum PP support (Figs. 3 and 4). The Dothideomycetes are monophyletic in both phylogenies and branch into two clades with high support under both ML and Bayesian reconstruction (Figs. 3 and 4). The Orbiliomycetes and Pezizomycetes are placed as the most basal Pezizomycotina classes, with strong support (94% BP/0.99 BP) for a sister relationship (Figs. 3 and 4). The Leotiomycetes and Sordariomycetes are also placed as a sister clades with maximum support in both phylogenies. The major difference in the resolution of the Sordariomycetes between the supermatrix phylogenies is the stronger branch supports within the order under Bayesian analysis (Figs. 3 and 4).

2.2 Parsimony Supertree Phylogenomic Analysis of Fungi

The most common supertree methods for reconstructing genome phylogenies are grounded in parsimony methods, in which changes to character states (i.e., evolutionary events such as presence of a given taxon in a tree or even a tree branch) are calculated and phylogeny is reconstructed using as little state changes as possible. The first supertree construction method to see widespread use in large-scale phylogenetic and phylogenomic analysis was the MRP method. MRP, which was developed independently by Baum (1992) and Ragan (1992), enables the use of source phylogenies with overlapping or missing taxa in generating a consensus phylogeny (Baum, 1992; Ragan, 1992). The method generates a matrix (referred to as a Baum–Ragan matrix) where each column represents one internal branch in each given source phylogeny such that the number of columns within the matrix is equal to the number of internal branches across all source phylogenies, and assigns a score of 1 to taxa from a given source phylogeny *P* which are present in the clade defined by internal branch *A*, 0 to taxa present in *P* but not within the clade defined by *A*, and ? to taxa that are not present in *P* (Creevey & McInerney, 2009). The Baum–Ragan matrix is then subject to parsimony analysis, with equal weighting given to each source phylogeny, and reconstructs the supertree phylogeny with the minimum of evolutionary changes required which includes all taxa represented across all source phylogenies. Similar parsimony methods, most notably gene tree parsimony (Slowinski & Page, 1999), extend MRP to include source phylogenies containing duplicated taxa; however, we do not cover such methods in this subsection. Parsimony-based supertree methods like MRP are generally quite accurate in reconstructing phylogeny for large datasets, although some issues have been observed (which we discuss in Section 2.3).

2.2.1 Matrix Representation With Parsimony Analysis in Fungal Phylogenomics

Many phylogenomic analyses of fungi have used parsimony methods. The first large-scale phylogenomic analysis of fungi to use MRP in supertree reconstruction was by Fitzpatrick et al. (2006), who carried out a phylogenomic reconstruction of fungi using 42 genomes from Dikarya and the zygomycete *Rhizopus oryzae* using both supertree and supermatrix methods (Fitzpatrick et al., 2006). Using a random BLASTp approach to identify homologous gene families, where randomly selected query sequences are sequentially searched against a full database and then both query sequences and homologs (if any) are sequentially removed from the database, Fitzpatrick et al. (2006) utilized 4805 single-copy gene phylogenies for MRP supertree reconstruction using the software package CLANN (Creevey & McInerney, 2005, 2009). The MRP phylogeny resolved the Pezizomycotina and Saccharomycotina subphyla within Ascomycota and inferred the Sordariomycetes and the Leotiomycetes as sister classes within Pezizomycotina. The MRP phylogeny also resolved two major clades within the Saccharomycotina: a monophyletic clade of species that translate the codon CTG as serine instead of leucine (the "CTG clade"), and a grouping of species that have undergone whole genome duplication (the "WGD clade") and their closest relatives. The authors compared the MRP phylogeny with a maximum-likelihood supermatrix phylogeny reconstructed using 38,000 characters from 153 gene families (as detailed in the previous subsection); both were highly congruent with conflict only in the placement of the sole Dothideomycetes species represented, *Stagonospora nodourum*. The authors also complemented their MRP phylogeny with two other supertree methods implemented in CLANN: a most similar supertree analysis (MSSA) method phylogeny which was identical to the MRP supertree (Creevey et al., 2004) and an average consensus (AV) method phylogeny based on branch lengths (Lapointe & Cucumel, 1997), which the authors believed to suffer from long-branch attraction in the erroneous placement of some species within the WGD clade in Saccharomycotina (Fitzpatrick et al., 2006). A follow-up analysis to Fitzpatrick et al. (2006) by Medina et al. (2011) using 103 genomes was extended to include multicopy gene families using the gene tree parsimony (GTP) method and successfully resolved the major groupings within the fungal kingdom (Medina et al., 2011). Using both a random BLASTp and a Markov Clustering Algorithm (MCL)-based approach with varying inflation values to identify orthologous gene families, the authors used as many as 30,012 single and paralogous gene phylogenies as input for supertree reconstruction.

As a follow-up to the supertree reconstructions of the fungal kingdom by Fitzpatrick et al. (2006) and Medina et al. (2011), we ran supertree analysis for 84 fungal species using MRP and AV methods and source phylogenies identified via a random BLASTp approach described later.

2.2.2 Phylogenomic Reconstruction of 84 Fungal Species From 8110 Source Phylogenies Using MRP and AV Supertree Methods

Following Fitzpatrick et al. (2006), families of homologous protein sequences within our 84-genome dataset were identified using BLASTp with an *e*-value cutoff of 10^{-20} by randomly selecting a query sequences from our database, finding all homologous sequences via BLASTp (Camacho et al., 2009), and removing the entire family from the database before reformatting and repeating. 12,964 single-copy gene families, which contained no more than one homolog from 4 or more taxa, were identified. Each single-copy gene family was aligned in MUSCLE, and conserved regions of each alignment were sampled using Gblocks with the default parameters (Castresana, 2000; Edgar, 2004). Sampled alignments were tested for phylogenetic signal using the PTP test as implemented in PAUP* with 100 replicates (Faith & Cranston, 1991; Swofford, 2002). 8110 sampled alignments which retained character data after Gblocks filtering and passed the PTP test were retained for phylogenomic reconstruction. 8110 approximately maximum-likelihood gene phylogenies were generated with FastTree, using the default JTT + CAT protein evolutionary model (Price, Dehal, & Arkin, 2010). All 8110 single-copy gene phylogenies were used to generate a matrix representation with parsimony (MRP) supertree using CLANN, with 100 bootstrap replicates (Creevey & McInerney, 2005). To complement the MRP supertree, an average consensus (AV) supertree was generated from the same input dataset in CLANN, with 100 bootstrap replicates. Both supertrees were visualized in iTOL and annotated according to the NCBI's taxonomy database. Both supertrees were rooted at *R. allomycis* (Figs. 5 and 6).

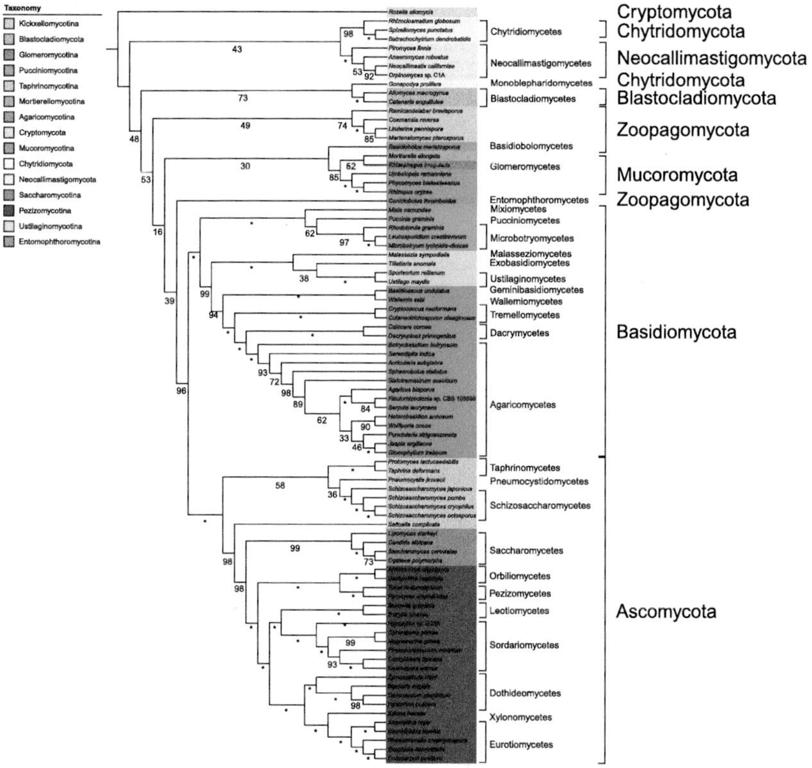

Fig. 5 Matrix representation with parsimony (MRP) phylogeny of 84 fungal species derived from 8110 source phylogenies. Bootstrap supports shown on branches. Maximum bootstrap support designated with an *asterisk* (*).

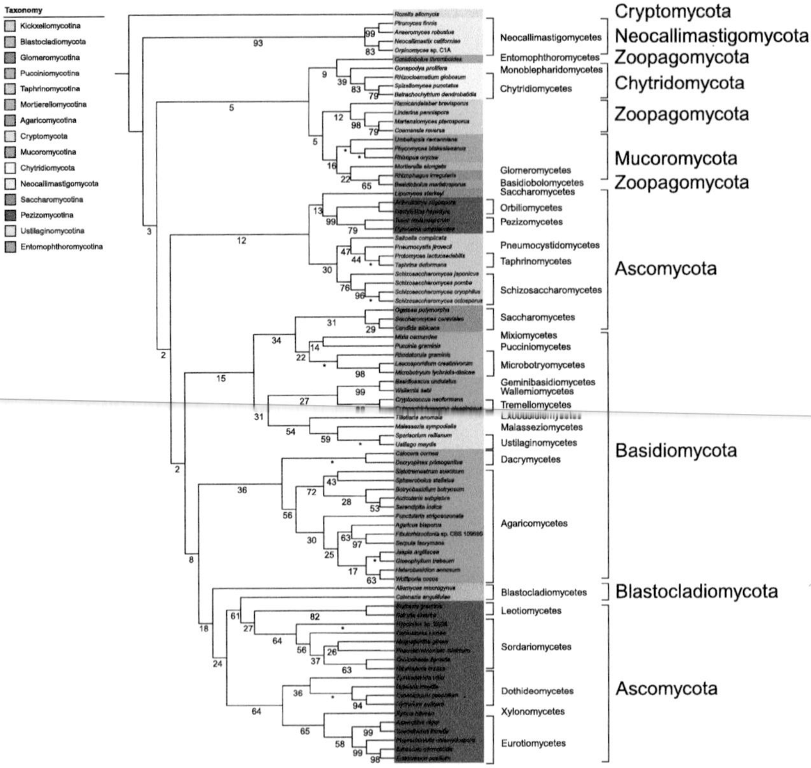

Fig. 6 Average consensus (AV) phylogeny of 84 fungal species derived from 8110 source phylogenies. Bootstrap supports shown on branches. Maximum bootstrap support designated with an *asterisk* (*).

2.2.3 MRP Phylogenomic Analysis of 84 Fungal Species Is Highly Congruent With Supermatrix Phylogenomic Analyses

We reconstructed the overall phylogeny of 8110 single-copy source phylogenies from our 84-genome dataset using an MRP supertree method analysis as implemented in CLANN (Fig. 5). MRP supertree reconstruction of the fungal kingdom recovers the majority of the eight fungal phyla in our dataset and is effective in resolving the Dikarya. However, there is poorer resolution of some of the basal phyla due to smaller taxon sampling perhaps having a negative influence on the distribution of basal taxa within our source phylogenies (we return to this in Section 3). Overall our MRP analysis is highly congruent with our supermatrix phylogenies detailed earlier, with some variation in the placement and resolution in some branches. We discuss the results of our MRP analysis for the basal fungal lineages and both Dikarya phyla and note some of the congruences and incongruences where noteworthy with our supermatrix phylogenies (Figs. 3–5).

2.2.3.1 Basal Fungi

After rooting at *R. allomycis*, the Neocallimastigomycota and Chytridiomycota (bar *G. prolifera*) emerge as the earliest-diverging fungal lineages. *G. prolifera* branches basal to the Blastocladiomycota with 73% BP (Fig. 5). This arrangement of the Neocallimastigomycota, Chytridiomycota, and Blastocladiomycota has poor support in general (43% BP for a sister relationship between Neocallimastigomycotina and 4 Chytridiomycota species); however with the exception of the aforementioned placement of *G. prolifera* the individual phyla receive maximum or near-maximum support as monophyletic (Fig. 5). Zoopagomycota is paraphyletic in our MRP phylogeny; a monophyletic Kicxellomycotina clade receives 74% BP support (Fig. 5), while as in the supermatrix phylogenies (Figs. 3 and 4) Entomophthoromycotina is paraphyletic. In our MRP analysis, *Basidiobolus meristosporus* branches at the base of Mucoromycota and *Conidiobolus thromboides* branches at the base of Dikarya, but those relationships are poorly supported (30% and 39% BP, respectively; Fig. 5). The Glomeromycotina species *R. irregularis* branches sister to the Mortierellomycota representative *M. elongata* with weak support (52% BP), but Murocomycota (the placement of Glomeromycotina, Mortierellomycota, and Mucoromycotina) receives higher support (85% BP). The monophyly of Mucoromycotina is also fully supported (Fig. 5). Overall many of the associations between basal phyla we observed in our supermatrix phylogenies are present in our MRP analysis as well; however, the overall placement of the basal fungal lineages varies between supermatrix and MRP analyses, such as the placement of Blastocladiomycota as a later-diverging clade than either Chytridiomycota or Neocallimastigomycota under MRP supertree analysis (Figs. 3–5).

2.2.3.2 Basidiomycota

The Basidiomycota are recovered with maximum support in our MRP phylogeny (Fig. 5). The Pucciniomycotina emerge as the most basal subphylum with maximum support, with *Mixia osmundae* branching at the base of the subphylum and *Puccinia graminis* placed as sister to the Microbotryomycetes (who are monophyletic with 97% BP). This reflects the topology of Pucciniomycotina seen in our supermatrix phylogenies (Figs. 3–5). The Ustilagomycotina and Agaricomycotina branch as sister subphyla with 99% BP and both are monophyletic; the former is fully supported at the branch level and the latter has 94% BP. *M. sympodialis* is placed at the base of Ustilagomycotina, reflecting the resolution of the Ustilagomycotina under ML supermatrix analysis (Figs. 3 and 5). In the Agaricomycotina, *W. sebi* and *B. undulatus* branch at the base of the subphylum with maximum support. The three larger classes from Agaricomycotina in our dataset (Agaricomycetes, Dacrymycetes, and Tremellomycetes) are all monophyletic and are recovered with maximum support (Fig. 5). The MRP phylogeny of the Basidiomycota is highly congruent overall with the supermatrix phylogenies, with comparable branch support (Figs. 3–5).

2.2.3.3 Ascomycota

Our MRP phylogeny supports the Ascomycota as a monophyletic group with maximum BP (Fig. 5). There is greater support along many deeper branches in the Ascomycota in our MRP phylogeny than in our ML supermatrix phylogeny and support is comparable with our Bayesian phylogeny; we ascribe this to a larger abundance of smaller source phylogenies containing closely related Ascomycotina species in our dataset (Figs. 3–5). Taphrinomycotina emerges as the earliest-diverging lineage but is paraphyletic; *Saitoella complicata* branches as an intermediate between Taphrinomycotina and a Saccharomycotina–Pezizomycotina clade with 98% BP, while the remaining members are monophyletic with weak support (58% BP). *Pneumocystis jirovecii* is placed as a sister taxon to Schizosaccharomycetes in our MRP analysis with weak support (36% BP); in the supermatrix phylogenies it was sister to Taphrinomycetes. The Taphrinomycetes and Schizosaccharomycetes themselves are monophyletic with maximum BP (Fig. 5). The Saccharomycotina are monophyletic with 99% BP (Fig. 5). The six larger classes (i.e., all bar Xylonomycetes) in our dataset from Pezizomycotina are all supported as monophyletic and receive maximum BP, with Pezizomycetes and Orbiliomycetes branching as the basal sister clades (Fig. 5). The MRP phylogeny mirrors Bayesian supermatrix reconstruction in placing a single origin for three classes (Xylonomycetes, Eurotiomycetes, and Dothideomycetes) with maximum support (Figs. 4 and 5). As in both supermatrix phylogenies, Dothideomycetes are split into two clades with high or maximum support. In the Sordariomycetes, MRP analysis reflects the ML supermatrix phylogeny in placing *Hypoxylon* sp. EC58 at the base of the class (Figs. 3 and 5). The MRP phylogeny of the Ascomycota is highly congruent with both of our supermatrix phylogenies with comparable branch supports, which is aided by the broad range of genomic data available for the phylum (Figs. 3–5).

2.2.4 *Average Consensus Phylogenomic Reconstruction of 84 Fungal Species Is Affected by Long-Branch Attraction Artifacts*

To complement our MRP phylogeny, we generated an average consensus (AV) method supertree phylogeny (Fig. 6) using the same set of input phylogenies as implemented in CLANN following Fitzpatrick et al. (2006). AV phylogeny infers phylogeny based on the branch lengths of source phylogenies, by computing the average value of the path-length matrices associated with said source phylogenies, and then using a least-squares method to find the source matrix closest to this average value (Lapointe & Cucumel, 1997). The tree that is associated with this source matrix is the average consensus phylogeny for the total set of source phylogenies, and the method is thought to work best with a set of source phylogenies of similar size (Lapointe & Cucumel, 1997). Our AV phylogeny was rooted at *R. allomycis* (Fig. 6). Given the results we obtained from our AV phylogeny, we believe that the method is susceptible to long-branch attraction (Felsenstein, 1978), as reported by Fitzpatrick et al. (2006). Long-branch attraction occurs when two very divergent taxa or clades with long branch lengths (i.e., many character changes occurring over time) are inferred as each other's closest relative due to convergent evolution of a given character (e.g., amino acid substitution), and is a common problem in parsimony and distance-based methods (Bergsten, 2005; Felsenstein, 1978). In the AV phylogeny, we recovered the two Blastocladiomycota species in our dataset within a large paraphyletic Pezizomycotina clade (Fig. 6). Additionally, the Ascomycota are paraphyletic: one clade containing two Pezizomycotina classes (Pezizomycetes and Orbiliomycetes), the Taphrinomycotina and the Saccharomycotina species *Lipomyces starkeyi* places at the base of Dikarya, while three Saccharomycotina species (including *S. cerevisiae*) appear as a sister clade to Pucciniomycotina (Fig. 6). The Agaricomycotina are also paraphyletic; Tremellomycetes and two basal Basidiomycota species (*B. undulatus* and *W. sebi*) appear closer to Ustilagomycota (Fig. 6). Many of the supports throughout the tree are extremely poor (almost all of the incongruences we highlighted all have <40% BP), which seems to be another effect of long-branch attraction (Fig. 6). Due to the breadth of fungal taxa, we have sampled for our multiple analyses, and the timescale of the evolution of the fungal kingdom being approximately 1 billion years old, it is unsurprising that a method using branch lengths to infer a close relationship between actually distantly related species that both have long branches, a classic example of the "Felsenstein Zone" (Bergsten, 2005; Huelsenbeck & Hillis, 1993). Ultimately, our AV phylogeny (Fig. 6) seems to confirm one of the concerns of Fitzpatrick et al. (2006) in a much more stark fashion that the AV method is not appropriate for large-scale phylogenomic reconstructions containing taxa sampled from across many phyla without prior predictive analysis of the potential for long branch attraction in such datasets (Su & Townsend, 2015).

2.3 Bayesian Supertree Phylogenomic Analysis of Fungi

While parsimony-based supertree reconstructions are generally reliable, concerns have been raised in the past as to some of the underlying methodology of MRP reconstruction and the effects that factors like input tree sizes (Pisani & Wilkinson, 2002; Wilkinson, Thorley, Pisani, Lapointe, & McInerney, 2004). There has long been the desire for a supertree method that infers phylogeny from source trees with more statistical rigor like Bayesian and maximum-likelihood inference methods. While Bayesian and ML analyses are the standard for supermatrix reconstruction, such methods have been difficult to implement in the past for supertree analysis due to computational limitations, most of which is down to the necessity of tree searching for the best supertree (i.e., calculating likelihoods for all possible supertrees given a set of source phylogenies).

It is only in recent years that phylogenomic inference based on ML and Bayesian methods has been implemented for supertree analysis; one such model for supertree likelihood estimation was first described by Steel and Rodrigo (2008) and then refined the following year (Bryant & Steel, 2009; Steel & Rodrigo, 2008). The Steel and Rodrigo method of likelihood estimation (henceforth referred to as ST-RF) is based on modeling the incongruences between input gene phylogenies and a corresponding unknown or provided supertree phylogeny. Two recent implementations of ST-RF ML analysis have been reported: the first a heuristic method of estimating approximate ML supertrees based on subtree pruning and regrafting implemented in the Python software L.U.St. by Akanni, Creevey, Wilkinson, and Pisani (2014), and the second a heuristic Bayesian MCMC criterion by Akanni, Wilkinson, Creevey, Foster, and Pisani (2015) implemented in the Python software package p4 (Akanni et al., 2014, 2015; Foster, 2004). Akanni et al. (2015) tested the Bayesian MCMC implementation on both a large kingdom-wide metazoan dataset and a smaller Carnivora dataset, notably the analysis produced a Bayesian supertree in full agreement with both the literature on metazoan relationships and a previous MRP supertree analysis on the same dataset (Holton & Pisani, 2010).

No parametric supertree reconstruction has been carried out for the fungal kingdom to date, and with that in mind we reconstructed the phylogeny of our 84-genome dataset with the MCMC Bayesian criterion developed by Akanni et al. (2015) using a slightly amended gene phylogeny dataset from our MRP and AV supertree phylogenies.

2.3.1 Heuristic MCMC Bayesian Supertree Reconstruction of 84 Fungal Genomes From 8050 Source Phylogenies

MCMC Bayesian supertree analysis was carried out on the single-copy phylogeny dataset using the ST-RF model as implemented in p4 (Akanni et al., 2015; Foster, 2004; Steel & Rodrigo, 2008). As ST-RF analysis is currently only implemented in p4 for fully bifurcating phylogenies, 60 phylogenies were removed from the total single-copy phylogeny dataset, for an input dataset of 8050 gene phylogenies. Two separate MCMC analyses with 4 chains each were ran for 30,000 generations with $\beta = 1$, sampling every 20 generations. The analyses converged after 30,000 generations, and a consensus phylogeny based on posterior probability of splits was generated from 150 supertrees sampled after convergence following Akanni et al. (2015). This consensus phylogeny was visualized in iTOL and annotated according to the NCBI's taxonomy database, and rooted at *R. allomycis* (Fig. 7).

2.3.2 Supertree Reconstruction With a Heuristic MCMC Bayesian Method Highly Congruent With MRP and Supermatrix Phylogenies

Using 8050 of the 8110 individual gene phylogenies which we identified in our MRP supertree analysis, we have reconstructed the first parametric supertree of the fungal kingdom (Fig. 7). We selected the ST-RF MCMC Bayesian supertree reconstruction method implemented in p4 for reconstruction over the heuristic method implemented in L.U.St. due to tractability issues regarding large datasets in the latter method (Akanni et al., 2014, 2015). Two ST-RF analyses were carried out for 30,000 generations, and the analyses were adjudged to have converged after 20,000 generations. To construct a phylogeny from our MCMC analysis, we sampled 150 trees generated after convergence and built a consensus tree in p4, where branch support values are the estimated posterior probabilities of a given split (i.e., bipartition) within a phylogeny (Fig. 7). Our ST-RF MCMC analysis is highly congruent with both our MRP supertree phylogeny and supermatrix phylogenies and supports the monophyly of the majority of the eight fungal phyla in our dataset (Fig. 7). Below, we detail the resolution of the basal and Dikarya lineages under ST-RF analysis.

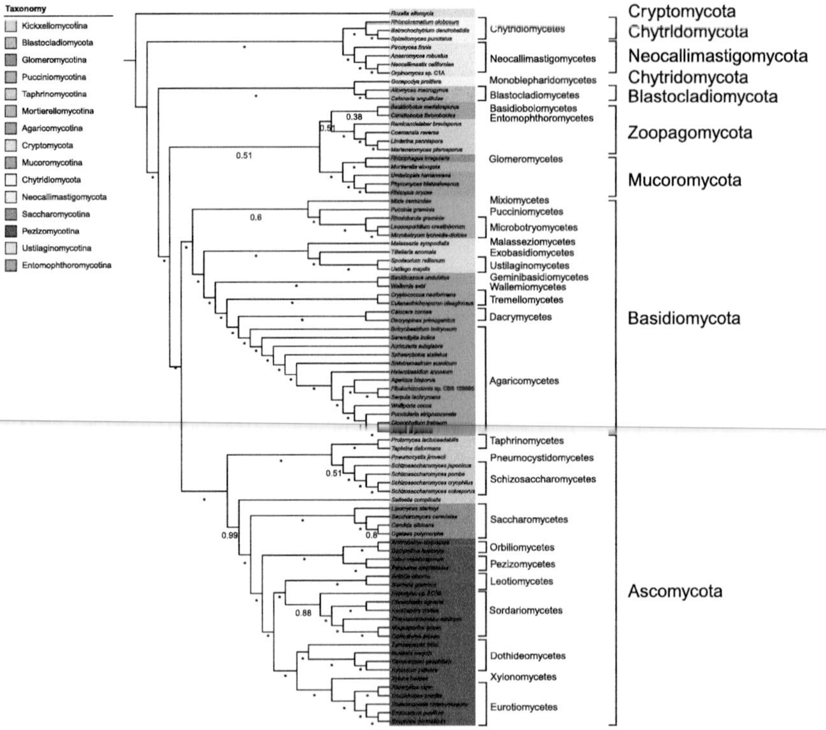

Fig. 7 MCMC Bayesian supertree phylogeny of 84 fungal species derived from 8050 fully bifurcating source phylogenies. Phylogeny generated in p4 using ST-RF model of maximum-likelihood supertree estimation running for 30,000 generations with $\beta = 1$. Posterior probabilities of bipartition(s) within 150 trees sampled after convergence shown on branches. Maximum posterior probability support designated with an *asterisk* (*).

2.3.2.1 Basal Fungi

After rooting at *R. allomycis*, the Neocallimastigomycota and Chytridiomycota (except *G. prolifera*) form a sister group relationship with maximum PP (Fig. 7). The Blastocladiomycota emerge after this branch, and the Chytridiomycota species *G. prolifera* branches as sister to the phylum with maximum PP (Fig. 7). There is weak support (0.51 PP) for a monophyletic clade containing both former zygomycetes phyla Zoopagomycota and Mucoromycota as sister clades (Fig. 7). Notably, unlike MRP and supermatrix analysis, ST-RF phylogeny places the Entomophthoromycotina as monophyletic but with very weak support (0.38 PP). There is also weak support for the placement the Entomophthoromycotina as basal within Zoopagomycota. Kickxellomycotina are monophyletic with maximum support. The monophyly of Mucoromycota is fully supported, with *R. irregularis* (Glomeromycotina) and *M. elongata* (Mortierellomycotina) branching as sister taxa (Fig. 7).

2.3.2.2 Basidiomycota

The Basidiomycota are supported as a monophyletic group with maximum PP (Fig. 7). There is weak support for the monophyly of Pucciniomycotina (0.6 PP); however, the deeper branches within the subphyla are all fully supported and their topology reflects both the MRP supertree and ML supermatrix phylogenies discussed earlier (Figs. 3, 5, and 7). There is full support for a sister relationship between Ustilaginomycotina and Agaricomycotina, and both these subphyla are fully supported. In Ustilaginomycotina, *M. sympodialis* is the basal species with maximum support (Fig. 7), as in our supermatrix and MRP supertree phylogenies. The topology of the Agaricomycotina is nearly identical on the class level to both the MRP and supermatrix phylogenies, with *B. undulatus* and *W. sebi* branching as basal species, the Tremellomycetes forming a monophyletic intermediate clade, and a fully supported sister relationship between the Dacrymycetes and the Agaricomycetes (Fig. 7).

2.3.2.3 Ascomycota

The monophyly of the Ascomycota is supported with maximum PP, as is the monophyly of two of the three subphyla in Ascomycota (Fig. 7). Taphrinomycotina is paraphyletic as in the MRP phylogeny, with *S. complicata* branching sister to Saccharomycota with near-maximum support (0.99 PP) and the remaining Taphrinomycotina species are placed as a monophyletic clade with maximum PP (Figs. 5 and 7). The Taphrinomycetes branch at the base of the Taphrinomycotina clade, and there is weak support (0.51 PP) for the placement of *P. jirovecii* as sister to the Schizosaccharomycotina (Fig. 7). The Saccharomycotina are fully supported as monophyletic (1.0 PP) with *L. starkeyi* placed at the base of the subphyla. The monophyly of the Pezizomycotina is also fully supported and there is maximum support for the monophyly of the six larger represented classes within the subphylum (Fig. 7). Additionally, the relationships between the individual classes within Pezizomycotina are identical to the topology seen in both the MRP supertree phylogeny and the ML supermatrix phylogeny (Figs. 3, 5, and 7). The Orbiliomycetes and Pezizomycetes branch as the earliest-diverging clades within Pezizomycotina with maximum PP, the Sordariomycetes and Leotiomycetes are sister classes with maximum PP and a monophyletic

Dothideiomycetes–Xylonomycetes–Eurotiomycetes clade receives maximum PP (Fig. 7).

2.4 Phylogenomics of Fungi Based on Gene Content

A common alternative to phylogenomic reconstruction using gene phylogenies is to take a "gene content" approach in which evolutionary relationships between species are derived from shared genomic content, such as the presence or absence of conserved orthologous genes (COGs) or the overall proportion of shared genes between two species, working under the assumption that species that share more of their genome are closely related (Snel, Bork, & Huynen, 1999; Snel, Huynen, & Dutilh, 2005). In the case of presence–absence analyses, a matrix can be constructed for the species under investigation, which can then have their phylogeny reconstructed via parsimony methods. Analyses based on proportions of shared genes can entail the construction of distance matrices for all input species, with values equal to the inverse ratio of shared genes (i.e., if two species share 75% of their genes, their distance is 0.25), which is then used to construct a neighbor-joining phylogeny. The advantages of such approaches are the relative tractability of parsimony or distance-based gene content methods, and their potential to use more information from genomes rather than the sourcing of data from smaller sets of gene families required by supertree or supermatrix approaches (Creevey & McInerney, 2009). However, the gene content approach is by its very nature a "broad strokes" approach and can ignore potentially important phylogenetic information from individual gene phylogenies such as HGT events, and assumes the same evolutionary history for missing orthologs or genomic content among species (Page & Holmes, 1998).

2.4.1 Gene Content Approaches to Phylogenomics in Fungi

Gene content approaches to phylogenomic reconstruction have seen application in a number of phylogenomics studies, although its greatest use predated many of the now common supertree and supermatrix methods. One of the earliest phylogenomic studies used a distance-based approach based on shared gene content to reconstruct the phylogeny of 13 unicellular species, including *S. cerevisiae* (Snel et al., 1999). Another study used a weighted distance matrix approach to reconstruct the phylogeny of 23 prokaryote and eukaryote species, including *S. cerevisiae* and partial genomic data from *S. pombe* (Tekaia, Lazcano, & Dujon, 1999). The most extensive gene content-based phylogenomic reconstruction of fungi was an analysis of 21 fungal genomes and 4 other eukaryote genomes in 2006 (Kuramae et al., 2006). In their analysis, the authors generated a presence–absence matrix (PAM) of 4852 COGs in fungal genomes as a complement to a supermatrix phylogeny using 531 concatenated proteins which was reconstructed using four different methods (MP, ML, neighbor-joining, and Bayesian inference). The authors reconstructed the phylogeny of all 25 genomes using this PAM and found that the PAM phylogeny differ most in the placement of *S. pombe* within Saccharomycetes as opposed to its basal position in Ascomycetes as seen in their supermatrix reconstructions (Kuramae et al., 2006).

To test the accuracy of inferring the phylogeny of a large genomic dataset using simple parsimony methods based on shared genomic content, we carried out a simple parsimony-based PAM phylogenomic reconstruction of 84 fungal species based on the presence of orthologs from single-copy gene families.

2.4.2 Phylogenomic Reconstruction of 84 Fungal Species Based on COG PAM

A simple PAM was generated for 84 fungal genomes based on their representation across 12,964 single-copy gene families identified via the random BLASTp approach detailed in Section 2.2. Parsimony analysis of this matrix was carried out using PAUP* with 100 bootstrap replicates. The resultant consensus phylogeny generated by PAUP* was visualized using iTOL and annotated according to the NCBI's taxonomy database. The phylogeny was rooted at *R. allomycis* (Fig. 8).

2.4.3 COG PAM Approach Displays Erroneous Placement of Branches Within Dikarya

We generated a simple PAM phylogeny for the 84 fungal genomes in our dataset by checking for the presence or absence of all 84 species across the 12,964 single-copy phylogenies we generated during our supertree analyses via random BLASTp searches and using the PAM as input for parsimony analysis (Fig. 8). The simple PAM phylogeny shows some level of congruence with the other phylogenomic analyses described here along certain branches (Fig. 8). The monophyly of Neocallimastigomycota, Chytridiomycota, and Blastocladiomycota all displays maximum or near-maximum BP, and there is 72% BP for a sister relationship between Chytridiomycota and Neocallimastigomycota (Fig. 8). The Zoopagomycota and Mucoromycota are placed in one monophyletic clade with 82% BP, with the two Entomophthoromycotina species in our dataset branching as closely related to the Mucoromycota (Fig. 8). However, some glaring conflicts with the

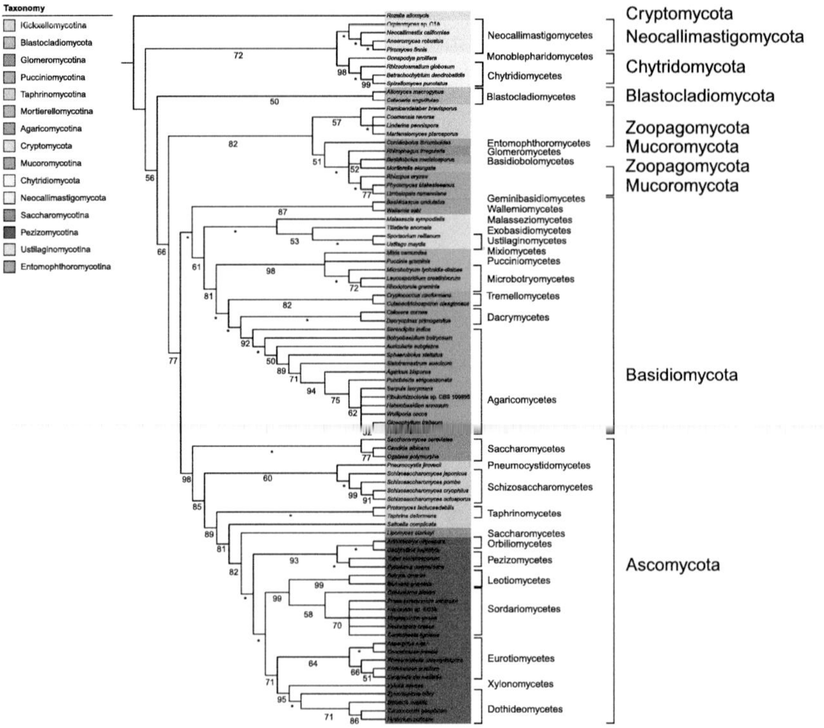

Fig. 8 Maximum parsimony (MP) phylogeny of 84 fungal species based on the presence of homologs from 12,964 single-copy gene families identified via random BLASTp searches. Bootstrap supports shown on branches. Maximum bootstrap support designated with an *asterisk* (*).

other phylogenomic methods we carried out can be observed within the Dikarya lineage. Most notably, the Agaricomycotina and Saccharomycotina are both paraphyletic in our single-copy PAM approach; for the former, *W. sebi* and *B. undulatus* branch at the base of the Basidiomycota adjacent to Ustilagomycotina, while in the latter three of the four Saccharomycotina (excluding *L. starkeyi*) species branch in our dataset at the base of the Ascomycota, implying that Taphrinomycotina diverged later than Saccharomycotina (Fig. 8). There is uncertain placement of clades within the Basidiomycota subphyla in particular. In the Ascomycota, the Taphrinomycotina are paraphyletic and *S. complicata* branches adjacent to *L. starkeyi*. The monophyly of all six larger Pezizomycotina classes are supported, many with relatively high or even maximum BP; however, there is poorer resolution of many relationships within these classes with the clearest examples being the Sordariomycetes and Eurotiomycetes (Fig. 8). In short, our PAM phylogeny is able to retrieve relationships with some level of accuracy within the fungal kingdom, but the method lacks the ability to resolve some of the more divergent relationships within fungi to the degree that some of our supermatrix or supertree phylogenies have illustrated.

2.5 Alignment-Free Phylogenomic Analysis of Fungi

Another alternative to the alignment-based methods of phylogenomic reconstruction we have detailed earlier is the use of a string-based comparison of genomes to infer phylogeny, based on the assumption that under such comparisons each species should have a characteristic genomic signature that can act as a phylogenetic marker (Delsuc, Brinkmann, & Philippe, 2005). Some analyses have thus used signatures such as distribution of protein folds or frequency of oligonucleotides from genetic and genomic data to infer phylogeny (Campbell, Mrázek, & Karlin, 1999; Lin & Gerstein, 2000; Pride, Meinersmann, Wassenaar, & Blaser, 2003). The most widely used alignment-free phylogenomic method, the composition vector (CV) approach, was first implemented by Qi, Luo, and Hao (2004) and by Qi, Wang, and Hao (2004), who used the approach to reconstruct the phylogeny of 87 prokaryote species from 11 bacterial and 2 archaeal phyla (Qi, Wang, et al., 2004). In their analysis, the authors detail the CV method for reconstructing phylogeny using genome-scale data, which we recount as follows:

1. Given a nucleic acid or amino acid sequence of length L in a genome, count the appearances of overlapping strings (i.e., oligonucleotides or oligopeptides) of a length K and construct a frequency vector of length 4^K for nucleic acid sequences and 20^K for amino acid sequences.
2. Subtract background noise, to account for random mutation at the molecular level, from each frequency vector to generate an overall composition vector for a given genome.
3. Calculate a distance matrix for the set of composition vectors corresponding to the set of input genomes.
4. Generate a neighbor-joining phylogeny from the distance matrix using software such as Neighbor or PAUP*.

The main advantages of the composition vector approach over traditional alignment-based methods of inferring phylogeny are the removal of artificial selection of phylogenetic markers from the process of reconstruction (the only variable in the method is K, the length of overlapping oligopeptides), and the relative speed with which the approach can infer phylogeny for large datasets over alignment-based supertree or supermatrix methods. Hence, it may be useful for quick phylogenomic identification of newly sequenced genomes against published data and as an independent verification step of previous alignment-based phylogenetic or phylogenomic analysis (Wang, Xu, Gao, & Hao, 2009). On that point however, interpreting the accuracy or otherwise of CV phylogenomic reconstructions is generally dependent on prior knowledge of the phylogeny of given taxa derived from alignment-based phylogenetic or phylogenomic analyses. An approach to inferring phylogeny based on nucleotide or amino acid composition may also be susceptible to compositional biases, and there has not been to the best of our knowledge a rigorous analysis of the potential effect these may have on accuracy of phylogenomic inference, as there have been for the supertree or supermatrix methods referred to earlier.

2.5.1 Composition Vector Method Phylogenomics of Fungi

Many of the phylogenomic analyses using the CV method have analyzed large prokaryotic datasets or broad global datasets sampled from many phyla or kingdoms across the three domains of life, whose phylogenies were recovered with quality comparative to alignment-based phylogenomic analyses. The most extensive application of the composition vector approach in fungal phylogenomics was an 85-genome analysis by Wang et al. (2009) using a CV implementation in the software program CVTree (Qi, Luo, et al., 2004; Wang et al., 2009). For their analysis, Wang et al. (2009) reconstructed the phylogeny of the fungal kingdom using 81 genomes from 4 fungal phyla (Basidiomycota, Ascomycota, Chytridiomycota, and Mucoromycota) as well as the microsporidian *Encephalitozoon cuniculi* and 3 eukaryotic outgroup taxa. The authors described the resolution of both the Basidiomycota and Ascomycota in detail in their analysis; the three subphyla within Basidiomycota were recovered but with poor bootstrap support due to issues with taxon sampling (only 12 Basiomycota species had genomic data at the time of the analysis), while the main focus of the authors analysis was on the resolution of 65 Ascomycota species. Within the Ascomycota, the Taphrinomycota (represented by three *Schizosaccharomyces* species) were fully resolved and in the Saccharomycotina the two clades described by Fitzpatrick et al. (2006), the CTG clade and the WGD clade, were also recovered. CV reconstruction recovered four classes within Pezizomycotina; the Dothideomycetes and Eurotiomycetes were placed as sister taxa with maximum support, as were the Sordariomycetes and Leotiomycetes.

To complement our phylogenomic analyses based on source gene phylogenies or identification of shared orthologs, we carried out alignment-free analysis of 84 fungal species using the composition vector method as implemented in CVTree.

2.5.2 Phylogenomic Reconstruction of 84 Fungal Species Using the CV Approach

Composition vector analysis was carried out on 84 genomes using CVTree with $K=5$ (Qi, Luo, et al., 2004). We selected $K=5$ as the best compromise of both computational requirements and resolution power. As the CV method does not generate bootstrapped phylogenies, we generated 100 bootstrap replicates of our 84-genome representative dataset using bespoke Python scripting and ran composition vector analysis on each replicate dataset (Zuo, Xu, Yu, & Hao, 2010). 100 replicate neighbor-joining phylogenies were calculated from their corresponding CVTree output distance matrices using Neighbor (Felsenstein, 1989). The majority-rule consensus phylogeny for all 100 composition vector replicate trees was generated using Consense (Felsenstein, 1989) and was visualized in iTOL, and annotated according to the NCBI's taxonomy database. The phylogeny was rooted at *R. allomycis* (Fig. 9).

2.5.3 Composition Vector Phylogenomic Reconstruction of 84 Fungal Species Is Congruent With Alignment-Based Methods

We carried out composition vector method phylogenomic reconstruction of our 84-genome dataset to complement the alignment-based and genomic content methods we detailed earlier (Fig. 9). Our composition vector analysis displays adequate levels of taxonomic congruence with our supermatrix and supertree analyses detailed in previous sections, supporting all the monophyly of each major fungal phyla and many of the subphyla within (Fig. 9). There are however some variations in topology and support between the basal lineages and within the Pezizomycotina subphylum in our CV phylogeny compared to our supermatrix and supertree phylogenies.

2.5.3.1 Basal Fungi

After rooting at *R. allomycis*, the Neocallimastigomycota emerge as the earliest-diverging fungal lineage (Fig. 9). The monophyly of Neocallimastigomycetes is also fully supported. Monophyletic Blastocladiomycota and Chytridiomycota clades branch as sister phyla with 62% BP. The monophyly of

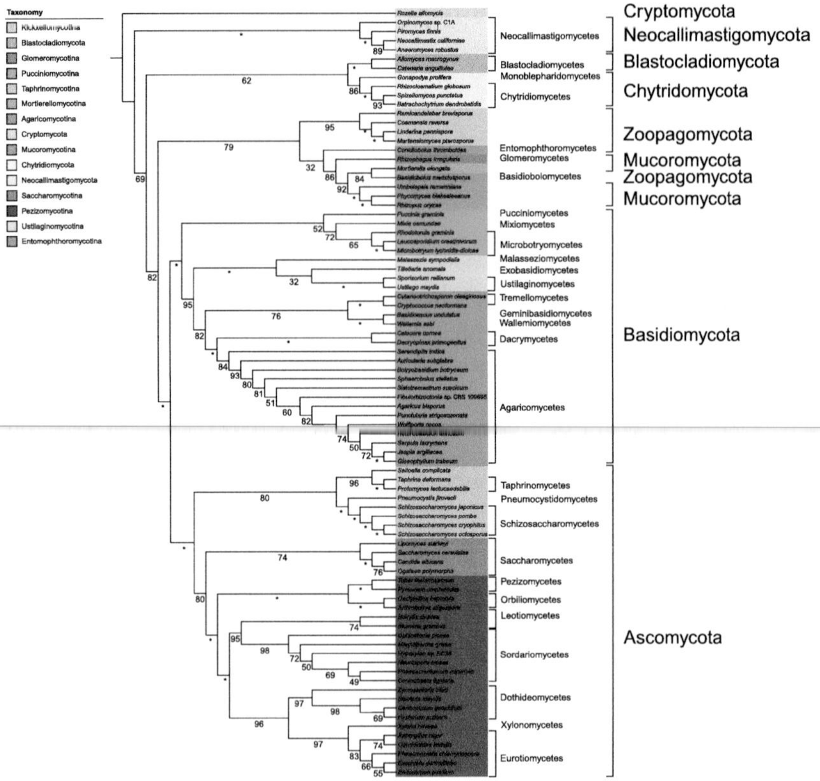

Fig. 9 Composition vector (CV) method phylogeny of 84 fungal species generated from 100 bootstrapped replicates of an 84-genome dataset. Bootstrap supports shown on branches. Maximum bootstrap support designated with an *asterisk* (*).

Blastocladiomycota receives maximum support, and notably unlike our MRP and supermatrix phylogenies *G. prolifera* branches within the Chytridiomycota with 86% BP (Figs. 3–5 and 9). In contrast to both supermatrix phylogenies and the MRP and ST-RF phylogenies, and like the AV and PAM phylogenies the two zygomycetes fungal phyla (Mucoromycota, Zoopagomycota) are placed within one monophyletic clade with 79% BP (Figs. 3–9). Kickxellomycotina are monophyletic with 95% BP and branch at the base of this Zoopagomycota–Mucoromycota clade. Resolution of the relationship between the rest of the former zygomycetes subphyla is harder to ascertain and has weaker support; the two Entomophthoromycotina species branch distant from each other with *B. meristosporus* branching within Mucoromycota adjacent to Mortierellomycotina and *C. thromboides* branching beside the Glomeromycotina species *R. irregularis*, similar to what is seen under PAM phylogenomic analysis (Figs. 8–9). Like the MRP phylogeny (Fig. 5), *R. irregularis* is within a paraphyletic Mucoromycota clade instead of at the base of the Dikarya as seen in the supermatrix phylogenies (Figs. 3, 4, and 9).

2.5.3.2 Basidiomycota

Pucciniomycotina is placed as the earliest-diverging subphylum within Basidiomycota with 52% BP, and the Ustilagomycotina and Agaricomycotina subphyla are sister clades with 95% BP (Fig. 9). The most-represented class within the Pucciniomycotina, the Microbotryomycetes, is monophyletic with 65% BP (Fig. 9), while unlike the rest of our phylogenies discussed earlier *P. graminis* is placed as the most basal species within Pucciniomycotina. Within the Ustilaginomycotina, *M. sympodialis* are placed as the basal lineage sister to the Exobasidiomycetes representative *T. anomala* similar to its position under ML supermatrix reconstruction and MRP reconstruction (Figs. 3, 5, and 9). The Agaricomycetes are monophyletic with 84% BP, with varying support for relationships within the class but a topology identical to both supermatrix phylogenies and MRP phylogeny with the exception of the placement of Tremellomycetes within a monophyletic ancestral branch adjacent to *B. undulatus* and *W. sebi* (Figs. 3–5 and 9).

2.5.3.3 Ascomycota

Within the Ascomycota, all three subphyla are resolved as monophyletic clades (Fig. 9). Taphrinomycotina is placed as the most basal subphylum within Ascomycota with maximum support, while the Pezizomycotina and Saccharomycotina are sister subphyla with 80% BP (Fig. 9). The Taphrinomycotina are monophyletic with 80% BP, and CV phylogeny displays maximum support for a sister relationship between *P. jirovecii* and the Schizosaccharomycetes and near-maximum (96% BP) support for a similar relationship between *S. complicata* and the two Taphrinomycetes representatives in our dataset (Fig. 9). The Saccharomycotina are monophyletic with 74% support (Fig. 9). All six larger classes from the Pezizomycotina represented in our dataset are resolved as monophyletic. The Orbiliomycetes and Pezizomycetes are placed as both sister subphyla and the earliest-diverging Pezizomycotina clades, both with maximum BP. The Leotiomycetes and Sordariomycetes are also sister clades with 95% BP. As our MRP phylogeny, the Eurotiomycetes are placed as sister to the Xylonomycetes species *X. heveae* with 97% BP (Figs. 5 and 9).

3. A GENOME-SCALE PHYLOGENY OF 84 FUNGAL SPECIES FROM SEVEN PHYLOGENOMIC METHODS

There is a large degree of congruence in the resolution of the fungal kingdom in most of the phylogenomic analyses we described in Section 2, which speaks to the quality of the genomic data we obtained from MycoCosm and the relative accuracy of the majority of the phylogenomic methods we utilized. In constructing a dataset for our analyses, we selected one representative from as many fungal orders as had been sequenced to date; this was to generate a phylogeny that was representative on the order level (though we do not focus on order phylogeny in this review) and to avoid overrepresentation of highly sampled taxa such as Eurotiomycetes or Saccharomycotina. Many of the best-known phylogenetic relationships within the fungal kingdom were recovered in our analyses, such as the monophyly of Dikarya as a whole (Hibbett et al., 2007). However, our analyses also supports more recent studies that have attempted to resolve outstanding branches of the fungal tree of life (Spatafora et al., 2016). In this section, we briefly describe the main trends seen across our seven phylogenomic reconstructions of the fungal kingdom and their congruence with previous studies and comment on the reconstructions of both the well-studied and highly represented Pezizomycotina subphylum and some of the newly circumscribed basal phyla. Finally, we discuss the suitability of the phylogenomic methods we have described and applied in this review for future fungal systematics studies.

3.1 Higher-Level Genome Phylogeny of the Fungal Kingdom

Despite variations in the resolution of some branches, there is a trend across the majority of phylogenies conducted of support or partial support for the eight phyla described in our dataset. Fig. 10 shows the congruence on the phylum level within the fungal kingdom in five of our seven phylogenetic reconstructions. We will refer to Fig. 10 and the subfigures (Figs. 10A–D) in Fig. 10 when comparing the different reconstructions on the phylum level and to the corresponding full phylogenies themselves for comparisons at lower levels here and elsewhere (average consensus and gene content phylogenies are omitted from Fig. 10 on the basis of erroneous placement of taxa). Beginning with the Cryptomycota species *R. allomycis*, the next-earliest-diverging clade within the fungal kingdom is the Blastocladiomycota under both supermatrix analyses followed by Neocallimastigomycota and Chytridiomycota

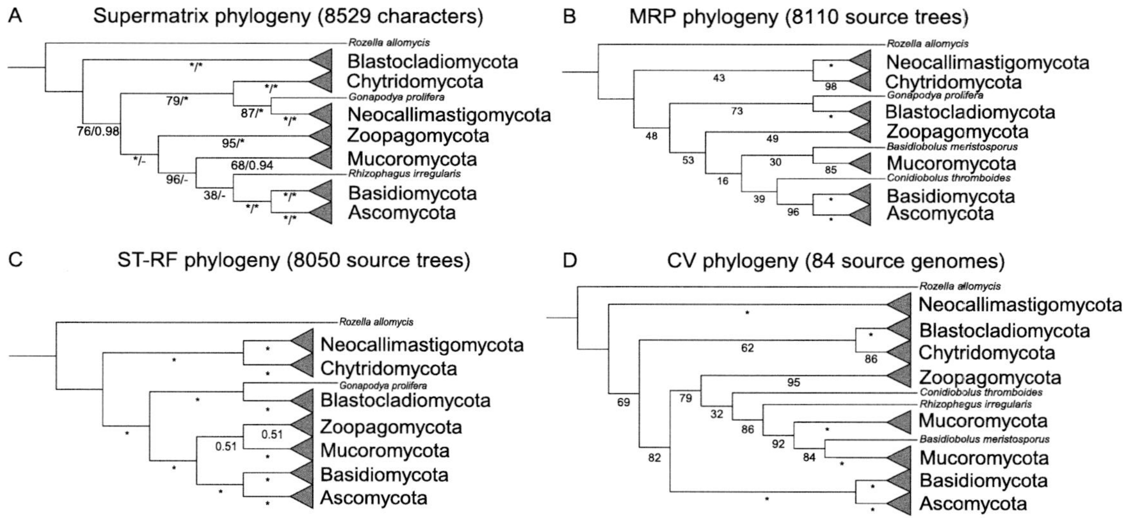

Fig. 10 Congruence of eight fungal phyla under five phylogenomic reconstructions. All clades bar Cryptomycota (represented *Rozella allomycis*) collapsed by phylum, paraphyletic species displayed as individual leaves. *Gonapodya prolifera* = Chytridiomycota, *Rhizophagus irregularis* = Mucoromycota, all other species except *R. allomycis* = Zoopagomycota. Refer to Figs. 3, 4, 5, 7, and 9, respectively, for original phylogenies. (A) ML and Bayesian supermatrix phylogenies. Branch supports given as ML bootstrap supports and, where topology is identical, Bayesian posterior probabilities. Maximum bootstrap or posterior probability support designated with an *asterisk* (*). (B) MRP supertree phylogeny. Branch supports given as bootstrap supports. Maximum bootstrap support designated with an *asterisk* (*). (C). MCMC Bayesian supertree phylogeny using ST-RF ML method. Branch supports given as posterior probabilities of bipartition(s). Maximum posterior probability support designated with an *asterisk* (*). (D) CV phylogeny. Branch supports given as bootstrap supports. Maximum bootstrap support designated with an *asterisk* (*).

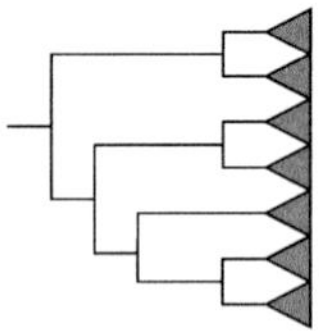

Pezizomycotina	Supermatrix		Supertree			PAM	CV
	ML	PB	MRP	AV	ST-RF		
Orbiliomycetes	✓	✓	✓	×	✓	✓	✓
Pezizomycetes	✓	✓	✓	×	✓	✓	✓
Leotiomycetes	✓	✓	✓	×	✓	✓	✓
Sordariomycetes	✓	✓	✓	×	✓	✓	✓
Dothideomycetes	✓	Xy	✓	×	✓	Eu	✓
Xylonomycetes	✓	Do	✓	×	✓	✓	✓
Eurotiomycetes	✓	✓	✓	×	✓	Do	✓

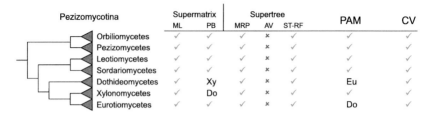

Fig. 11 Congruence of Pezizomycotina under seven phylogenomic methods. Placement of classes identical to topology on the left (see text) indicated with a *tick*, varying placement of classes indicated by the first two letters of a class. Average consensus (AV) phylogeny produced paraphyletic Pezizomycotina and so entire column labeled with *crosses*. Refer to text for discussion of topology of Pezizomycotina under AV phylogeny. Refer to Figs. 3–9 for original phylogenies.

(Fig. 10A). Other analyses place Neocallimastigomycota and Chytridiomycota (except *G. prolifera*) as closest to *R. allomycis* (Fig. 10B–D).

We describe the resolution of the former zygomycetes in greater detail later, but in the five phylogenies in Fig. 10 all support at least a sister relationship between the two zygomycetes phyla Zoopagomycota and Mucoromycota. The placement of the Glomeromycotina species *R. irregularis* varies, but Mucoromycota is generally placed as sister to the Dikarya (Fig. 10). The Basidiomycota are fully supported as monophyletic in each of the five phylogenies represented in Fig. 10, and all bar ML supermatrix reconstruction is in exact agreement with the two most extensive fungal genome phylogenies containing all three Basidiomycota subphyla (Medina et al., 2011; Wang et al., 2009). The Ascomycota are also fully supported as monophyletic in each of the five phylogenies represented in Fig. 10, with the only major variation being the placement of *S. complicata* within (or paraphyletic to) Taphrinomycotina (Fig. 10). The Saccharomycotina are monophyletic in all five phylogenies (Fig. 10). We discuss the class-level phylogeny within Pezizomycotina in greater detail in Section 3.3 and Fig. 11, but to briefly summarize here we see strong-to-maximum support for all six of the larger classes that were present in our dataset, and support for the two unofficial "Sordariomyceta" and "Dothideomyceta" groupings within Pezizomycotina (Schoch et al., 2009).

3.2 Multiple Phylogenomic Methods Show Moderate Support for the Modern Designations of Mucoromycota and Zoopagomycota

There is moderate support for the recent designation of the zygomycetes phyla Zoopagomycota and Mucoromycota by Spatafora et al. (2016) across most of our phylogenomic methods (Fig. 10). Previously the species within these two phyla were classified within Zygomycota, a phylum-level classification that had dated back to the 1950s until it was formally disputed by Hibbett et al. (2007). Six *incertae sedis* zygomycetes subphyla were later circumscribed (Hoffmann, Voigt, & Kirk, 2011), and subsequent phylogenetic analyses informally classified the zygomycetes subphyla into two groups, which were later established as Mucoromycota and Zoopagomycota (Chang et al., 2015; Spatafora et al., 2016).

Our phylogenomic analyses included 11 species from the 2 zygomycetes phyla, with the best resolution found in the ST-RF phylogeny where Zoopagomycota and Mucoromycota are placed as sister phyla with 0.51 PP and branch sister to Dikarya (Fig. 10C). Notably, our ST-RF phylogeny is the only phylogeny that resolves Entomophthoromycotina as a monophyletic clade (Fig. 7), albeit with extremely weak posterior probability support (0.38 PP). Within Zoopagomycota in our ST-RF phylogeny, Entomophthoromycotina branch as the basal clade with 0.51 PP, sister to Kickxellomycotina (Fig. 7). Our ST-RF phylogeny also places *R. irregularis* (Glomeromycotina) adjacent to *M. elongata* (Mortierellomycotina) within the Mucoromycota (Fig. 7). Within Mucoromycota, Mortiellomycotina and Mucoromycotina are supported as sister subphyla throughout the majority of our phylogenies (e.g., Bayesian supermatrix analysis, Fig. 4), with high to maximum support. Both of these phylum-level topologies are in agreement with Spatafora et al. (2016), though their phylogeny does not support a distinctive monophyletic branch containing both Zoopagomycota and Mucoromycota (Fig. 10C). The majority of our remaining phylogenomic analysis all shows some degree of support for both Zoopagomycota and Mucoromycota in relative agreement with Spatafora et al. (2016); however, in each of these phylogenies there is some conflict in either subphylum-level topology or lower BP/PP support due to issues of taxon sampling or low gene tree coverage in our dataset (of our 8110 source phylogenies for MRP analysis over 3500 contain 7 taxa or less; Fig. 10). With greater sampling of species from these lineages, we hope to see more consistent support of both the Zoopagomycota and Mucoromycota in future genome phylogenies using these methods, in line with what appears to be moderate-to-strong support for the new classification in our analyses based on total evidence (Kluge, 1989).

3.3 Pezizomycotina as a Benchmark for Phylogenomic Methodologies

The Pezizomycotina are by far the most sampled subphylum within the fungal kingdom in terms of genome sequencing (375 Pezizomycotina species have genomic data available from MycoCosm as of May 2017). Reflecting this, 22 Pezizomycotina species representing 7 classes are present in our 84-genome dataset (>25% of our final dataset). As a well-represented clade within our dataset at both the subphylum and individual class level, we are able to see how multiple phylogenomic analyses conducted in a total evidence approach (Kluge, 1989) are able to resolve a single clade of closely related classes containing some important ecological and pathogenic fungi. In every phylogenomic reconstruction, we attempted bar average consensus (AV) phylogeny, Pezizomycotina were monophyletic with maximum bootstrap or posterior probability branch support, and every class within Pezizomycotina is monophyletic with high or maximum BP or PP support (Figs. 3–5 and 7–9). There is a consistent trend within each of these phylogenies in the resolution of relationships between Pezizomycetes classes:

1. The Orbiliomycetes and Pezizomycetes always branch as the basal classes within Pezizomycotina and are always sister taxa (Figs. 3–5 and 7–9).
2. The relationship between Sordariomycetes and Leotiomycetes (within "Sordariomyceta" sensu Schoch et al., 2009) is always present and is fully supported in each phylogeny (Figs. 3–5 and 7–9).
3. The relationship between Dothideomycetes, Xylonomycetes, and Eurotiomycetes (within "Dothideomyceta" sensu Schoch et al., 2009) is always present and is fully supported in each phylogeny (Figs. 3–5 and 7–9).

Fig. 11 displays on the left the topology of the Pezizomycotina classes supported under ML supermatrix reconstruction, MRP supertree reconstruction, and ST-RF supertree reconstruction (Figs. 3, 5, and 7) and indicates the congruence (or otherwise) of Pezizomycotina under every phylogenomic analysis we attempted (Figs. 3–9). All methods bar AV are highly congruent in their resolution of the Pezizomycotina subphylum, with placement of the Xylonomycetes class the most notable variation. Even within the highly aberrant AV phylogeny, sister relationships such as those between Orbiliomycetes and Pezizomycetes or the association of classes within Sordariomyceta or Dothideomyceta can still be observed, though with lower resolution and support (Fig. 6). There is a high degree of congruence between our genome phylogenies of Pezizomycotina (Fig. 11) and the most extensive molecular phylogenies of Pezizomycotina that we could find in the literature derived from either small concatenated sets or whole genomes (Medina et al., 2011; Spatafora et al., 2006; Wang et al., 2009). The relative consistency of our analyses both with each other and with previous literature suggests that the resolution of Pezizomycotina could be considered a good benchmark for the accuracy of novel or existing phylogenomic methods (e.g., ST-RF analysis) when incorporated into a total evidence analysis, as the subphylum is large and diverse (the 10th edition of Ainsworth & Bisby's Dictionary of the Fungi estimates close to 70,000 Pezizomycetes species) but also densely sampled in genomic terms and containing a number of genomes of reference quality (Kirk, Cannon, Minter, & Stalpers, 2008).

3.4 The Use of Phylogenomics Methods in Fungal Systematics

Phylogenomic analyses with larger datasets across a wider spectrum of taxa are becoming more and more computationally tractable as methods of identifying potential phylogenetic markers on a genome-wide scale (e.g., identification and reconstruction of orthologous gene phylogenies in supertree analysis) and genome-scale reconstruction improve. In as much as the majority of our multiple analyses strongly support the major phyla of the fungal kingdom, we can also treat our analyses as measures of the accuracy of each of these phylogenomic methods in the reconstruction of large datasets. Supermatrix, MRP and ST-RF supertree, and CV method reconstructions all appear to arrive at relatively congruent results and may be useful for approximating a total evidence style approach for phylogenomic analyses of fungi. Simplified parsimony methods like our PAM phylogeny or branch length-based methods like our average consensus phylogeny may be useful for the reconstruction of smaller but well-represented datasets (for example, our PAM phylogeny does reconstruct the Pezizomycotina with support and topology close to supertree and supermatrix phylogenies) but for phylum or kingdom-wide analyses issues such as long-branch attraction begin to emerge (Bergsten, 2005). Long-branch attraction is thought to be an issue with MRP reconstruction as well, and while it is likely a factor in the weaker supports in some of the ancestral branches in our MRP phylogeny (for example, the weak supports in some of the internal branches grouping the basal phyla together), the MRP phylogeny seems to have been relatively immune to the topological effects of long-branch attraction that are very apparent in our branch length-dependent average consensus method phylogeny (Pisani & Wilkinson, 2002).

For our supertree analyses, we identified groups of orthologous proteins using a sequential random BLASTp approach as implemented by Fitzpatrick et al. (2006), where a random sequence from a given database is searched against that entire database, and then the sequence and its homologs (if any) are removed and the database reformatted (Fitzpatrick et al., 2006). Overall, this ad hoc approach to identifying orthology within our dataset seems to have been sufficient as a first step to generating source gene phylogenies; however, it may have had an impact downstream on resolution of internal branches within our MRP analysis. It is possible that a random BLASTp approach is too conservative, in that the orthologous families it identifies are missing members or that two "separate" orthologous families may in fact be one large orthologous family. Other established methods of identifying orthologous families, such as the OrthoMCL pipeline, have been used in phylogenomic analyses and can be tuned for granularity (i.e., orthologous cluster size) which may produce broader source phylogenies (Li, Stoeckert, & Roos, 2003). However, the large SQL-dependent computational overhead required for the current implementation of OrthoMCL was not considered suitable for an analysis of this scale.

Most of the phylogenomic methods we attempted are relatively tractable even for a dataset as large as ours. Depending on computational resources and available data, some of the methods we have discussed may be more appropriate for future fungal phylogenomic analyses than others. The most common techniques like MRP analysis and both ML and Bayesian supermatrix analysis were both tractable and produced phylogenies with largely congruent topologies and supports on most branches (although we should note that we utilized the parallelized version of PhyloBayes for our Bayesian analysis). The heuristic MCMC Bayesian supertree reconstruction we attempted using the ST-RF model as implemented in p4 was also relatively tractable despite not being parallelized, and Akanni et al. (2015) note that the method is far more efficient than the approximate ML reconstruction implemented in L.U.St. (Akanni et al., 2015). However, ST-RF analysis using either p4 or L.U.St. is currently only able to use fully resolved input phylogenies. While in our case this meant only 60 single-copy phylogenies (<1% of our total dataset) had to be removed before carrying out analysis, this may cause issues for more polytomous datasets. Bayesian and ML supertree reconstruction is certainly a promising development for phylogenomics, and hopefully methods like ST-RF should see more widespread use in future phylogenomic analysis as they mature.

Phylogenomic reconstruction using average consensus as implemented in CLANN was extremely inefficient time-wise and returned a severely erroneous phylogeny, so while it is certainly desirable for branch lengths to be incorporated in supertree reconstruction, a branch length-based method like AV is not appropriate for this kind of large-scale analysis. While PAM method reconstruction was straightforward to carry out, as we state earlier there were issues with erroneous placement of taxa and as such we do not recommend the method for large-scale datasets. Finally, composition vector method analysis produced a phylogeny relatively congruent to our alignment-based methods at $K=5$. Other CV method analyses have recommended K-values between 5 and 7 for most datasets (Zuo, Li, & Hao, 2014), however with the size of our dataset and the increase in computational resources required for generating distance matrices for eukaryotic genomes at $K>5$ in CVTree we felt that $K=5$ was the best compromise between accuracy and computational tractability. We would recommend however as in Section 2.5 that CV analysis should be used in conjunction with alignment-based methods for eukaryotic datasets, as interpretation of CV analysis requires a priori knowledge of the phylogeny of a given dataset.

4. CONCLUDING REMARKS

Fungi make up one of the major eukaryotic kingdoms, with an estimated 1.5 million member species inhabiting a diverse variety of ecological niches and an evolutionary history dating back over a billion years. It is imperative that evolutionary relationships within the fungal kingdom are well understood by analysis of as much quality phylogenetic data as is available with the most accurate methodologies possible. In this chapter, we discussed the evolutionary diversity of the fungal kingdom and the important role that fungi have had in the area of genomic and phylogenomics. We have reviewed previous phylogenomic analyses of the fungal kingdom over the last decade, and using seven phylogenomic methods, we have reconstructed the phylogeny of 84 fungal species across 8 fungal phyla. We found that established supermatrix and supertree methods produced relatively congruent phylogenies that were in large agreement with the literature. We also conducted the first analysis of the fungal kingdom using a heuristic MCMC Bayesian approach to supertree reconstruction previously used in Metazoa and found that this novel supertree approach resolves the fungal kingdom with a high degree of accuracy. The majority of our analyses overall show moderate-to-strong support of the newly assigned zygomycete phyla Mucoromycota and Zoopagomycota and strongly support the monophyly of Dikarya, while within the highly sampled Pezizomycotina subphylum there is a large amount of congruence between different phylogenomic methods as to the resolution of class relationships within the subphylum. We also conclude that supermatrix and supertree analyses remain the exemplar methods of phylogenomic reconstruction for fungi, based on their accuracy and computational tractability. We believe through both our discussion of the ecological diversity of the fungal kingdom and the history of its study on the genomic level we have demonstrated the need for a robust fungal tree of life with a broad representation, and that through our multiple phylogenomic analysis we have generated an important backbone for future comparative genomic analysis of fungi, particularly with the constantly increasing amount of quality genomic data arising from the 1000 Fungal Genomes Project and its certain use in future studies.

ACKNOWLEDGMENTS

We wish to acknowledge the JGI and all individual contributors to the 1000 Fungal Genomes Project for both the sheer scope of their undertaking and the quantity and quality of genomic data that they have made publically available and that we were able to use in this chapter. We also wish to acknowledge the DJEI/DES/SFI/HEA Irish Centre for High-End Computing (ICHEC) for the provision of computational facilities and support. C.G.P.M. is funded by an Irish Research Council Government of Ireland Postgraduate Scholarship (Grant No. GOIPG/2015/2242).

REFERENCES

Akanni, W. A., Creevey, C. J., Wilkinson, M., & Pisani, D. (2014). L.U.St: A tool for approximated maximum likelihood supertree reconstruction. *BMC Bioinformatics*, *15*(1), 183. https://doi.org/10.1186/1471-2105-15-183.

Akanni, W. A., Wilkinson, M., Creevey, C. J., Foster, P. G., & Pisani, D. (2015). Implementing and testing Bayesian and maximum-likelihood supertree methods in phylogenetics. *Royal Society Open Science*, *2*(8), 140436. https://doi.org/10.1098/rsos.140436.

Annaluru, N., Muller, H., Mitchell, L. A., Ramalingam, S., Stracquadanio, G., Richardson, S. M., et al. (2014). Total synthesis of a functional designer eukaryotic chromosome. *Science*, *344*(6179), 55–58. https://doi.org/10.1126/science.1249252.

Baldauf, S. L., & Palmer, J. D. (1993). Animals and fungi are each other's closest relatives: Congruent evidence from multiple proteins. *Proceedings of the National Academy of Sciences of the United States of America*, *90*(24), 11558–11562. https://doi.org/10.1073/pnas.90.24.11558.

Baum, B. R. (1992). Combining trees as a way of combining data sets for phylogenetic inference, and the desirability of combining gene trees. *Taxon*, *41*(1), 3–10. https://doi.org/10.2307/1222480.

Benson, D. A., Cavanaugh, M., Clark, K., Karsch-Mizrachi, I., Lipman, D. J., Ostell, J., et al. (2013). GenBank. *Nucleic Acids Research*, *41*(D1), D36–42. https://doi.org/10.1093/nar/gks1195.

Berbee, M. L., & Taylor, J. W. (1992). Detecting morphological convergence in true fungi, using 18S ribosomal RNA gene sequence data. *Biosystems*, *28*(1–3), 117–125.

Berbee, M. L., & Taylor, J. W. (2010). Dating the molecular clock in fungi—How close are we? *Fungal Biology Reviews*, *24*(1–2), 1–16. https://doi.org/10.1016/j.fbr.2010.03.001.

Bergsten, J. (2005). A review of long-branch attraction. *Cladistics*, *21*(2), 163–193. https://doi.org/10.1111/j.1096-0031.2005.00059.x.

Bininda-Emonds, O. R. P. (2004). The evolution of supertrees. *Trends in Ecology and Evolution*, *19*(6), 315–322. https://doi.org/10.1016/j.tree.2004.03.015.

Bryant, D., & Steel, M. (2009). Computing the distribution of a tree metric. *IEEE/ACM Transactions on Computational Biology and Bioinformatics*, *6*(3), 420–426. https://doi.org/10.1109/TCBB.2009.32.

Butler, G., Rasmussen, M. D., Lin, M. F., Santos, M. C. M. A. S., Sakthikumar, S., Munro, C. A., et al. (2009). Evolution of pathogenicity and sexual reproduction in eight Candida genomes. *Nature*, *459*(7247), 657–662. https://doi.org/10.1038/nature08064.

Byrne, K. P., & Wolfe, K. H. (2005). The yeast gene order browser: Combining curated homology and syntenic context reveals gene fate in polyploid species. *Genome Research*, *15*(10), 1456–1461. https://doi.org/10.1101/gr.3672305.

Camacho, C., Coulouris, G., Avagyan, V., Ma, N., Papadopoulos, J., Bealer, K., et al. (2009). BLAST+: Architecture and applications. *BMC Bioinformatics*, *10*(1), 421. https://doi.org/10.1186/1471-2105-10-421.

Campbell, A., Mrázek, J., & Karlin, S. (1999). Genome signature comparisons among prokaryote, plasmid, and mitochondrial DNA. *Proceedings of the National Academy of Sciences of the United States of America*, *96*(16), 9184–9189. https://doi.org/10.1073/pnas.96.16.9184.

Castresana, J. (2000). Selection of conserved blocks from multiple alignments for their use in phylogenetic analysis. *Molecular Biology and Evolution*, *17*(4), 540–552. https://doi.org/10.1093/oxfordjournals.molbev.a026334.

Cavalier-Smith, T. (1998). A revised six-kingdom system of life. *Biological Reviews of the Cambridge Philosophical Society*, *73*(3), 203–266.

Chang, Y., Wang, S., Sekimoto, S., Aerts, A. L., Choi, C., Clum, A., et al. (2015). Phylogenomic analyses indicate that early fungi evolved digesting cell walls of algal ancestors of land plants. *Genome Biology and Evolution*, 7(6), 1590–1601. https://doi.org/10.1093/gbe/evv090.

Creevey, C. J., Fitzpatrick, D. A., Philip, G. K., Kinsella, R. J., O'Connell, M. J., Pentony, M. M., et al. (2004). Does a tree-like phylogeny only exist at the tips in the prokaryotes? *Proceedings Biological Sciences/The Royal Society*, *271*(1557), 2551–2558. https://doi.org/10.1098/rspb.2004.2864.

Creevey, C. J., & McInerney, J. O. (2005). Clann: Investigating phylogenetic information through supertree analyses. *Bioinformatics*, *21*(3), 390–392. https://doi.org/10.1093/bioinformatics/bti020.

Creevey, C. J., & McInerney, J. O. (2009). Trees from trees: Construction of phylogenetic supertrees using CLANN. *Methods in Molecular Biology*, *537*, 139–161. https://doi.org/10.1007/978-1-59745-251-9_7.

Cuomo, C. A., & Birren, B. W. (2010). The fungal genome initiative and lessons learned from genome sequencing. *Methods in Enzymology*, *470*(C), 833–855. https://doi.org/10.1016/S0076-6879(10)70034-3.

Darriba, D., Taboada, G. L., Doallo, R., & Posada, D. (2011). ProtTest 3: Fast selection of best-fit models of protein evolution. *Bioinformatics*, *27*(8), 1164–1165. https://doi.org/10.1093/bioinformatics/btr088.

De Barros Lopes, M., Bellon, J. R., Shirley, N. J., & Ganter, P. F. (2002). Evidence for multiple interspecific hybridization in Saccharomyces sensu stricto species. *FEMS Yeast Research*, *1*(4), 323–331. https://doi.org/10.1016/S1567-1356(01)00051-4.

Delsuc, F., Brinkmann, H., & Philippe, H. (2005). Phylogenomics and the reconstruction of the tree of life. *Nature Reviews Genetics*, *6*(5), 361–375. https://doi.org/10.1038/nrg1603.

de Queiroz, A., & Gatesy, J. (2007). The supermatrix approach to systematics. *Trends in Ecology and Evolution*, *22*(1), 34–41. https://doi.org/10.1016/j.tree.2006.10.002.

Edgar, R. C. (2004). MUSCLE: Multiple sequence alignment with high accuracy and high throughput. *Nucleic Acids Research*, *32*(5), 1792–1797. https://doi.org/10.1093/nar/gkh340.

Engel, S. R., Dietrich, F. S., Fisk, D. G., Binkley, G., Balakrishnan, R., Costanzo, M. C., et al. (2014). The reference genome sequence of Saccharomyces cerevisiae: Then and now. *G3 (Bethesda)*, *4*(3), 389–398. https://doi.org/10.1534/g3.113.008995.

Faith, D. P., & Cranston, P. S. (1991). Could a cladogram this short have arisen by chance alone? On permutation tests for cladistic structure. *Cladistics*, 7(1), 1–28. https://doi.org/10.1111/j.1096-0031.1991.tb00020.x.

Federhen, S. (2012). The NCBI taxonomy database. *Nucleic Acids Research*, *40*(D1), D136–D143. https://doi.org/10.1093/nar/gkr1178.

Felsenstein, J. (1978). Cases in which parsimony or compatibility methods will be positively misleading. *Systematic Zoology*, *27*(4), 401. https://doi.org/10.2307/2412923.

Felsenstein, J. (1989). PHYLIP—Phylogeny inference package—v3.2. *Cladistics*, *5*(2), 164–166. https://doi.org/10.1111/j.1096-0031.1989.tb00562.x.

Fisk, D. G., Ball, C. A., Dolinski, K., Engel, S. R., Hong, E. L., Issel-Tarver, L., et al. (2006). Saccharomyces cerevisiae S288C genome annotation: A working hypothesis. *Yeast*, *23*(12), 857–865. https://doi.org/10.1002/yea.1400.

Fitzpatrick, D. A., Logue, M. E., & Butler, G. (2008). Evidence of recent interkingdom horizontal gene transfer between bacteria and Candida parapsilosis. *BMC Evolutionary Biology*, *8*(1), 181. https://doi.org/10.1186/1471-2148-8-181.

Fitzpatrick, D. A., Logue, M. E., Stajich, J. E., & Butler, G. (2006). A fungal phylogeny based on 42 complete genomes derived from supertree and combined gene analysis. *BMC Evolutionary Biology*, *6*(1), 99. https://doi.org/10.1186/1471-2148-6-99.

Fitzpatrick, D. A., O'Gaora, P., Byrne, K. P., & Butler, G. (2010). Analysis of gene evolution and metabolic pathways using the Candida Gene Order Browser. *BMC Genomics*, *11*(1), 290. https://doi.org/10.1186/1471-2164-11-290.

Foster, P. G. (2004). Modeling compositional heterogeneity. *Systematic Biology*, *53*(3), 485–495. https://doi.org/10.1080/10635150490445779.

Galagan, J. E., Calvo, S. E., Borkovich, K. A., Selker, E. U., Read, N. D., Jaffe, D., et al. (2003). The genome sequence of the filamentous fungus Neurospora crassa. *Nature*, *422*(6934), 859–868. https://doi.org/10.1038/nature01554.

Galagan, J. E., Calvo, S. E., Cuomo, C., Ma, L.-J., Wortman, J. R., Batzoglou, S., et al. (2005). Sequencing of Aspergillus nidulans and comparative analysis with A. fumigatus and A. oryzae. *Nature*, *438*(7071), 1105–1115. https://doi.org/10.1038/nature04341.

Galagan, J. E., Henn, M. R., Ma, L. J., Cuomo, C. A., & Birren, B. (2005). Genomics of the fungal kingdom: Insights into eukaryotic biology. *Genome Research*, *15*(12), 1620–1631. https://doi.org/10.1101/gr.3767105.

Goffeau, A., Barrell, B. G., Bussey, H., Davis, R. W., Dujon, B., Feldmann, H., et al. (1996). Life with 6000 genes. *Science*, *274*(5287), 546–567. https://doi.org/10.1126/science.274.5287.546.

Goffeau, A., & Vassarotti, A. (1991). The European project for sequencing the yeast genome. *Research in Microbiology*, *142*(7–8), 901–903. https://doi.org/10.1016/0923-2508(91)90071-H.

Grigoriev, I. V., Cullen, D., Goodwin, S. B., Hibbett, D., Jeffries, T. W., Kubicek, C. P., et al. (2011). Fueling the future with fungal genomics. *Mycology*, *2*(3), 192–209. https://doi.org/10.1080/21501203.2011.584577.

Grigoriev, I. V., Nikitin, R., Haridas, S., Kuo, A., Ohm, R., Otillar, R., et al. (2014). MycoCosm portal: Gearing up for 1000 fungal genomes. *Nucleic Acids Research*, *42*(D1), D699–704. https://doi.org/10.1093/nar/gkt1183.

Grigoriev, I. V., Nordberg, H., Shabalov, I., Aerts, A., Cantor, M., Goodstein, D., et al. (2011). The genome portal of the department of energy joint genome institute. *Nucleic Acids Research*, *40*(D1), 1–7. https://doi.org/10.1093/nar/gkr947.

Guarro, J., Gené, J., & Stchigel, A. M. (1999). Developments in fungal taxonomy. *Clinical Microbiology Reviews*, *12*(3), 454–500. https://doi.org/0893-8512/99/$04.00?0.

Guindon, S., Dufayard, J.-F., Lefort, V., Anisimova, M., Hordijk, W., & Gascuel, O. (2010). New algorithms and methods to estimate maximum-likelihood phylogenies: Assessing the performance of PhyML 3.0. *Systematic Biology*, *59*(3), 307–321. https://doi.org/10.1093/sysbio/syq010.

Hall, C., & Dietrich, F. S. (2007). The reacquisition of biotin prototrophy in Saccharomyces cerevisiae involved horizontal gene transfer, gene duplication and gene clustering. *Genetics*, *177*(4), 2293–2307. https://doi.org/10.1534/genetics.107.074963.

Hawksworth, D. L. (2001). The magnitude of fungal diversity: The 1.5 million species estimate revisited. *Mycological Research*, *105*(12), 1422–1432. https://doi.org/10.1017/S0953756201004725.

Heath, I. B. (1980). Variant mitoses in lower eukaryotes: Indicators of the evolution of mitosis? *International Review of Cytology*, *64*(C), 1–80. https://doi.org/10.1016/S0074-7696(08)60235-1.

Hibbett, D. S., Binder, M., Bischoff, J. F., Blackwell, M., Cannon, P. F., Eriksson, O. E., et al. (2007). A higher-level phylogenetic classification of the fungi. *Mycological Research*, *111*(5), 509–547. https://doi.org/10.1016/j.mycres.2007.03.004.

Hoffmann, K., Voigt, K., & Kirk, P. M. (2011). Mortierellomycotina subphyl. nov., based on multi-gene genealogies. *Mycotaxon*, *115*(1), 353–363. https://doi.org/10.5248/115.353.

Holley, R. W., Apgar, J., Everett, G. A., Madison, J. T., Marquisee, M., Merrill, S. H., et al. (1965). Structure of a ribonucleic acid. *Science (New York, N.Y.)*, *147*(3664), 1462–1465. https://doi.org/10.1126/science.147.3664.1462.

Holton, T. A., & Pisani, D. (2010). Deep genomic-scale analyses of the metazoa reject coelomata: Evidence from single-and multigene families analyzed under a supertree and supermatrix paradigm. *Genome Biology and Evolution*, *2*(1), 310–324. https://doi.org/10.1093/gbe/evq016.

Huelsenbeck, J. P., & Hillis, D. M. (1993). Success of phylogenetic methods in the four taxon case. *Systematic Biology*, *42*(3), 247–264. https://doi.org/10.1093/sysbio/42.3.247.

Huelsenbeck, J. P., Ronquist, F., Nielsen, R., & Bollback, J. P. (2001). Bayesian inference of phylogeny and its impact on evolutionary biology. *Science*, *294*(5550), 2310–2314. https://doi.org/10.1126/science.1065889.

Jackson, A. P., Gamble, J. A., Yeomans, T., Moran, G. P., Saunders, D., Harris, D., et al. (2009). Comparative genomics of the fungal pathogens Candida dubliniensis and Candida albicans. *Genome Research*, *19*(12), 2231–2244. https://doi.org/10.1101/gr.097501.109.

James, T. Y., Kauff, F., Schoch, C. L., Matheny, P. B., Hofstetter, V., Cox, C. J., et al. (2006). Reconstructing the early evolution of fungi using a six-gene phylogeny. *Nature*, *443*(7113), 818–822. https://doi.org/10.1038/nature05110.

Jones, M. D. M., Forn, I., Gadelha, C., Egan, M. J., Bass, D., Massana, R., et al. (2011). Discovery of novel intermediate forms redefines the fungal tree of life. *Nature*, *474*(7350), 200–203. https://doi.org/10.1038/nature09984.

Jones, M. D. M., Richards, T. A., Hawksworth, D. L., & Bass, D. (2011). Validation and justification of the phylum name Cryptomycota phyl. nov. *IMA Fungus*, *2*(2), 173–175. https://doi.org/10.5598/imafungus.2011.02.02.08.

Keller, N. P., Turner, G., & Bennett, J. W. (2005). Fungal secondary metabolism—From biochemistry to genomics. *Nature Reviews Microbiology*, *3*(12), 937–947. https://doi.org/10.1038/nrmicro1286.

Kellis, M., Birren, B. W., & Lander, E. S. (2004). Proof and evolutionary analysis of ancient genome duplication in the yeast Saccharomyces cerevisiae. *Nature*, *428*(VN-(6983)), 617–624. https://doi.org/10.1038/nature02424.

Khaldi, N., Seifuddin, F. T., Turner, G., Haft, D., Nierman, W. C., Wolfe, K. H., et al. (2010). SMURF: Genomic mapping of fungal secondary metabolite clusters. *Fungal Genetics and Biology*, *47*(9), 736–741. https://doi.org/10.1016/j.fgb.2010.06.003.

Kirk, P. M., Cannon, P. F., Minter, D. W., & Stalpers, J. A. (2008). *Ainsworth & Bisby's dictionary of the fungi* (10th ed.). Wallingford, UK: CABI.

Kluge, A. G. (1989). A concern for evidence and a phylogenetic hypothesis of relationships among epicrates (Boidae, serpentes). *Systematic Biology*, *38*(1), 7–25. https://doi.org/10.1093/sysbio/38.1.7.

Koonin, E. V., Fedorova, N. D., Jackson, J. D., Jacobs, A. R., Krylov, D. M., Makarova, K. S., et al. (2004). A comprehensive evolutionary classification of proteins encoded in complete eukaryotic genomes. *Genome Biology*, *5*(2), R7. https://doi.org/10.1186/gb-2004-5-2-r7.

Kück, P., & Meusemann, K. (2010). FASconCAT: Convenient handling of data matrices. *Molecular Phylogenetics and Evolution*, *56*(3), 1115–1118. https://doi.org/10.1016/j.ympev.2010.04.024.

Kuramae, E. E., Robert, V., Snel, B., Weiß, M., & Boekhout, T. (2006). Phylogenomics reveal a robust fungal tree of life. *FEMS Yeast Research*, *6*(8), 1213–1220. https://doi.org/10.1111/j.1567-1364.2006.00119.x.

Lapointe, F. J., & Cucumel, G. (1997). The average consensus procedure: Combination of weighted trees containing identical or overlapping sets of taxa. *Systematic Biology*, *46*(2), 306–312. https://doi.org/10.1093/sysbio/46.2.306.

Lartillot, N., Brinkmann, H., & Philippe, H. (2007). Suppression of long-branch attraction artefacts in the animal phylogeny using a site-heterogeneous model. *BMC Evolutionary Biology*, 7(Suppl. 1), S4. https://doi.org/10.1186/1471-2148-7-S1-S4.

Lartillot, N., & Philippe, H. (2004). A Bayesian mixture model for across-site heterogeneities in the amino-acid replacement process. *Molecular Biology and Evolution*, *21*(6), 1095–1109. https://doi.org/10.1093/molbev/msh112.

Lartillot, N., Rodrigue, N., Stubbs, D., & Richer, J. (2013). PhyloBayes MPI: Phylogenetic reconstruction with infinite mixtures of profiles in a parallel environment. *Systematic Biology*, *62*(4), 611–615. https://doi.org/10.1093/sysbio/syt022.

Léjohn, H. B. (1974). *Biochemical parameters of fungal phylogenetics. In T. Dobzhansky, M. K. - Hecht, & W. C. Steere (Eds.), Evolutionary biology* (pp. 79–125). Boston, MA: Springer. *https://doi.org/10.1007/978-1-4615-6944-2_3.*

Letunic, I., & Bork, P. (2016). Interactive tree of life (iTOL) v3: An online tool for the display and annotation of phylogenetic and other trees. *Nucleic Acids Research*, *44*(W1), W242–W245. https://doi.org/10.1093/nar/gkw290.

Li, L., Stoeckert, C. J., & Roos, D. S. (2003). OrthoMCL: Identification of ortholog groups for eukaryotic genomes. *Genome Research*, *13*(9), 2178–2189. https://doi.org/10.1101/gr.1224503.

Lin, J., & Gerstein, M. (2000). Whole-genome trees based on the occurrence of folds and orthologs: Implications for comparing genomes on different levels. *Genome Research*, *10*(6), 808–818. https://doi.org/10.1101/gr.10.6.808.

Marcet-Houben, M., & Gabaldón, T. (2009). The tree versus the forest: The fungal tree of life and the topological diversity within the yeast phylome. *PLoS One*, *4*(2). e4357 https://doi.org/10.1371/journal.pone.0004357.

Marcet-Houben, M., & Gabaldón, T. (2010). Acquisition of prokaryotic genes by fungal genomes. *Trends in Genetics: TIG*, *26*(1), 5–8. https://doi.org/10.1016/j.tig.2009.11.007.

Marcet-Houben, M., Marceddu, G., & Gabaldón, T. (2009). Phylogenomics of the oxidative phosphorylation in fungi reveals extensive gene duplication followed by functional divergence. *BMC Evolutionary Biology*, *9*(1), 295. https://doi.org/10.1186/1471-2148-9-295.

Medina, E. M., Jones, G. W., & Fitzpatrick, D. A. (2011). Reconstructing the fungal tree of life using phylogenomics and a preliminary investigation of the distribution of yeast prion-like proteins in the fungal kingdom. *Journal of Molecular Evolution*, *73*(3–4), 116–133. https://doi.org/10.1007/s00239-011-9461-4.

Nikoh, N., Hayase, N., Iwabe, N., Kuma, K., & Miyata, T. (1994). Phylogenetic relationship of the kingdoms Animalia, Plantae, and Fungi, inferred from 23 different protein species. *Molecular Biology and Evolution*, *11*(5), 762–768. https://doi.org/0737-4038/94f I lOS-0005$02.

Oliver, S. G., Van Der Aart, Q. J., Agostoni-Carbone, M. L., Aigle, M., Alberghina, L., Alexandraki, D., et al. (1992). The complete DNA sequence of yeast chromosome III. *Nature*, *357*(6373), 38–46. https://doi.org/10.1038/357038a0.

Page, R. D. M., & Holmes, E. C. (1998). *Molecular evolution: A phylogenetic approach*. Oxford, UK: Blackwell Science.

Parra, G., Bradnam, K., & Korf, I. (2007). CEGMA: A pipeline to accurately annotate core genes in eukaryotic genomes. *Bioinformatics*, *23*(9), 1061–1067. https://doi.org/10.1093/bioinformatics/btm071.

Pisani, D., & Wilkinson, M. (2002). Matrix representation with parsimony, taxonomic congruence, and total evidence. *Systematic Biology*, *51*(1), 151–155. https://doi.org/10.1080/106351502753475925.

Price, M. N., Dehal, P. S., & Arkin, A. P. (2010). FastTree 2—Approximately maximum-likelihood trees for large alignments. *PLoS One*, *5*(3). e9490 https://doi.org/10.1371/journal.pone.0009490.

Pride, D. T., Meinersmann, R. J., Wassenaar, T. M., & Blaser, M. J. (2003). Evolutionary implications of microbial genome tetranucleotide frequency biases. *Genome Research*, *13*(2), 145–156. https://doi.org/10.1101/gr.335003.

Qi, J., Luo, H., & Hao, B. (2004). CVTree: A phylogenetic tree reconstruction tool based on whole genomes. *Nucleic Acids Research*, *32*(Suppl. 2), W45–7. https://doi.org/10.1093/nar/gkh362.

Qi, J., Wang, B., & Hao, B. I. (2004). Whole proteome prokaryote phylogeny without sequence alignment: A K-string composition approach. *Journal of Molecular Evolution*, *58*(1), 1–11. https://doi.org/10.1007/s00239-003-2493-7.

Ragan, M. A. (1992). Phylogenetic inference based on matrix representation of trees. *Molecular Phylogenetics and Evolution*, *1*(1), 53–58. https://doi.org/10.1016/1055-7903(92)90035-F.

Redecker, D. (2000). Glomalean fungi from the ordovician. *Science*, *289*(5486), 1920–1921. https://doi.org/10.1126/science.289.5486.1920.

Richards, T. A., Soanes, D. M., Jones, M. D. M., Vasieva, O., Leonard, G., Paszkiewicz, K., et al. (2011). Horizontal gene transfer facilitated the evolution of plant parasitic mechanisms in the oomycetes. *Proceedings of the National Academy of Sciences of the United States of America*, *108*(37), 15258–15263. https://doi.org/10.1073/pnas.1105100108.

Robbertse, B., Reeves, J. B., Schoch, C. L., & Spatafora, J. W. (2006). A phylogenomic analysis of the Ascomycota. *Fungal Genetics and Biology*, *43*(10), 715–725. https://doi.org/10.1016/j.fgb.2006.05.001.

Rokas, A., Williams, B. L., King, N., & Carroll, S. B. (2003). Genome-scale approaches to resolving incongruence in molecular phylogenies. *Nature*, *425*(6960), 798–804. https://doi.org/10.1038/nature02053.

Schoch, C. L., Sung, G. H., López-Giráldez, F., Townsend, J. P., Miadlikowska, J., Hofstetter, V., et al. (2009). The ascomycota tree of life: A phylum-wide phylogeny clarifies the origin and evolution of fundamental reproductive and ecological traits. *Systematic Biology*, *58*(2), 224–239. https://doi.org/10.1093/sysbio/syp020.

Simão, F. A., Waterhouse, R. M., Ioannidis, P., Kriventseva, E. V., & Zdobnov, E. M. (2015). BUSCO: Assessing genome assembly and annotation completeness with single-copy orthologs. *Bioinformatics*, *31*(19), 3210–3212. https://doi.org/10.1093/bioinformatics/btv351.

Slowinski, J. B., & Page, R. D. (1999). How should species phylogenies be inferred from sequence data? *Systematic Biology*, *48*(4), 814–825. https://doi.org/10.1080/106351599260030.
Snel, B., Bork, P., & Huynen, M. a. (1999). Genome phylogeny based on gene content. *Nature Genetics*, *21*(1), 108–110. https://doi.org/10.1038/5052.
Snel, B., Huynen, M. A., & Dutilh, B. E. (2005). Genome trees and the nature of genome evolution. *Annual Review of Microbiology*, *59*(1), 191–209. https://doi.org/10.1146/annurev.micro.59.030804.121233.
Spatafora, J. W., Chang, Y., Benny, G. L., Lazarus, K., Smith, M. E., Berbee, M. L., et al. (2016). A phylum-level phylogenetic classification of zygomycete fungi based on genome-scale data. *Mycologia*, *108*(5), 1028–1046. https://doi.org/10.3852/16-042.
Spatafora, J., Sung, G., Johnson, D., Hesse, C., O'Rourke, B., Serdani, M., et al. (2006). A five-gene phylogeny of Pezizomycotina. *Mycologia*, *98*(6), 1018–1028. https://doi.org/10.3852/mycologia.98.6.1018.
Stanke, M., Steinkamp, R., Waack, S., & Morgenstern, B. (2004). AUGUSTUS: A web server for gene finding in eukaryotes. *Nucleic Acids Research*, *32*(Suppl. 2) W309-12https://doi.org/10.1093/nar/gkh379.
Steel, M., & Rodrigo, A. (2008). Maximum likelihood supertrees. *Systematic Biology*, *57*(2), 243–250. https://doi.org/10.1080/10635150802033014.
Su, Z., & Townsend, J. P. (2015). Utility of characters evolving at diverse rates of evolution to resolve quartet trees with unequal branch lengths: Analytical predictions of long-branch effects. *BMC Evolutionary Biology*, *15*(86), 86. https://doi.org/10.1186/s12862-015-0364-7.
Swofford, L. D. (2002). *PAUP*: Phylogenetic analysis using parsimony (* and other methods). Version 4.0 beta.* Sunderland, MA: Sinauer.
Szöllősi, G. J., Davín, A. A., Tannier, E., Daubin, V., & Boussau, B. (2015). Genome-scale phylogenetic analysis finds extensive gene transfer among fungi. *Philosophical Transactions of the Royal Society of London. Series B, Biological Sciences*, *370*(1678), 20140335. https://doi.org/10.1098/rstb.2014.0335.
Taylor, F. J. R. (1978). Problems in the development of an explicit hypothetical phylogeny of the lower eukaryotes. *Biosystems*, *10*(1–2), 67–89. https://doi.org/10.1016/0303-2647(78)90031-X.
Tekaia, F., Lazcano, A., & Dujon, B. (1999). The genomic tree as revealed from whole proteome comparisons. *Genome Research*, *9*(6), 550–557. https://doi.org/10.1101/gr.9.6.550.
Wang, H., Xu, Z., Gao, L., & Hao, B. (2009). A fungal phylogeny based on 82 complete genomes using the composition vector method. *BMC Evolutionary Biology*, *9*(1), 195. https://doi.org/10.1186/1471-2148-9-195.
Wilkinson, M., Thorley, J. L., Pisani, D. E., Lapointe, F.-J., & McInerney, J. O. (2004). *Some desiderata for liberal supertrees. In O. R. P. Bininda-Emonds (Ed.), Vol. 3. Phylogenetic supertrees: Combining information to reveal the Tree of Life* (pp. 227–246). Dordrecht, The Netherlands: Springer. https://doi.org/10.1007/978-1-4020-2330-9_11.
Wolfe, K. H., & Shields, D. C. (1997). Molecular evidence for an ancient duplication of the entire yeast genome. *Nature*, *387*(6634), 708–713. https://doi.org/10.1038/42711.
Wood, V., Gwilliam, R., Rajandream, M. A., Lyne, M., Lyne, R., Stewart, A., et al. (2002). The genome sequence of Schizosaccharomyces pombe. *Nature*, *415*(6874), 871–880. https://doi.org/10.1038/nature724.
Zuo, G., Li, Q., & Hao, B. (2014). On K-peptide length in composition vector phylogeny of prokaryotes. *Computational Biology and Chemistry*, *53*(Part A), 166–173. https://doi.org/10.1016/j.compbiolchem.2014.08.021.
Zuo, G., Xu, Z., Yu, H., & Hao, B. (2010). Jackknife and bootstrap tests of the composition vector trees. *Genomics, Proteomics and Bioinformatics*, *8*(4), 262–267. https://doi.org/10.1016/S1672-0229(10)60028-9.

CHAPTER SEVEN

Phylogenetics and Phylogenomics of Rust Fungi

M. Catherine Aime*, Alistair R. McTaggart†, Stephen J. Mondo‡, Sébastien Duplessis§,1

*Purdue University, West Lafayette, IN, United States
†University of Pretoria, Pretoria, South Africa
‡US Department of Energy Joint Genome Institute, Walnut Creek, CA, United States
§IUnité Mixte de Recherche INRA/Université de Lorraine, Champenoux, France
[1]Corresponding author: e-mail address: sebastien.duplessis@inra.fr

Contents

Abstract

Rust fungi (Pucciniales) are the most speciose and the most complex group of plant pathogens. Historically, rust taxonomy was largely influenced by host and phenotypic characters, which are potentially plastic. Molecular systematic studies suggest that the extant diversity of this group was largely shaped by host jumps and subsequent shifts. However, it has been challenging to reconstruct the evolutionary history for the order, especially at deeper (family-level) nodes. Phylogenomics offer a potentially powerful tool to reconstruct the Pucciniales tree of life, although researchers working at this vanguard still face unprecedented challenges working with nonculturable organisms that possess some of the largest and most repetitive genomes now known in kingdom fungi. In this chapter, we provide an overview of the current status and

Advances in Genetics, Volume 100
ISSN 0065-2660
https://doi.org/10.1016/bs.adgen.2017.09.011

special challenges of rust genomics, and we highlight how phylogenomics may provide new perspectives and answer long-standing questions regarding the biology of rust fungi.

1. RUST PHYLOGENETICS

1.1 Introduction

Rust fungi (Pucciniales) are an important order of obligate pathogens in the Basidiomycota with complex life cycles, the members of which are usually host specific to species, genera, or families of vascular plants. There are ca. 8000 described species, which makes rust fungi the largest natural group of plant pathogens, and one of the more speciose orders of fungi. Even though rust fungi represent one of the better characterized fungal groups, discovery and description of new taxa is still on the rise, with 10 new genera and ca. 250 new species described since the turn of the century (Toome-Heller, 2016). Cryptic diversity is also believed to be high within the order: detailed studies of what were once considered single broadly distributed species demonstrated that they comprised numerous cryptic species (Beenken, Zoller, & Berndt, 2012; Bennett, Aime, & Newcombe, 2011; Doungsa-ard et al., 2015; Liu & Hambleton, 2013; McTaggart, Doungsa-ard, Geering, Aime, & Shivas, 2015). Their diversity is presently classified into approximately 125 genera and 11–15 families (e.g., Cummins & Hiratsuka, 2003; Kirk, Cannon, Minter, & Stalpers, 2008). Rusts have been the causal agents of devastating disease epidemics (e.g., coffee rust; Cressey, 2013), have been investigated as biological warfare agents (e.g., wheat stem rust; Kortepeter & Parker, 1999), and are used successfully for biological control of invasive plants in countries such as Australia and South Africa (e.g., *Uromycladium tepperianum*; Wood & Morris, 2007). The Pucciniales contain fungal species with the largest known genomes, averaging 380 Mb (Tavares et al., 2014), and are estimated to have evolved ca. 235 Mya (Aime, M.C., & Wilson, A., unpublished data). Each of these topics is discussed in more detail later.

1.2 Life Cycle and Evolution

1.2.1 Life Cycle

Rust fungi have complex and variable life cycles that include up to five different spore stages (spermatia, aeciospores, urediniospores, teliospores, and basidiospores) produced in structures called sori (singular: sorus). This classic life cycle, typified by *Puccinia graminis*, is termed macrocyclic, and in some cases two unrelated hosts may be required for production of different spore stages, termed heteroecious. However, life cycles of rust fungi are plastic, and many species have reduced life cycles that lack one or more of these spore stages (e.g., Tranzschel, 1904; Fig. 1). Also, as a result of the overall complexity of rust life cycles and the fact that spore stages are spatially and temporally separated, entire life cycles of the majority of described species are yet to be elucidated. Traditionally, spore stages of species were linked via painstaking inoculation experiments (Arthur, 1903). Molecular systematics has simplified linkage of life cycle stages via DNA sequences (e.g., barcodes). For example, the aecial and telial stages of wheat stripe rust (*Puccinia striiformis*) were linked by observation of a common molecular barcode more than a century after the rust was first described (Jin, Szabo, & Carson, 2010).

In heteroecious species, spermogonia and aecia (the sori that produce spermatia and aeciospores, respectively) are produced on a separate and unrelated host than that on which uredinia and telia (the sori that produce urediniospores and teliospores, respectively) are produced. The aecial hosts

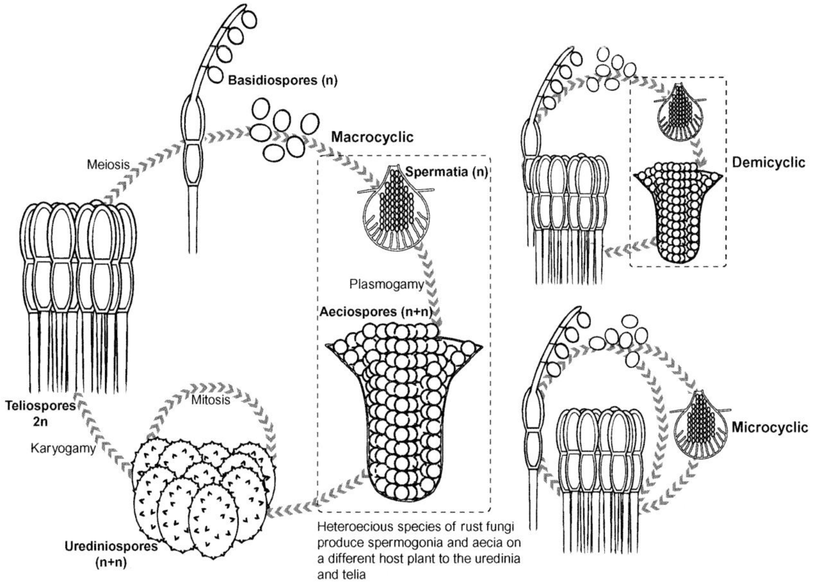

Fig. 1 Stylized spore stages for macrocyclic, demicyclic, and microcyclic life cycles of *Puccinia* spp. Life cycle stages that are heteroecious in some taxa are outlined. Hemicyclic life cycles (not pictured) do not form spermogonia or aecia, and occur on one host species.

of heteroecious species were thought to be pathways for speciation of rust fungi under an observed phenomenon known as Tranzschel's Law, in which a microcyclic species would be correlated with the aecial host of a macrocyclic, heteroecious species (Arthur, 1934; Shattock & Preece, 2000).

Rust fungi that complete their life cycle on a single host are called autoecious. Microcyclic rusts (those that lack the uredinial and aecial stages, and in some instances the spermogonial stage as well) are all autoecious; macro- (all five stages) and demicyclic (no uredinial stage) rusts can be either autoecious or heteroecious (Fig. 1). A final variation, termed hemicyclic, is used to describe species for which spermogonial and aecial stages are not known. In true hemicyclic rusts, basidiospores are homothallic and capable of reinfecting the uredinial/telial hosts (e.g., Anikster et al., 2004); other species—sometimes erroneously termed hemicyclic—can actually be heteroecious and macrocyclic rusts with unknown alternate hosts. While the complete life cycles of many rust fungi, including important species such as soybean rust (*Phakopsora pachyrhizi*), are unknown, the presence of different life cycle stages is believed to be homoplasious in the Pucciniales (Aime, 2006; McTaggart, Geering, & Shivas, 2014).

1.2.2 Evolution and Diversification

One hypothesis regarding the evolution of rust fungi was that speciation has occurred primarily through coevolution with their host plants, beginning with ferns and coevolving to gymnosperms and angiosperms (e.g., Savile, 1971). However, early cladistic and phylogenetic research challenged this hypothesis (Hart, 1988; Sjamsuridzal, Nishida, Ogawa, Kakishima, & Sugiyama, 1999) and subsequent researchers relied more heavily on morphological characters for derivation of phylogenetic hypotheses for the rusts. Nonetheless, early attempts to reconstruct evolutionary relationships for the rusts based on molecular characters have provided evidence, at least at small scales, for divergence by coevolution or from taxonomically small host shifts. Such examples include genera in Phragmidiaceae and Raveneliaceae on species of Rosaceae and Fabaceae (Aime, 2006), and species of *Endoraecium* and *Uromycladium* on *Acacia* in Australia (McTaggart et al., 2015). At larger scales, McTaggart, Shivas, Doungsa-ard, et al. (2016) and McTaggart, Shivas, van der Nest, et al. (2016) implicate large host jumps as drivers for diversification in this lineage. Inferring the ancestral species to the rust lineage has also been rigorously debated, but molecular data suggest that the most recent common ancestor of extant rust fungi might have diversified on hosts in early gymnosperm lineages such as Araucariaceae and Taxales (Aime, 2006;

Aime, M.C. and Wilson, A., unpublished data). Taken together these studies indicate a strong dependency on host and coevolution punctuated by jumps to unrelated hosts as drivers for diversification of rust fungi (e.g., van der Merwe, Walker, Ericson, & Burdon, 2008).

1.3 Taxonomy and Systematics

The Pucciniales are members of the Pucciniomycetes and have a sister relationship with the Platyglocales, Helicobasidiales, Pachnocybales, and Septobasidiales (Aime et al., 2006). Other orders of the Pucciniomycetes mainly germinate from simple teliospores with phragmobasidia and include pathogens of mosses, scale insects, and the spermogonia of rust fungi (Aime et al., 2006), and the character of simple teliospores with auricularioid basidia is considered plesiomorphic for Pucciniales (Aime, 2006).

Rust fungi were first classified based on the behavior of their basidium and/or whether teliospores were pedicillate (Cunningham, 1931). This classification generally divided rusts into three or four families, for example, the Pucciniaceae, Melampsoraceae, and Zaghouaniaceae (Cunningham, 1931; Sydow & Sydow, 1915). This early three-family system is reflected in the current classification of rust fungi sensu (Aime, 2006) that found support for three suborders: Uredinineae, Melampsorineae, and Mikronegeriineae.

Cummins and Hiratsuka (2003) provided the most widely used familial classification of rust fungi, which included 13 families. This classification differed from earlier works in emphasizing phenotypic characters of the teliospores and structure of the spermogonia rather than host associations. Aime (2006) investigated these families within a systematic framework and determined that several were polyphyletic or redundant. The current familial classification of rust fungi based on phylogenetic studies by Aime (2006), McTaggart, Shivas, Doungsa-ard, et al. (2016), McTaggart, Shivas, van der Nest, et al. (2016), and Beenken (2017) supports 11 families, namely Mikronegeriaceae, Coleosporiaceae (including Cronartiaceae), Melampsoraceae s.l., Phakopsoraceae p.p., Phragmidiaceae, Pileolariaceae, Pucciniaceae, Pucciniastraceae, Raveneliaceae, Sphaerophragmiaceae, and Uropyxidaceae.

Many genera and families of rust fungi have not been included in phylogenetic analyses to date, or do not fit into resolved monophyletic groups. This is likely to change when more taxa are included within modern analyses. This may include splitting polyphyletic genera and families to reflect systematic relationships. For example, the species of *Gymnosporangium* are not confamilial with other members of Pucciniaceae, as previously thought, and likely represent a currently undescribed family-level lineage (Aime, 2006; Maier, Begerow, Weiss, & Oberwinkler, 2003). Other genera, such as *Allodus*, remain unplaceable with currently available markers (Minnis, McTaggart, Rossman, & Aime, 2012). Genera in polyphyletic families, such as Chaconiaceae and Uropyxidaceae sensu (Cummins & Hiratsuka, 2003), will ultimately be placed in different families, whether this is by new combinations, descriptions, or divisions. The current understanding of relationships between genera and families of rust fungi is shown in Fig. 2.

A stable familial classification of rust fungi will hinge on the reconstruction of deeper level nodes in the Pucciniales tree of life, the resolution of polyphyletic and uncertain families, and the robust placement of orphaned genera in a systematic context. This kind of robust resolution is dependent on two factors. First, taxonomic resolution for higher-level taxa will depend on the incorporation of type species of families and genera. However, many keystone rust fungi are rarely collected and, for reasons elaborated later, accommodating these taxa in a resolved phylogeny will hinge on the collection of fresh material. A second challenge in determining the phylogenetic relationships between rust fungi is that molecular data are not readily available or easily obtained. Rust fungi are obligate biotrophs that for the most part cannot be maintained in pure culture on artificial media. For this reason, molecular systematics studies of rust fungi depend on DNA extracted from minute sori rendering the amount of extractable rust DNA a limiting factor for molecular work. A confounding complication is that DNA extracted from leaf material will contain the DNA of other organisms, such as phylloplane yeasts, endophytic fungi of the host, and the host plant itself. Thus, molecular work depends on the development of specific primers in addition to the extraction of adequate material (Aime, 2006; van der Merwe et al., 2008). Because single-copy DNA markers are not easily amplified, which is further discussed later, the most robust phylogenetic studies have relied at most on data from two to three genes.

1.4 Phylogenetic Analyses and Molecular Barcoding of Rust Fungi

A major advancement in phylogenetic studies for rust fungi was the development of rust-specific primers for the internal transcribed spacer (ITS), large subunit (LSU/28S), and small subunit (SSU/18S) regions of ribosomal DNA (rDNA) (Aime, 2006; Beenken et al., 2012; Pfunder, Schurch, & Roy, 2001). Because rDNA regions contain hundreds of copies within a

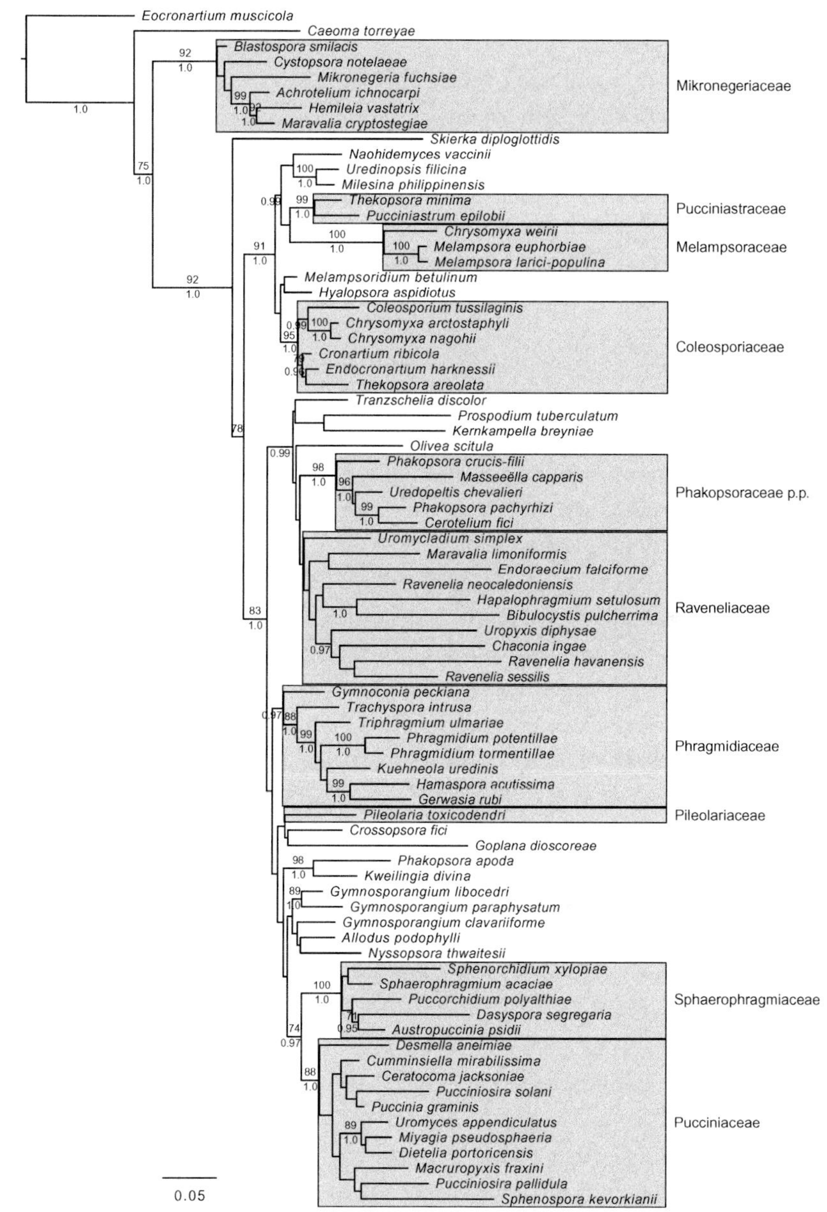

Fig. 2 Phylogram of genera and currently accepted families of Pucciniales. The phylogram was obtained from a maximum likelihood search of large subunit and small subunit regions of ribosomal DNA and cytochrome *c* oxidase subunit 3 of mitochondrial DNA, partitioned as three separate genes in RAxML v.8.2 (Stamatakis, 2014). Bootstrap values ($\geq$70%) from 1000 replicates shown above nodes. Posterior probability values ($\geq$0.95) summarized from 30,000 trees in a Bayesian search in MrBayes (Ronquist & Huelsenbeck, 2003) shown below nodes. GTRGAMMA was used as a model of evolution for both phylogenetic criteria. Monophyletic families of rust fungi are *shaded*; however, the topology and taxonomy of genera and families used in the present tree are expected to change when more taxa and gene regions are sampled.

genome, the small quantities of rust DNA in a typical extraction are not as limiting when working with these loci. The LSU and SSU regions are most often used in phylogenetic studies at the infrageneric and infrafamilial level in rust fungi (Aime, 2006; Beenken, 2017; Maier et al., 2003; Scholler & Aime, 2006; Wingfield, Ericson, Szaro, & Burdon, 2004; Yun, Minnis, Kim, Castlebury, & Aime, 2011), whereas ITS, discussed in more detail later, is more often employed for delimitation of closely related species or species complexes (e.g., Alaei et al., 2009; Barilli, Satovic, Sillero, Rubiales, & Torres, 2010). None of the rDNA regions have yet provided robust resolution of deeper, e.g., interfamilial, phylogenetic nodes.

Single-copy nuclear genes are used less for phylogenetic studies of rust fungi than in Ascomycota and other orders of Basidiomycota because of the limiting amounts of DNA and the high frequency of amplification of other fungi (such as epi- and endophytes) with nonspecific primers. However, translation elongation factor 1 alpha (TEF), β-tubulin (B-tub), and RNA polymerase II second largest subunit (RPB2) have been successfully applied at the species level for discerning relationships within specific genera of Pucciniaceae (the largest family of rust fungi) (Liu & Hambleton, 2010, 2013; van der Merwe, Ericson, Walker, Thrall, & Burdon, 2007; van der Merwe et al., 2008). The cytochrome *c* oxidase subunit 3 (CO3) gene of mitochondrial DNA was investigated as a barcode for rust fungi (Vialle et al., 2009) and has been subsequently used in several phylogenetic studies (Beenken, 2014; Doungsa-ard et al., 2015; Feau, Vialle, Allaire, Maier, & Hamelin, 2011; McTaggart et al., 2015; McTaggart, Shivas, Doungsa-ard, et al., 2016; McTaggart, Shivas, van der Nest, et al., 2016).

As already mentioned, relationships between closely related species of rust fungi have relied heavily on the ITS region in combination with the intergenic spacer or LSU regions of rDNA (Beenken, 2014; Beenken & Wood, 2015; Beenken et al., 2012; Demers, Liu, Hambleton, & Castlebury, 2017; McTaggart et al., 2014; McTaggart, Shivas, Doungsa-ard, et al., 2016; McTaggart, Shivas, van der Nest, et al., 2016). However, the ITS region is not a reliable marker for molecular barcoding and direct Sanger sequencing in Puccinales because intraspecific and intraisolate diversity has been observed within a wide range of rust fungi species (Alaei et al., 2009; Virtudazo, Nakamura, & Kakishima, 2001). Intraisolate diversity of haplotypes of the ITS region has been discovered in genera such as *Austropuccinia*, *Coleosporium*, *Phakopsora*, and *Puccinia*; paralogous/redundant copies of the ITS region have been found in *Gymnosporangium* and *Puccinia* (e.g., McTaggart & Aime, 2012; Novick, 2008; Rush, 2012).

To summarize, there remain multiple evolutionary questions regarding the diversification of Pucciniales, and the adaptations that allowed them to become the largest yet most complex group of plant pathogens. A well-resolved phylogenetic hypothesis of the Pucciniales has yet to be developed and will be critical for answering many or these long-standing questions. These questions include, for instance, (i) when, how, and why did heteroecism evolve; (ii) how have closely related rust fungi evolved to infect distantly related host plants; (iii) how and why have different life cycle stages evolved; and (iv) what makes rust fungi such effective biotrophic pathogens? Phylogenomics offer real promise for overcoming most of the shortcomings of traditional phylogenetics studies in rusts. For instance, a phylogenomic approach was recently used on fungal herbarium material to finally resolve intraordinal relationships for another previously intractable order, the Agaricales (mushroom-forming fungi), by generating hundreds of single-copy orthologs (Dentinger et al., 2016). The identification of reliable species-specific alternative barcode loci, such as those that regulate sexual compatibility, is also possible through comparative genomics and phylogenomics. Although rust fungus genomics has its own special set of problems, discussed in detail later, strides are being made in this area such that the resolution of the rust tree of life now appears more feasible than ever before.

2. RUST PHYLOGENOMICS

In the past few years, a revolution in biology has driven a transition from sequencing single genomes of model eukaryotes to the systematic sequencing of genomes from different individuals and taxa within a given kingdom. The speed and depth of progression have shifted genomics from an emerging field to an established discipline and a very active front of science. Progress in the ever-expanding field of bioinformatics fostered this revolution. High-quality genomes can now be sequenced and assembled, and projects that build on genetic information carried by the DNA molecule, such as the encyclopedia of DNA elements (ENCODE), become more robust as more genomes are available for major branches of life (Hug et al., 2016). However, it is not trivial to obtain complete reference genomes for given species; problems in assembly and annotation are encountered depending on the complexity and composition of the chromosomes and the available technologies. The cost of sequencing genomes was reduced with the emergence of next-generation sequencing technologies. These different technologies can generate either huge amounts of high-fidelity short sequences or very long sequences that may have a higher error rate.

The major bottlenecks to the study of eukaryotic genomes in general are at the stages of (1) DNA isolation and (2) assembly of complex genomes that contain a high number of repetitive regions. These obstacles are forefront in rust fungal genomics, which have the added complication of sheer size in Mbs and dikaryotic source material, and the field is still developing when compared to other fungi because of these challenges. In this section, we will present an update on rust genomics and how advances and current knowledge can support future progress in phylogenomics.

2.1 Rust Fungal Genomics Projects

Collaborative international efforts pioneered the sequencing of rust fungi genomes in the mid-2000s. The *Puccinia* comparative genome project conducted at the Broad Institute of MIT and Harvard targeted the genomes of three cereal rust fungi: *P. graminis* f. sp. *tritici*, *P. striiformis* f. sp. *tritici*, and *Puccinia triticina*. These pathogens are responsible for major diseases of wheat, particularly *P. graminis* f. sp. *tritici*, the causal agent of wheat stem rust. Another project conducted at the US Department of Energy Joint Genome Institute (US DOE JGI) sequenced the genome of poplar rust, *Melampsora larici-populina*, in line with previous efforts to sequence microbes associated with the model poplar tree *Populus trichocarpa* (Martin et al., 2008; Tisserant et al., 2013; Tuskan et al., 2006). These parallel efforts were combined to establish a joint comparative genomic study, and analysis of the wheat stem rust and poplar rust genomes has already helped to unravel general features of biotrophy in rust fungi (Duplessis et al., 2011). Concomitantly in 2011, the genome of the yellow stripe rust fungus *P. striiformis* f. sp. *tritici* was sequenced and assembled from short reads of Illumina sequencing, and its release provided raw but pertinent genomic insights (Cantu et al., 2011).

In a second phase, several genomes of cereal rusts and other important rust fungi were sequenced and published in parallel and collaborative efforts. Diverse sequencing strategies were applied—mostly exhaustive assembly based on short-read sequencing—and a few species were sequenced several times, with different target isolates. So far, 20 genomes have been reported, representing a total of 11 species. However, only five of these available genomes are of adequate quality, in terms of completeness and number of scaffolds (see Table 1). In this second phase, different species and isolates of rust fungi were resequenced by Illumina sequencing, and these data have

Table 1 Genomic Resources for the Order Pucciniales

Species Name	Family	Common Disease Name	Genome Features: Total Assembly Size	Contigs/Scaffolds No	Transposable and Repetitive Elements (% of Genome)	Predicted Gene Number	Notes	Publication
Melampsora larici-populina	Melampsoraceae	Poplar rust	101.1 Mb	3264/462	45%	16,399	Sanger sequencing	Duplessis et al. (2011)
Melampsora lini	Melampsoraceae	Flax rust	220 Mb[a]	NA/21,310	45.9%	16,271	Illumina sequencing	Nemri et al. (2014)
Puccinia graminis f. sp. *tritici* isolate CDL 75-36-700-3	Pucciniaceae	Wheat stem (black) rust	88.6 Mb	4557/392	43.7% (updated to 36.5%[b])	17,773 (updated to 15,800[b])	Sanger sequencing	Duplessis et al. (2011)
Puccinia graminis f. sp. *tritici* isolate Pgt 21-0			92 Mb	21,517/NA	NA	21,874 (annotated from pan-genome)	Illumina sequencing, pan-genome from five Australian isolates	Upadhyaya et al. (2015)
Puccinia triticina race 77	Pucciniaceae	Wheat leaf rust	100.6 Mb	44,586/2651	37.5%	27,678	454 sequencing, Sanger sequencing of >800 fosmids	Kiran et al. (2016)
Puccinia triticina race 106			106.5 Mb	67,044/7448	40%	26,384	454 sequencing	Kiran et al. (2016)

Continued

Table 1 Genomic Resources for the Order Pucciniales—cont'd

			Genome Features					
Species Name	Family	Common Disease Name	Total Assembly Size	Contigs/Scaffolds No	Transposable and Repetitive Elements (% of Genome)	Predicted Gene Number	Notes	Publication
Puccinia triticina isolate BBBD race 1			135.3 Mb	24,838/14,818	50.9%	14,880	Illumina, 454 and Sanger sequencing, Sanger sequencing of fosmids and BACs	Cuomo et al. (2017)
Puccinia striiformis f. sp. *tritici* isolate PST-130	Pucciniaceae	Wheat stripe (yellow) rust	68.2–78.8 Mb[a]	29,178/NA	17.8%	20,423 (updated to 18,149[c])	Illumina sequencing	Cantu et al. (2011)
Puccinia striiformis f. sp. *tritici* isolate PST-21			73 Mb	43,106/NA	NA	20,653	Illumina sequencing	Cantu et al. (2013)
Puccinia striiformis f. sp. *tritici* isolate PST-43			71 Mb	49,784/NA	NA	21,036	Illumina sequencing	Cantu et al. (2013)

Puccinia striiformis f. sp. *tritici* isolate PST-87/7	53 Mb	55,502/NA	NA	20,688	Illumina sequencing	Cantu et al. (2013)
Puccinia striiformis f. sp. *tritici* isolate PST-08/21	56 Mb	50,898/NA	NA	20,875	Illumina sequencing	Cantu et al. (2013)
Puccinia striiformis f. sp. *tritici* isolate PST-CY32	110 Mb[a]	12,833/4283	48.9%	25,288	Fosmid-to-fosmid Illumina sequencing, Sanger sequencing of 10 fosmids	Zheng et al. (2013)
Puccinia striiformis f. sp. *tritici* isolate PST-78	117.3 Mb	17,295/9715	31.4%	19,542	454 and Illumina sequencing, Illumina sequencing of fosmids	Cuomo et al. (2017)
Puccinia striiformis f. sp. *tritici* isolate PST-31	66.3 Mb	30,066/NA	36.8%	18,362	Illumina sequencing, assembly derived from PST-78	Kiran et al. (2017)
Puccinia striiformis f. sp. *tritici* isolate PST-K	69.77 Mb	32,818/NA	36.3%	18,880	Illumina sequencing, assembly derived from PST-78	Kiran et al. (2017)

Continued

Table 1 Genomic Resources for the Order Pucciniales—cont'd

Species Name	Family	Common Disease Name	Genome Features: Total Assembly Size	Genome Features: Contigs/Scaffolds No	Genome Features: Transposable and Repetitive Elements (% of Genome)	Genome Features: Predicted Gene Number	Notes	Publication
Puccinia striiformis f. sp. *tritici* isolate PST-46S 119			70.2 Mb	24,737/NA	35.2%	19,795	Illumina sequencing, assembly derived from PST-78	Kiran et al. (2017)
Puccinia sorghi	Pucciniaceae	Maize common rust	102 Mb[a]	28,773/15,722	33%	21,087	Illumina sequencing	Rochi et al. (2016)
Uromyces fabae	Pucciniaceae	Broad bean rust	330–379 Mb[a]	NA/59,735	NA	NA	Illumina sequencing, raw assembly	Link et al. (2014)
Hemileia vastatrix	Mikronegeriaceae	Coffee rust	333 Mb	396,264/302,466	74.4%	14,445	Illumina and 454 sequencing, raw and partial assembly, hybrid genome from eight different isolates	Cristancho et al. (2014)

Austropuccinia psidii	Sphaerophragmiaceae	Myrtle rust	103–145 Mb[a]	37,605/NA	27%	>19,000	Illumina sequencing, raw assembly	Tan, Collins, Chen, Englezou, and Wilkins (2014)
Phakopsora pachyrhizi	Phakosporaceae *sensu stricto*	Asian soybean rust	850 Mb–1 Gb[a]	NA	NA	NA	Illumina sequencing, not assembled	Loehrer et al. (2014)

[a]Estimated total size.
[b]Updated in Cuomo et al. (2017).
[c]Updated in Cantu et al. (2013).

provided valuable information about intraspecific variation and impact of selection on genes related to host infection (see details in Duplessis, Bakkeren, & Hamelin, 2014).

Based on our knowledge of the current efforts ongoing in the rust community, we have entered a third phase of rust fungal genomics that will benefit from new sequencing technologies. These new technologies, such as PacBio and Nanopore sequencing, can generate longer sequences that improve genome assembly and enable sorting of haplotypes for dikaryotic rust genomes (e.g., preprint by Miller et al., 2017 in bioRxiv). Due to their highly repetitive nature, exploitation of these long-read sequencing technologies will be important as we move forward in the field of rust genomics.

Most reported genome sequences of rust fungi are hosted at the JGI Mycocosm webportal (http://genome.jgi.doe.gov/programs/fungi). The tools provided by this platform enable cross-comparison between rust genomes. Deposition of new published rust genomes is encouraged to foster comparative studies. Genomes that are not yet published but that have been sequenced in the framework of community sequencing projects (CSPs; e.g., *Cronartium quercuum* f. sp. *fusiforme* and *Melampsora allii-populina*) are also hosted in Mycocosm, and the growing list of accepted and ongoing CSPs at the JGI show an increasing number of projects to sequence rust fungi are underway.

2.1.1 Super-Size Genomes of Rust Fungi

So far, the size reported for genomes of rust fungi varies from ~60 to >300 Mb (Table 1). The upper limit is expected to be much larger, up to >2 Gbp, according to estimates based on flow cytometry, with the average Pucciniales genome now estimated at 380 Mb (Ramos et al., 2015; Tavares et al., 2014). Compared to the genome sizes of other basidiomycetes (range = 7.6–176.4 Mb; mean size = 40.2 Mb; based on published genomes available in Mycocosm in July 2017), genomes of rust fungi average nearly one order of magnitude larger than other Basidiomycota. Different isolates of the same species have varied in their reported genome sizes, which may be a result of the sequencing technologies used that vary in their quality of assembly (e.g., 100–135 Mb for *P. triticina*; 53–117 Mb for *P. striiformis*; Table 1). Remarkable differences exist within rust families, such as the Melampsoraceae or the Pucciniaceae (Table 1). For example, in a single genus, *Melampsora*, genome size is highly variable. The genome of *M. larici-populina* (poplar rust) is close to 100 Mb (101 Mb in Duplessis et al., 2011; revised at 110 Mb in the unpublished version 2, available at the JGI Mycocosm), whereas the genome of *Melampsora lini* (flax rust) is estimated at 220 Mb (Nemri et al., 2014), and the genome of *M. allii-populina* (poplar rust) is estimated at 336 Mb (JGI Mycocosm, not yet published).

A couple of very important and damaging rust fungi such as coffee rust, *Hemileia vastatrix*, and soybean rust, *Ph. pachyrhizi*, exhibit very large genome sizes, estimated between 300 and >850 Mb (Table 1). A raw genome assembly from different isolates of *H. vastatrix* was produced using 454 pyrosequencing and Illumina sequencing technologies—but with unsatisfying assembly metrics (>300,000 scaffolds), and a sequenced genome size of 333 Mb that contrasts with the nearly 800 Mb estimate based on flow cytometry (Cristancho et al., 2014; Tavares et al., 2014). The size and sequencing approach combined have been problematic for assembly of this important fungal genome. Similarly, several attempts to sequence the *Ph. pachyrhizi* genome were conducted using different sequencing strategies, but all have proven unsuccessful due to the genome size and the level of heterozygosity from dikaryotic starting material used for DNA extraction (see Loehrer et al., 2014 for details).

2.1.2 Challenges for Assembly of Rust Genomes

The step of assembling sequence reads is critical in any genomic project targeting species with large genomes, such as rust fungi. Aside from their large genome size, the Pucciniales have several other features that are problematic for genome assembly. The most important of these are a high level of heterozygosity, which is primarily a result of the dikaryotic spore states used as source material for DNA isolation (Duplessis et al., 2014), and the high content of repetitive elements in the genomes of rust fungi. These issues in assembly of rust genomes preclude sequencing strategies based only on short reads. So far, only the pioneer rust genome projects that mostly relied on Sanger sequencing technology featured acceptable assembly metrics, with less than 500 assembled scaffolds (Duplessis et al., 2011).

Approaches that have since used short-read sequencing techniques have produced lower-quality assemblies characterized by very large numbers of scaffolds (Table 1). A few intermediate-quality genomes have been assembled into less than 10,000 scaffolds, but most assemblies are highly fragmented, comprising over 20,000 assembled scaffolds. Although differences in genome assemblies can result from intraspecific genetic diversity, in the examples listed in Table 1, the divergence in genome size and quality of assemblies observed for some species is most likely caused by the technologies and assembly strategies used (e.g., *P. striiformis* f. sp. *tritici*). The quality of the genes annotated is dependent on the quality of the assembled genome, thus, a robust assembly is critical for phylogenomic approaches based on gene order.

2.1.3 The Repetitive Nature of Rust Genomes

The identification of repetitive elements in the genome and the classification of transposable elements (TEs) are essential for masking these sequences and performing subsequent de novo gene prediction and annotation. TE annotation is based on preexisting databases and tools that allow recognition of highly repetitive elements of unknown nature but with typical features, e.g., terminal repeats and known TE-related coding sequences (Flutre, Duprat, Feuillet, & Quesneville, 2011). Depending on the database(s) used to search for TEs or the tools employed for a proper annotation, the results can differ strikingly. For instance, the original annotation of TEs in the poplar rust and wheat stem rust fungal genomes identified a 43%–45% total TE coverage (Duplessis et al., 2011). Recently, estimates of the TE content of the wheat stem rust *P. graminis* f. sp. *tritici* were revised to 36.5% (Cuomo et al., 2017). However, the approaches differed between these analyses and the discrepancies may simply reflect the level of accuracy of TE annotation. For instance, a reassessment of the TE content of the poplar rust genome that used the original tools from Duplessis et al. (2011) and updated TE-reference databases showed a similar estimate of 45% TE content (Lorrain, C., & Duplessis, S., unpublished data). TE content in rust genomes is very high and accounts for 17.8%–74.4% of whole-genome assemblies, according to published reports (Table 1). Other fungal genomes in the Basidiomycota exhibit a much lower coverage with 6% TE content on average (Min–Max: 0.15%–71%; based on published genomes available in Mycocosm in July 2017). TEs account for a large proportion of the size of rust genomes, and it is expected that other rust species, such as soybean rust, with very large expected genome sizes may harbor an even higher percentage of TE content.

In the past two years, rust genomes and transcriptomes with detailed annotations have been reported (Cuomo et al., 2017; de Carvalho et al., 2017; Dobon, Bunting, Cabrera-Quio, Uauy, & Saunders, 2016; Jing, Guo, Hu, & Niu, 2017; Kiran et al., 2017, 2016; Liu et al., 2015; Rochi et al., 2016; Rutter et al., 2017; Upadhyaya et al., 2015). However, no specific standards have been clearly established in the field, which makes assembly and comparison of TE content difficult. Some basic standards could be followed to allow for future use of such data in comparative genomics and phylogenomics that would facilitate sequence quality summarization regardless of the technology in use and the scope of the desired end product (complete or draft genome, coding space capture-only). For any comparison of genome content/gene complements, authors should make sure that the same tools with comparable parameters were applied to other rust genomes before any conclusions are made. Similarly, genome-scale synteny or gene order should not be reported or compared when the assembly is highly fragmented. TEs represent a considerable part of rust genomes, and their identification and annotation is critical for users to make conclusions from comparative studies of gene family expansions or contractions. The community should aim for the best annotation possible and use dedicated de novo TE detection and annotation tools, such as the REPET pipeline (Hoen et al., 2015; Jamilloux, Daron, Choulet, & Quesneville, 2017).

2.1.4 A Landscape of Unknown: The Gene Complement of Rust Fungi

Rust genomes have a higher average range of predicted genes than those reported for other basidiomycetes. There are between 15,000 and 20,000 genes per genome in rust fungi (Table 1), whereas there are ~13,500 genes on average for basidiomycetes (Min–Max: 3517–35,274; based on published genomes available in Mycocosm in July 2017). Here again, large intraspecific differences are noticeable, such as *P. triticina* for which 14,880–27,678 genes have been reported from different isolates (Cuomo et al., 2017; Kiran et al., 2016).

The quality of both the assembly and annotation of TEs impacts the potential for overprediction of genes. A fragmented assembly can generate truncated genes and an artificial inflation of the gene complement, which is critical for comparisons that stem from these predictions. Additionally, if not properly masked, transposons can be falsely identified as nontransposon coding genes, further inflating the total number of genes reported for a given genome. As this inflation has become apparent, only a few genomes have been revised since their original publication. In the case of sequencing *P. graminis* f. sp. *tritici*, it was first estimated to have at least 17,773 genes (Duplessis et al., 2011), then was revised to 15,800 (Cuomo et al., 2017). In *P. striiformis* f. sp. *tritici* isolate PST-130, gene counts were revised from 20,423 (Cantu et al., 2011) to 18,149 (Cantu et al., 2013). In contrast, the 16,399 genes of the *M. larici-populina* genome published in 2011 have been revised to a higher number in its latest version in the JGI Mycocosm, thanks to the inclusion of new RNA-sequencing data related to new stages of the rust fungus life cycle (Duplessis, S., et al., unpublished data). In any revised genome annotation, a table system that allows changes to be easily tracked between versions should be provided to the community.

The large gene complements of rust fungi contain a very important fraction of genes with an unknown function that are not represented outside the Pucciniales. Some of these are assignable to small and large gene families, and many of them are specific at the rust family or species levels. The absence of functional annotation for many of those expanded gene families specific to the Pucciniales precludes speculation regarding their possible involvement in the biology of these fungi. However, comparative genomic studies taking into consideration the expression of these genes in regard to the ecology and the biology of rust fungi, including the use of coregulation networks, may help to identify their involvement in some biological processes (Kohler et al., 2015; Rutter et al., 2017).

In 2011, the gene complements of poplar rust and wheat stem rust were scrutinized in detail by international consortia that assessed their overall quality, including through manual expert gene curation (see supplemental data of Duplessis et al., 2011). However, the manual curation step is often omitted and recent comparative genomic studies have relied solely on automatic ab initio annotation without any posterior validation. However, the quality and the completeness of the resulting gene complements and annotations are particularly important for phylogenomic approaches. Once more, standard requirements must be observed in any lab-level genome sequencing effort, so those data can be included in further comparative and phylogenomics studies. Transcript support-based annotation through RNA sequencing should be a minimal prerequisite for annotation. An ideal situation would be inclusion of several stages of the rust life cycle to maximize the opportunity to capture expression of most if not all genes. However, sequencing of multiple spore stages is rarely met or even possible for rust fungal species in which some spore stages are difficult to obtain or are still unknown. Predefined sets of core orthologous and conserved fungal or eukaryotic genes are often used to demonstrate completeness of a new genome annotation (e.g., tools such as CEGMA or BUSCO were applied in almost all genomics studies listed in Table 1, see Section 2.2.3) (Parra, Bradnam, & Korf, 2007; Simão, Waterhouse, Ioannidis, Kriventseva, & Zdobnov, 2015). In the case of rust fungi that display super-size genomes, the capture of only the coding space can be an approach for obtaining gene information. Similarly, assembly and annotation of RNA-seq transcriptomic data can provide valuable insights. For instance, several RNA-seq efforts targeting expression profiling of purified infection structures such as haustoria or of infected plant tissues during time course studies have been reported for soybean rust (de Carvalho et al., 2017; Link et al., 2014; Tremblay, Hosseini, Li, Alkharouf, & Matthews, 2012, 2013). These studies of the transcriptome may help in making substantial progress to understand crucial cellular functions for this fungus. However, the lack of a backbone genome prevents in-depth comparative studies between other rust fungi.

2.2 Composition and Organization of Rust Fungal Genomes: Steps Toward Phylogenomics

The identification of homologous characters between different organisms is crucial for comparative and phylogenomic analyses. The assessment of gene homology and orthology is central to all current methods (Delsuc, Brinkmann, & Herve, 2005; Emms & Kelly, 2015). In the present section, we summarize studies on the composition and organization of published rust genomes that provided useful information on gene homology/orthology and gene order, how these may form the basis for future phylogenomics studies with this group.

2.2.1 Synteny in Rust Fungal Genomes

There is little evidence of synteny in rust fungi (Duplessis et al., 2014). However, this lack of evidence may be largely attributable to the fragmented assemblies of rust genomes, and perhaps the large phylogenetic distance between sampled taxa. The wheat stem rust fungus, *P. graminis* f. sp. *tritici*, and the poplar rust fungus, *M. larici-populina*, exhibit the best genome assemblies so far with 392 and 462 scaffolds, respectively (Table 1; Duplessis et al., 2011). In the pioneer study comparing these two relatively well-assembled genomes, only a few syntenic blocks could be revealed. These regions contained very few common genes and the largest reported block contained six orthologous gene pairs on a 281-Kb genomic sequence. Many different genes and numerous TEs spanned the corresponding regions in each genome.

The divergence between the Melampsoraceae and Pucciniaceae is estimated to have occurred approximately 67–77 Mya according to McTaggart, Shivas, Doungsa-ard, et al. (2016) and McTaggart, Shivas, van der Nest, et al. (2016), and the absence of synteny could reflect massive genomic rearrangements via past TE activity. The fragmented nature of the flax rust genome, *M. lini*, prevented accurate estimates of synteny with *M. larici-populina* (Nemri et al., 2014). Comparisons drawn for several genomes of closely related species in the Pucciniaceae, i.e., *P. graminis* f. sp. *tritici*, *P. striiformis* f. sp. *tritici*, *P. triticina*, and *P. sorghi* which have diverged less than

10 Mya (McTaggart, Shivas, Doungsa-ard, et al., 2016; McTaggart, Shivas, van der Nest, et al., 2016), indicate extensive microsynteny. With the exception of *P. graminis* f. sp. *tritici*, these genomes show fragmented assemblies, which also prevent the identification of very large syntenic blocks.

Analysis of microsynteny between *P. graminis* f. sp. *tritici* and the raw assembly of *P. striiformis* f. sp. *tritici* isolate PST-130 revealed colinearity for many contigs. The interruption of contigs by unshared genes between the two genomes illustrated that multiple rearrangements may have occurred since the most recent common ancestor of the two wheat rust species (Cantu et al., 2011). To test synteny between genomes of different rust fungi, three full bacterial artificial chromosomes (BACs) containing a total of 381 Kb from the *P. triticina* genome were sequenced and compared to the *P. graminis* f. sp. *tritici* genome (Fellers et al., 2013). This comparison revealed local conservation of gene order, as well as gene shuffling and inversions, moderate to important sequence divergence between genes from the two species, and the presence of different insertion loci for various types of TEs (Fellers et al., 2013). This more detailed analysis for conservation of gene order confirmed microsynteny in closely related species of Pucciniaceae. Interestingly, one *P. triticina* BAC was syntenic with two scaffolds from the *P. graminis* f. sp. *tritici* genome, possibly representing the two haplotypes for this rust fungus (Fellers et al., 2013). Analysis of microsynteny was also performed between the genomes of *P. sorghi* and *P. striiformis* f. sp. *tritici* isolate PST-78, and more than 300 syntenic blocks with a conserved gene order were revealed that represented 11% of the assembled genome of *P. sorghi*. However, the number of conserved genes within a given block remained low, even if a few examples of true colinear genomic segments could be identified (Rochi et al., 2016).

The sequencing of the *P. triticina* genome and comparison with other cereal rust fungi have revealed the presence of many syntenic blocks accounting for large portions of these genomes (Cuomo et al., 2017). Strikingly, these blocks are 30% larger overall in *P. triticina* than in *P. graminis* f. sp. *tritici*. However, the poor assembly for the genome of *P. striiformis* f. sp. *tritici* isolate PST-78, compared with those of the two other cereal rust fungi, does not allow definitive conclusions to be made. Nevertheless, the analysis of the syntenic regions between *P. triticina* and *P. graminis* f. sp. *tritici* clearly shows that genome expansion in the former is due to larger insertions of repetitive elements between conserved genes in syntenic blocks. This difference confirmed the intuitive hypothesis that expansion of rust fungal genomes likely occurs through extensive TE integration (Cuomo et al., 2017). These different examples indicate that beyond the obvious problem of assembly quality, the past action of TEs has shuffled genes during the evolution of rust genomes, rendering a phylogenomics approach based on conservation of gene order very difficult.

Mitochondrial genomes have been identified and reported in a couple of rust genomic studies, not only for high-quality draft genomes but also for rust fungi with no reference genomes or with low quality assembly, e.g., the coffee rust fungus (*H. vastatrix*) and the soybean rust fungi (*Ph. pachyrhizi* and *Ph. meibomiae*) (Cristancho et al., 2014; Duplessis et al., 2011; Rochi et al., 2016; Stone, Buitrago, Boore, & Frederick, 2010). These assemblies were possible based on the higher sequencing depth associated with mitochondrial scaffolds. The gene content of mitochondrial DNA was assessed in the soybean rust fungi, and gene order was highly conserved. Using a selection of these conserved essential mitochondrial genes—some being classical markers used in phylogenetic studies—the systematic placement of these rust species was supported among a set of selected fungi (Stone et al., 2010). Surprisingly, the two soybean rust fungi exhibited rather compact mitochondrial genomes close to 32 Kb, whereas other reports for rust fungi are closer to 70–80 Kb (Cristancho et al., 2014; Duplessis et al., 2011; Rochi et al., 2016). Analysis of Illumina-based resequencing data from rust isolates can also help to identify missed mitochondrial scaffolds in previous assemblies (e.g., in the poplar rust genome; Persoons et al., 2014). Extensive data are available for studying mitochondrial genomes of rust fungi encouraging further assessment of their organization and evolution.

The potentially low level of synteny in rust fungi nuclear genomes—in part arising from the specific challenges in assembly already discussed—makes the study of gene order conservation at a higher level more difficult. One solution for improved assemblies is the construction of genetic maps, which is underway for different rust fungi (Kolmer, 1996; Kubisiak et al., 2011; Pernaci et al., 2014; Zambino, Kubelik, & Szabo, 2000). Recently, version 2 of the *M. larici-populina* genome was anchored onto linkage groups for an greatly improved assembly (Frey, P., Hellsten, U., Duplessis, S., Grigoriev, I., et al., unpublished data). Such an advance may help to reconcile synteny at a higher scale, and to better understand through paleogenomic approaches how past TE activity may have participated in reshaping the genomes of rust fungi into the mosaics we now observe.

Another shortcoming of short-read sequencing technologies (e.g., next- or second-generation technologies) is the limited capacity to resolve complex regions with repetitive or heterozygous sequences. However, new sequencing technologies (i.e., third-generation technologies) are being developed that provide longer reads (e.g., PacBio and Nanopore sequencing technologies) that have the potential to separate individual haplotypes of dikaryotic rust fungi as well as assemble through TE-rich regions for improved assemblies (Miller et al., 2017). Altogether, current approaches now applied to rust fungal genome sequencing hold promise for improvement of the analysis of the genomic landscape of rusts.

2.2.2 Paralogs and Orthologs in Rust Fungi

Various bioinformatics tools are used to study the extent of duplications within fungal genomes and to unravel the presence of orthologs between species at different taxonomic levels. Homology-based tools starting from all-against-all BLAST (Altschul, Gish, Miller, Myers, & Lipman, 1990) similarity searches and running sequence comparisons with different methods are used to deduce relationships in proteomes, to identify putative paralogs within a species, or to infer orthology between several species. For instance, Markov-clustering tools (e.g., Tribe-MCL, ortho-MCL, Enright, Van Dongen, & Ouzounis, 2002; Li, Stoeckert, & Roos, 2003) are the methods of choice to identify gene clusters that are further considered as gene families among fungal genomes (Grigoriev et al., 2014). Such approaches also allow the identification of orphans that are unique to a given species. Tools such as i-ADHoRE (Proost et al., 2015; Simillion, Janssens, Sterck, & Van de Peer, 2008) were applied to study duplications in rust fungal genomes (Duplessis et al., 2011; Rochi et al., 2016). No whole-genome duplications or large-scale dispersed segmental duplications have been observed for the best genome assemblies of rust fungi (Duplessis et al., 2011). A pair of studies reported low to large amounts of segmental duplications in the genomes of *P. striiformis* f. sp. *tritici* and *P. triticina*, respectively (Kiran et al., 2017, 2016). However, these conclusions are dependent on the quality of the assembly, which is complicated by the large amount of TEs in the case of *P. triticina*. Further confirmation or improved analyses of gene or genome duplications will be needed. Interestingly, studies on rust genomes all reported high proportions of orphan genes and expanding specific gene families (Duplessis et al., 2014). Paralogous and orthologous gene pairs were considered between and within the poplar and the wheat stem rust fungal genomes, and their respective rates of synonymous substitution indicate an older date of duplication between than within genomes (Duplessis et al., 2011).

Gene family expansions are noticeable at a range of taxonomic levels, i.e., order, families, species, or between populations, and were particularly scrutinized for predicted secreted proteins (Cantu et al., 2011, 2013; Cuomo et al., 2017; Duplessis et al., 2011; Kiran et al., 2016; Nemri et al., 2014; Rochi et al., 2016; Zheng et al., 2013). The pioneer genomic study that compared *M. larici-populina* and *P. graminis* f. sp. *tritici* revealed the acquisition of many gene families at the species and ordinal levels compared to lower loss of gene families, and the presence of an important proportion of specific orphan genes (see fig. 1 in Duplessis et al., 2011). A similar analysis of gene conservation between basidiomycetes including more rust genomes depicted similar results and identified the presence of clusters shared between cereal rusts of the Pucciniaceae (Cuomo et al., 2017).

We have collected 16 published genomes (Cuomo et al., 2017; Duplessis et al., 2011; Firrincieli et al., 2015; Kamper et al., 2006; Mondo et al., 2017; Morin et al., 2012; Nemri et al., 2014; Padamsee et al., 2012; Perlin et al., 2015; Riley et al., 2014; Schirawski et al., 2010; Toome, Kuo, et al., 2014; Toome, Ohm, et al., 2014; Zhang et al., 2016) from the JGI Mycocosm database, including five rust fungi and five nonrust species belonging to the Pucciniomycotina, and performed MCL clustering analysis (Figs. 3 and 4) to illustrate the dramatic level of specific gene families conserved in the order Pucciniales compared to other basidiomycetes. Our sampling includes different examplars in two different rust families, allowing the detection of important expansions at the genus level. Overall almost half of the gene complement in each rust species is conserved at the ordinal level. Indeed, some large expansions of gene families occurred at the species, family, or order levels (Fig. 4). Annotation of expanded gene families in rust fungi showed that most proteins had an unknown function. These gene categories included small-secreted proteins that could represent putative virulence effectors (see later), and different types of transporters, transcription factors, and those with functions related to interaction and modification of nucleic acid such as zinc-finger proteins or helicases (Cuomo et al., 2017; Duplessis et al., 2011; Rochi et al., 2016). It has been speculated that expanded families associated with interaction with nucleic acid or DNA repair and maintenance may relate to the consequent proportion of TEs invading rust genomes; however, such a link remains to be established. With more rust fungal genomes in hand, systematic comparison within the order Pucciniales and with sister branches in the Pucciniomycotina showing very distinct genomic profiles (small genome size, few TEs, few genes) may help to address such a link in the future.

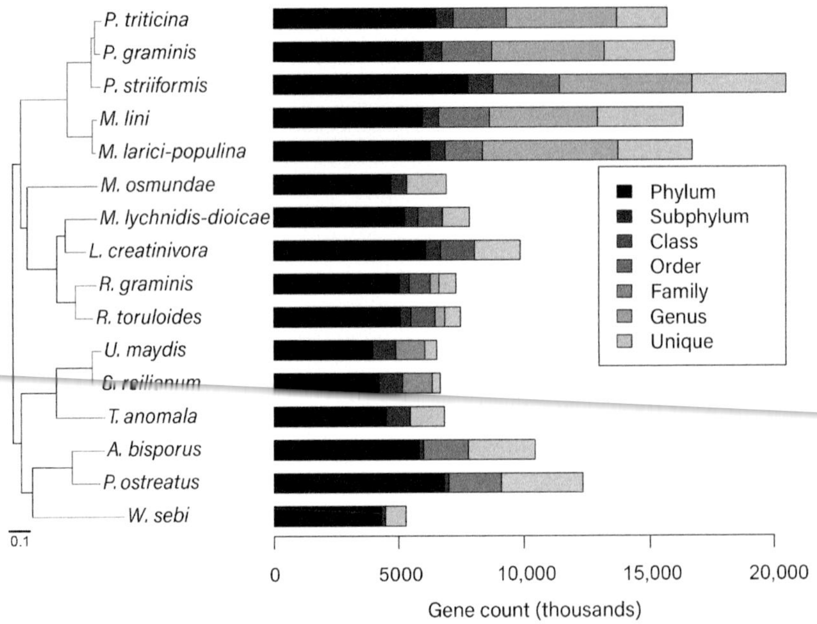

Fig. 3 Gene content and conservation across select basidiomycetes. RAxML (Stamatakis, 2014) phylogeny using 1259 single-copy gene orthologs. Gene orthologs were identified using MCL (Enright et al., 2002), then separated by conservation at various taxonomic levels according to the NCBI Taxonomy database. Genomes used in this analysis include: *Puccinia triticina* 1-1 BBBD race 1 (Cuomo et al., 2017), *Puccinia graminis* f. sp. *tritici* CRL 75-36-700-3 (Cuomo et al., 2017; Duplessis et al., 2011), *Puccinia striiformis* f. sp. *tritici* PST-78 (Cuomo et al., 2017), *Melampsora lini* CH5 (Nemri et al., 2014), *Melampsora larici-populina* 98AG31 (Duplessis et al., 2011), *Mixia osmundae* IAM 14324 (Toome, Ohm, et al., 2014), *Microbotryum lychnidis-dioicae* p1A1 Lamole (Perlin et al., 2015), *Leucosporidiella creatinivora* 62-1032 (Mondo et al., 2017), *Rhodotorula graminis* WP1 (Firrincieli et al., 2015), *Rhodosporidium toruloides* IFO0880 (Zhang et al., 2016), *Ustilago maydis* 521 (Kamper et al., 2006), *Sporisorium reilianum* SRZ2 (Schirawski et al., 2010), *Tilletiaria anomala* UBC 951 (Toome, Kuo, et al., 2014), *Agaricus bisporus* var. *bisporus* H97 (Morin et al., 2012), *Pleurotus ostreatus* PC15 (Riley et al., 2014), and *Wallemia sebi* CBS 633.66 (Padamsee et al., 2012).

2.2.3 Core Ortholog Genes in Rust Fungi: Building Robust Phylogenomic Trees

A great benefit of MCL approaches for phylogenomics is the identification of core orthologous genes. Such genes that are identified as unique copy and conserved across all species at different taxonomical levels can be used to resolve or to ascertain phylogenetic positions.

The quality of a genome assembly can be assessed through the completeness of genome annotation by verifying the minimal gene set conserved

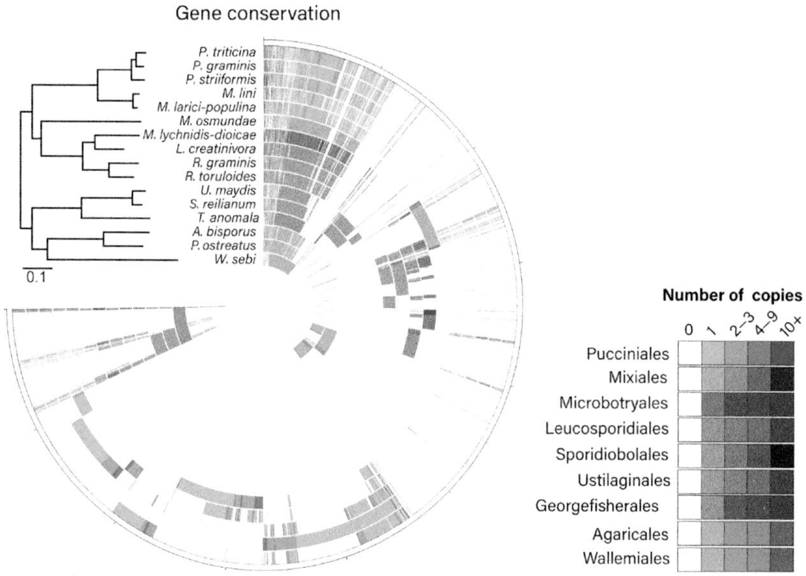

Fig. 4 Expansion of unique genes in the Pucciniales. Each track represents a genome, and columns represent individual orthologous gene clusters (identified using mcl; Enright et al., 2002). Tracks are colored based on their taxonomy, at the order level, and shading within track indicates gene copy number (shown in the *bottom right*). *Blues*, Pucciniales; *purples*, Mixiales; *reds*, Microbotryales; *oranges*, Leucosporidiales; *grays*, Sporidiobolales; *orange-brown*, Ustilaginales; *yellow-red*, Georgefisherales; *greens*, Agaricales; *blue-green*, Wallemiales. For full lineage information, see Fig. 3.

across eukaryotes or fungi. A number of databases and pipelines have been erected through the years that perform the isolation of a maximized set of unique or low copy and conserved genes, such as CEGMA (Core Eukaryotic Genes Mapping Approach; Parra et al., 2007), FUNNYBASE (Fungal Phylogenomic database; Marthey et al., 2008), or BUSCO (Benchmarking Universal Single-Copy Orthologs; Simão et al., 2015). Similar sets of conserved and unique genes can also be drawn out of any MCL analysis. Such tools and approaches have been used in all rust genomic studies to either assess completeness of annotation (Cantu et al., 2011, 2013; Cristancho et al., 2014; Cuomo et al., 2017; Kiran et al., 2016; Nemri et al., 2014; Rochi et al., 2016; Zheng et al., 2013) or perform phylogenomic analyses (Cuomo et al., 2017; Duplessis et al., 2011). For example, analyses of assemblies and annotations were conducted with CEGMA for most rust fungal genomes and demonstrated very high conservation of 248 selected core eukaryotic genes (72%–97%). Notably, several rust genomes had over 94% complete CEGMA genes (Cuomo et al., 2017; Nemri et al., 2014;

Zheng et al., 2013). This high level of completeness indicates that despite the large number of repetitive elements, high-quality assemblies with most of the coding space may be captured even when short-read sequencing approaches are used, for example, *H. vastatrix* (Cristancho et al., 2014) and *Uromyces fabae* (Link et al., 2014). This finding is promising for progress in the field of rust genomics with species exhibiting very large genome sizes.

Capturing most of the coding space from very large genomes may be possible and sufficient to fuel future phylogenomic analyses. MCL analyses run from fungal genomes stored in the Mycocosm at the JGI can help to identify sets of unique genes and perform phylogenomic analyses. The results of such analyses can be visible by using the "Tree" tool implemented in the Mycocosm that can be called by users when a given taxonomic level is selected, i.e., end branches or intermediate nodes (Grigoriev et al., 2014). This process was applied to a selection of 16 published genomes, including five rust fungi (Figs. 3 and 4), and the result is congruent with species placement in classic phylogenetic studies of the Pucciniales (see also Fig. 2), as well as with previous phylogenomic trees (e.g., the most recent published by Cuomo et al., 2017). Most analyses of this kind are derived from MCL clustering, which relies on gene sets that are often annotated by automated approaches. This approach can induce false signals in the phylogeny through wrong alignment of the selected core genes (e.g., Tan et al., 2015) Even if time consuming, particularly when large sets of genomes are taken under consideration, expert curation of such core gene sets can deliver phylogenies of greater quality and should be recommended.

2.2.4 Marked Expansion of Gene Families in Rust Genomes

All MCL analyses and detailed studies of paralogous and orthologous genes in rust fungi have unraveled specific expansions of gene families encoding secreted proteins of unknown function, including small and cysteine-rich secreted proteins. These features are typical of virulence effector genes (Duplessis, Joly, & Dodds, 2012; Lo Presti et al., 2015; Petre, Joly, & Duplessis, 2014), and they are typically used to define pipelines that will identify subsets of so-called candidate secreted effector proteins (CSEPs) (Lorrain, Hecker, & Duplessis, 2015; Saunders et al., 2012; Sperschneider, Taylor, Dodds, & Duplessis, 2017).

Previous comparative reports showed that rust fungi, as is true for other obligate biotrophs among filamentous plant pathogens, exhibit larger secretomes in proportion to the whole predicted proteome (Lo Presti et al., 2015). Rust fungi have fewer plant cell wall-degrading enzymes and a much larger proportion of secreted proteins of unknown function compared with other fungi with different trophic modes (Lo Presti et al., 2015). Identification, classification, and comparison of CSEPs between genomes have been central to almost all genomic studies conducted with rust fungi (Cantu et al., 2011, 2013; Cristancho et al., 2014; Cuomo et al., 2017; Duplessis et al., 2011; Kiran et al., 2017, 2016; Link et al., 2014; Nemri et al., 2014; Rochi et al., 2016; Upadhyaya et al., 2015; Zheng et al., 2013). CSEPs have also been investigated in transcriptomic studies on in planta gene expression and spore germination of rust fungi (de Carvalho et al., 2017; Dobon et al., 2016; Jing et al., 2017; Liu et al., 2015; Rutter et al., 2017; Talhinhas et al., 2014; see Duplessis et al., 2014, 2012, for extensive lists of earlier transcriptome studies focusing on CSEPs).

In all cases of genomic and transcriptomic analyses, very similar conclusions were drawn: genes encoding secreted proteins are very important in proportion, many are specific at the order level but more importantly at the family and the species levels, almost all lack functional annotation, and for a very large proportion they show specific or induced gene expression during host infection. Also, a variable portion of the secretome, depending on the species, is organized in large to very large gene families. Resequencing of isolates and analyses of genomic variants conducted in different rust fungi also showed that some of the CSEP genes accumulated nonsynonymous mutations and are marked by diversifying selection (Cantu et al., 2011; Kiran et al., 2017, 2016; Persoons et al., 2014; Upadhyaya et al., 2015; Zheng et al., 2013). Selection analysis of clusters of paralogous genes encoding CSEPs in the poplar rust genome has similarly revealed evidence of diversifying selection for some families (Hacquard et al., 2012). These observations indicate diversification within some CSEP families after expansion in rust fungal genomes are likely to face corresponding resistance genes in host plant genomes (Duplessis et al., 2012).

Genomic studies of other filamentous plant pathogens have shown that some virulence effector genes preferentially reside in gene sparse regions of the genome where repetitive elements and TEs are prominent (Raffaele & Kamoun, 2012). In a two-speed genome model, CSEP genes in repeat-rich compartments show accelerated evolution rates, which can be of benefit to the pathogen to face the arsenal of recognition receptors in respective host plants (Dong, Raffaele, & Kamoun, 2015). Powdery mildews, like the rusts, are biotrophic fungi belonging to the Ascomycota that share some genomic features with rust fungi, such as a large genome size and a high content of TEs (Duplessis, Spanu, & Schirawski, 2013; Spanu et al., 2010). Expansion of CSEP and virulence effector gene families in the genome of *Blumeria graminis* f. sp. *tritici* has been associated with DNA derived from retrotransposons (Pedersen et al., 2012; Sacristán et al., 2009). It is tempting to speculate activity of TEs has played a role in the expansion of secreted protein gene families in the Pucciniales; however, so far, no clear systematic association between rust CSEPs and TE families has been demonstrated.

CSEPs are not particularly overrepresented in TE or repeat-rich regions, likely due to the fact that repetitive elements are dispersed everywhere in rust genomes (Duplessis et al., 2011). Only a few cases of association have been reported for members of CSEPs gene families with TIR type and non-categorized TEs in the poplar rust genome (Duplessis et al., 2011). The drivers of the expansions observed for CSEP gene families and whether TEs are an active part of a process leading to expansion and diversification of CSEP families remains unanswered. Improved genome assemblies, as well as improved TE annotation may help to establish clearer links between TE invasion of rust genomes and evolution of virulence-related CSEP families. For example, the reassessment of TE annotation in the second version of the poplar rust genome found less uncategorized elements and more DNA transposons (Lorrain, C., & Duplessis, S., unpublished data). In the secretome of rust fungi, only a limited fraction showed evidence of functional annotation, including proteases, lipases, and carbohydrate active enzymes. Overall, the total number of potential effectors is rather large, comprising several hundreds to over a thousand genes and suggests a complex and diversified weaponry to infect host plants (Dong et al., 2015; Duplessis et al., 2012; Petre et al., 2014). It is also tempting to consider that some CSEPs of unknown function are specialized subsets required to achieve infection in different host plants of heteroecious rust fungi (Duplessis et al., 2014; Lo Presti et al., 2015; Schulze-Lefert & Panstruga, 2011). These gene families may shed light on a long-standing biological and evolutionary question in the host specificity of rust fungi.

2.2.5 Gene Loss and Convergent Evolution in Obligate Plant Pathogens

Finally, a striking feature of rust fungal genomes is the loss of, or strong reduction in, notable gene categories, such as those involved in the generation of secondary metabolites or carbohydrate active enzymes, that are understood as a means for remaining undetected by plant recognition machinery during infection (Collemare et al., 2014; Duplessis et al., 2011). Similar observations were made for unrelated obligate biotrophs in three independent lineages (ascomycetes, basidiomycetes, and oomycetes), indicating convergent evolution to adapt to the plant host environment (Hacquard, 2014; McDowell, 2011; Spanu, 2011). A few genes that are essential for sulfur and nitrogen acquisition were lost in these distinct biotrophic lineages (Baxter et al., 2010; Duplessis et al., 2011; Kemen et al., 2011; Spanu et al., 2010). However, reanalysis of core genes in larger sets of biotrophic and nonbiotrophic pathogens suggested that the status of obligate biotrophy is more related to overlapping lineage-specific gene losses (Hacquard, 2014). The recent sequencing and reanalysis of cereal rust genomes confirmed gene losses initially reported in the comparative analysis of the wheat stem rust fungus and the poplar rust fungus (Cuomo et al., 2017). The knowledge of gene loss across major branches of the Pucciniales and beyond in the Pucciniomycotina may also inform phylogenomic studies and help to better understand how and when these losses have happened and if it may relate to host specialization (e.g., plant cell wall-degrading enzymes).

3. BEYOND SEQUENCES AND ASSEMBLY: THE FUTURE FOR RUST GENOMICS AND PHYLOGENOMICS

The field of rust genomics still faces many issues related to the nature of these complex organisms. As biotrophic pathogens, they cannot be cultivated and acquisition of primary material for DNA isolation can still be challenging. By nature, the genomes of rust fungi exhibit an intrinsic complexity due to their large size and high number of repeat regions, which pose serious problems for obtaining high-quality genome assemblies. Currently, several rust fungal genomes with sizes larger than those published so far are being sequenced within the frame of international CSP projects at the JGI or through individual lab sequencing efforts. Systematic inclusion of long-read sequencing approaches as well as improved assembly methods (e.g., PacBio, Nanopore technologies, 10× genomics) may help to achieve accurate assembly of a reference genome of several hundred megabases and to improve the quality of rust genomes already available. As discussed in this chapter, the quality of genome assembly is central and critical for gene and TE annotation. The heterogeneity of rust fungal genomic reports argues in favor of standard practices shared at the community level to ensure accuracy of future comparative genomic analyses. Community guidelines will be particularly important for MCL clustering approaches that will support phylogenomic tree reconstruction. Considering their relative importance in rust genomes, annotation of TEs should not be overlooked as they may be crucial to better understand the evolution of rust lineages at the family level. Finally, expert curation of genes is also recommended, particularly when given genes are selected for the purpose of phylogenomic studies.

Despite the relative heterogeneity of genomes reported for a single Pucciniales species, the overall picture is very stable for all rust fungi. Also it is possible to capture large fractions of the coding space by short-read sequencing, even in the absence of a neat assembly. Sequencing many more rust fungal genomes should help in defining what are the minimal sets of genes specific at the family levels, and how conserved gene sets have evolved from basal lineages to the more recent ones. On the path to understand the processes underlying host specificity, RNA sequencing holds promise in pinpointing the minimal sets of determinants required at the different steps of the complex rust life cycles. Particularly, the characterization and the study of evolution of virulence effector gene families are of the utmost importance.

The order Pucciniales contain more than 8000 described species; of these, genomes for less than 10, or 0.1%, have thus far been released. This number is a stark underrepresentation of one of the major orders of Dikarya. Yet, only by generation and analyses of rust genomic data from across the Pucciniales can many of the key questions or knowledge gaps regarding the biology and life histories of these unique organisms be resolved.

ACKNOWLEDGMENTS

S.D. would like to acknowledge the "Investissements d'Avenir" program (ANR-11-LABX-0002-01, Lab of Excellence ARBRE); the Joint Genome Institute (Office of Science of the U.S. Department of Energy under contract no. DE-Ac02-05cH11231) for the sequencing of the genome of the poplar rust fungus *Melampsora larici-populina* and other rust-related ongoing CSP projects; and his colleagues Francis Martin, Pascal Frey, Fabien Halkett, and Stéphane De Mita at INRA Nancy for discussions on rust genomics and comparative genomics. M.C.A. is grateful to the students and postdocs who have pioneered genomics and rust phylogenetics studies in her lab, in particular Merje Toome-Heller, Maj Padamsee, Andy Wilson, and Tas Kijpornyongpan.

REFERENCES

Aime, M. C. (2006). Toward resolving family-level relationships in rust fungi (Uredinales). *Mycoscience*, *47*(3), 112–122. https://doi.org/10.1007/s10267-006-0281-0.

Aime, M. C., Matheny, P. B., Henk, D. A., Frieders, E. M., Nilsson, R. H., Piepenbring, M., et al. (2006). An overview of the higher level classification of Pucciniomycotina based on combined analyses of nuclear large and small subunit rDNA sequences. *Mycologia*, *98*(6), 896–905.

Alaei, H., De Backer, M., Nuytinck, J., Maes, M., Höfte, M., & Heungens, K. (2009). Phylogenetic relationships of *Puccinia horiana* and other rust pathogens of *Chrysanthemum* x *morifolium* based on rDNA ITS sequence analysis. *Mycological Research, 113*, 668–683.

Altschul, S. F., Gish, W., Miller, W., Myers, E. W., & Lipman, D. J. (1990). Basic local alignment search tool. *Journal of Molecular Biology, 215*(3), 403–410. https://doi.org/10.1016/S0022-2836(05)80360-2.

Anikster, Y., Szabo, L. J., Eilam, T., Manisterski, J., Koike, S. T., & Bushnell, W. R. (2004). Morphology, life cycle biology, and DNA sequence analysis of rust fungi on garlic and chives from California. *Phytopathology, 94*(6), 569–577. https://doi.org/10.1094/phyto.2004.94.6.569.

Arthur, J. C. (1903). Problems in the study of plant rusts. *Bulletin of the Torrey Botanical Club, 30*(1), 1–18. https://doi.org/10.2307/2478644.

Arthur, J. C. (1934). *Manual of the rusts in the United States and Canada.* Lafayette, Indiana: Purdue Research Foundation.

Barilli, E., Satovic, Z., Sillero, J. C., Rubiales, D., & Torres, A. M. (2010). Phylogenetic analysis of *Uromyces* species infecting grain and forage legumes by sequence analysis of nuclear ribosomal internal transcribed spacer region. *Journal of Phytopathology, 159*, 137–145.

Baxter, L., Tripathy, S., Ishaque, N., Boot, N., Cabral, A., Kemen, E., et al. (2010). Signatures of adaptation to obligate biotrophy in the *Hyaloperonospora arabidopsidis* genome. *Science, 330*(6010), 1549.

Beenken, L. (2014). Pucciniales on *Annona* (Annonaceae) with special focus on the genus *Phakopsora. Mycological Progress, 13*(3), 791–809. https://doi.org/10.1007/s11557-014-0963-5.

Beenken, L. (2017). *Austropuccinia*: A new genus name for the myrtle rust *Puccinia psidii* placed within the redefined family Sphaerophragmiaceae (Pucciniales). *Phytotaxa, 297*, 53–61.

Beenken, L., & Wood, A. R. (2015). *Puccorchidium* and *Sphenorchidium*, two new genera of Pucciniales on Annonaceae related to *Puccinia psidii* and the genus *Dasyspora. Mycological Progress, 14*(7), 1–13. https://doi.org/10.1007/s11557-015-1073-8.

Beenken, L., Zoller, S., & Berndt, R. (2012). Rust fungi on Annonaceae II: The genus *Dasyspora* Berk. & M.A. Curtis. *Mycologia, 104*(3), 659–681. https://doi.org/10.3852/11-068.

Bennett, C., Aime, M. C., & Newcombe, G. (2011). Molecular and pathogenic variation within *Melampsora* on *Salix* in western North America reveals numerous cryptic species. *Mycologia, 103*(5), 1004–1018. https://doi.org/10.3852/10-289.

Cantu, D., Govindarajulu, M., Kozik, A., Wang, M., Chen, X., Kojima, K. K., et al. (2011). Next generation sequencing provides rapid access to the genome of *Puccinia striiformis* f. sp. *tritici*, the causal agent of wheat stripe rust. *PLoS One, 6*(8), e24230. https://doi.org/10.1371/journal.pone.0024230.

Cantu, D., Segovia, V., MacLean, D., Bayles, R., Chen, X., Kamoun, S., et al. (2013). Genome analyses of the wheat yellow (stripe) rust pathogen *Puccinia striiformis* f. sp. *tritici* reveal polymorphic and haustorial expressed secreted proteins as candidate effectors. *BMC Genomics, 14*(1), 270. https://doi.org/10.1186/1471-2164-14-270.

Collemare, J., Griffiths, S., Iida, Y., Karimi Jashni, M., Battaglia, E., Cox, R. J., et al. (2014). Secondary metabolism and biotrophic lifestyle in the tomato pathogen *Cladosporium fulvum. PLoS One, 9*(1), e85877. https://doi.org/10.1371/journal.pone.0085877.

Cressey, D. (2013). Coffee rust regains foothold: Researchers marshal technology in bid to thwart fungal outbreak in Central America. *Nature, 493*(7434), 587–588.

Cristancho, M. A., Botero Rozo, D. O., Giraldo, W., Tabima, J., Riaño, D. M., Escobar, C., et al. (2014). Annotation of a hybrid partial genome of the coffee rust (*Hemileia vastatrix*) contributes to the gene repertoire catalogue of the Pucciniales. *Frontiers in Plant Science, 5*, Article No. 594. https://doi.org/10.3389/fpls.2014.00594.

Cummins, G. B., & Hiratsuka, Y. (2003). *Illustrated genera of rust fungi* (3rd ed.). St. Paul, Minnesota, U.S.A: American Phytopathological Society.

Cunningham, G. H. (1931). *The rust fungi of New Zealand: Together with the biology, cytology and therapeutics of the Uredinales*. Dunedin, New Zealand: John McIndoe.

Cuomo, C. A., Bakkeren, G., Khalil, H. B., Panwar, V., Joly, D., Linning, R., et al. (2017). Comparative analysis highlights variable genome content of wheat rusts and divergence of the mating loci. *G3: Genes-Genomes-Genetics*, 7(2), 361–376.

de Carvalho, M. C. D. C. G., Costa Nascimento, L., Darben, L. M., Polizel-Podanosqui, A. M., Lopes-Caitar, V. S., Qi, M., et al. (2017). Prediction of the in planta *Phakopsora pachyrhizi* secretome and potential effector families. *Molecular Plant Pathology*, *18*(3), 363–377. https://doi.org/10.1111/mpp.12405.

Delsuc, F., Brinkmann, H., & Herve, P. (2005). Phylogenomics and the reconstruction of the tree of life. *Nature Reviews Genetics*, *6*(5), 361.

Demers, J. E., Liu, M., Hambleton, S., & Castlebury, L. A. (2017). Rust fungi on *Panicum*. *Mycologia*, 1–17. https://doi.org/10.1080/00275514.2016.1262656.

Dentinger, B. T. M., Gaya, E., O'Brien, H., Suz, L. M., Lachlan, R., Díaz-Valderrama, J. R., et al. (2016). Tales from the crypt: Genome mining from fungarium specimens improves resolution of the mushroom tree of life. *Biological Journal of the Linnean Society*, *117*, 11–32. https://doi.org/10.1111/bij.12553.

Dobon, A., Bunting, D. C. E., Cabrera-Quio, L. E., Uauy, C., & Saunders, D. G. O. (2016). The host-pathogen interaction between wheat and yellow rust induces temporally coordinated waves of gene expression. *BMC Genomics*, 17(1), 380. https://doi.org/10.1186/s12864-016-2684-4.

Dong, S., Raffaele, S., & Kamoun, S. (2015). The two-speed genomes of filamentous pathogens: Waltz with plants. *Current Opinion in Genetics & Development*, *35*, 57–65. https://doi.org/10.1016/j.gde.2015.09.001.

Doungsa-ard, C., McTaggart, A. R., Geering, A. D. W., Dalisay, T. U., Ray, J., & Shivas, R. G. (2015). *Uromycladium falcatarium* sp. nov., the cause of gall rust on *Paraserianthes falcataria* in south-east Asia. *Australasian Plant Pathology*, *44*, 25–30. https://doi.org/10.1007/s13313-014-0301-z.

Duplessis, S., Bakkeren, G., & Hamelin, R. (2014). Advancing knowledge on biology of rust fungi through genomics. In F. Martin (Ed.), *Advances in botanical research: Vol. 70*. (pp. 173–209). Oxford, United Kingdom: Elsevier Ltd. Academic Press.

Duplessis, S., Cuomo, C. A., Lin, Y.-C., Aerts, A., Tisserant, E., Veneault-Fourrey, C., et al. (2011). Obligate biotrophy features unraveled by the genomic analysis of rust fungi. *Proceedings of the National Academy of Sciences of the United States of America*, *108*(22), 9166–9171. https://doi.org/10.1073/pnas.1019315108.

Duplessis, S., Joly, D. L., & Dodds, P. N. (2012). Rust effectors. In F. Martin & S. Kamoun (Eds.), *Effectors in plant-microbe interactions* (pp. 155–193). Oxford, United Kingdom: Wiley-Blackwell.

Duplessis, S., Spanu, P. D., & Schirawski, J. (2013). Biotrophic fungi (powdery mildews, rusts, and smuts). In *The ecological genomics of fungi* (pp. 149–168). Oxford, United Kingdom: John Wiley & Sons, Inc.

Emms, D. M., & Kelly, S. (2015). OrthoFinder: Solving fundamental biases in whole genome comparisons dramatically improves orthogroup inference accuracy. *Genome Biology*, *16*, 157.

Enright, A. J., Van Dongen, S., & Ouzounis, C. A. (2002). An efficient algorithm for large-scale detection of protein families. *Nucleic Acids Research*, *30*(7), 1575–1584.

Feau, N., Vialle, A., Allaire, M., Maier, W., & Hamelin, R. C. (2011). DNA barcoding in the rust genus *Chrysomyxa* and its implications for the phylogeny of the genus. *Mycologia*, *103*(6), 1250–1266. https://doi.org/10.3852/10-426.

Fellers, J. P., Soltani, B. M., Bruce, M., Linning, R., Cuomo, C. A., Szabo, L. J., et al. (2013). Conserved loci of leaf and stem rust fungi of wheat share synteny interrupted by lineage-specific influx of repeat elements. *BMC Genomics*, *14*(1), 60. https://doi.org/10.1186/1471-2164-14-60.

Firrincieli, A., Otillar, R., Salamov, A., Schmutz, J., Khan, Z., Redman, R. S., et al. (2015). Genome sequence of the plant growth promoting endophytic yeast *Rhodotorula graminis* WP1. *Frontiers in Microbiology*, *6*, 978. https://doi.org/10.3389/fmicb.2015.00978.

Flutre, T., Duprat, E., Feuillet, C., & Quesneville, H. (2011). Considering transposable element diversification in de novo annotation approaches. *PLoS One*, *6*(1), e16526.

Grigoriev, I. V., Nikitin, R., Haridas, S., Kuo, A., Ohm, R., Otillar, R., et al. (2014). MycoCosm portal: Gearing up for 1000 fungal genomes. *Nucleic Acids Research*, *42*(D1), D699–D704. https://doi.org/10.1093/nar/gkt1183.

Hacquard, S. (2014). The genomics of powdery mildew fungi. *Advances in Botanical Research*, *70*, 109–142. https://doi.org/10.1016/B978-0-12-397940-7.00004-5.

Hacquard, S., Joly, D. L., Lin, Y. C., Tisserant, E., Feau, N., Delaruelle, C., et al. (2012). A comprehensive analysis of genes encoding small secreted proteins identifies candidate effectors in *Melampsora larici-populina* (poplar leaf rust). *Molecular Plant-Microbe Interactions*, *25*(3), 279–293. https://doi.org/10.1094/mpmi-09-11-0238.

Hart, J. A. (1988). Rust fungi and host plant coevolution: Do primitive hosts harbor primitive parasites? *Cladistics*, *4*(4), 339–366. https://doi.org/10.1111/j.1096-0031.1988.tb00519.x.

Hoen, D. R., Hickey, G., Bourque, G., Casacuberta, J., Cordaux, R., Feschotte, C., et al. (2015). A call for benchmarking transposable element annotation methods. *Mobile DNA*, *6*, 13.

Hug, L. A., Baker, B. J., Anantharaman, K., Brown, C. T., Probst, A. J., Castelle, C. J., et al. (2016). A new view of the tree of life. *Nature Microbiology*, *1*, 16048.

Jamilloux, V., Daron, J., Choulet, F., & Quesneville, H. (2017). De Novo annotation of transposable elements: Tackling the fat genome issue. *Proceedings of the IEEE*, *105*(3), 474–481.

Jin, Y., Szabo, L. J., & Carson, M. (2010). Century-old mystery of *Puccinia striiformis* life history solved with the identification of *Berberis* as an alternate host. *Phytopathology*, *100*(5), 432–435. https://doi.org/10.1094/PHYTO-100-5-0432.

Jing, L., Guo, D., Hu, W., & Niu, X. (2017). The prediction of a pathogenesis-related secretome of *Puccinia helianthi* through high-throughput transcriptome analysis. *BMC Bioinformatics*, *18*(1), 166. https://doi.org/10.1186/s12859-017-1577-0.

Kamper, J., Kahmann, R., Bolker, M., Ma, L.-J., Brefort, T., Saville, B. J., et al. (2006). Insights from the genome of the biotrophic fungal plant pathogen *Ustilago maydis*. *Nature*, *444*(7115), 97–101.

Kemen, E., Gardiner, A., Schultz-Larsen, T., Kemen, A. C., Balmuth, A. L., Robert-Seilaniantz, A., et al. (2011). Gene gain and loss during evolution of obligate parasitism in the white rust pathogen of *Arabidopsis thaliana*. *PLoS Biology*, *9*(7), e1001094. https://doi.org/10.1371/journal.pbio.1001094.

Kiran, K., Rawal, H. C., Dubey, H., Jaswal, R., Bhardwaj, S. C., Prasad, P., et al. (2017). Dissection of genomic features and variations of three pathotypes of *Puccinia striiformis* through whole genome sequencing. *Scientific Reports*, 7, 42419. https://doi.org/10.1038/srep42419.

Kiran, K., Rawal, H. C., Dubey, H., Jaswal, R., Devanna, B. N., Gupta, D. K., et al. (2016). Draft genome of the wheat rust pathogen (*Puccinia triticina*) unravels genome-wide structural variations during evolution. *Genome Biology and Evolution*, *8*, 2702–2721. https://doi.org/10.1093/gbe/evw197.

Kirk, P. M., Cannon, P. F., Minter, D. W., & Stalpers, J. A. (2008). *Dictionary of the fungi* (10th ed.). Wallingford, UK: CABI.

Kohler, A., Kuo, A., Nagy, L. G., Morin, E., Barry, K. W., Buscot, F., et al. (2015). Convergent losses of decay mechanisms and rapid turnover of symbiosis genes in mycorrhizal mutualists. *Nature Genetics*, *47*, 410–415. https://doi.org/10.1038/ng.3223.

Kolmer, J. A. (1996). Genetics of resistance to wheat leaf rust. *Annual Review of Phytopathology*, *34*(1), 435–455. https://doi.org/10.1146/annurev.phyto.34.1.435.

Kortepeter, M. G., & Parker, G. W. (1999). Potential biological weapons threats. *Emerging Infectious Diseases*, *5*(4), 523.

Kubisiak, T. L., Anderson, C. L., Amerson, H. V., Smith, J. A., Davis, J. M., & Nelson, C. D. (2011). A genomic map enriched for markers linked to Avr1 in *Cronartium quercuum* f.sp. *fusiforme*. *Fungal Genetics and Biology*, *48*(3), 266–274. https://doi.org/10.1016/j.fgb.2010.09.008.

Li, L., Stoeckert, C. J., & Roos, D. S. (2003). OrthoMCL: Identification of ortholog groups for eukaryotic genomes. *Genome Research*, *13*(9), 2178–2189. https://doi.org/10.1101/gr.1224503.

Link, T. I., Lang, P., Scheffler, B. E., Duke, M. V., Graham, M. A., Cooper, B., et al. (2014). The haustorial transcriptomes of *Uromyces appendiculatus* and *Phakopsora pachyrhizi* and their candidate effector families. *Molecular Plant Pathology*, *15*(4), 379–393. https://doi.org/10.1111/mpp.12099.

Liu, M., & Hambleton, S. (2010). Taxonomic study of stripe rust, *Puccinia striiformis* sensu lato, based on molecular and morphological evidence. *Fungal Biology*, *114*(10), 881–899. https://doi.org/10.1016/j.funbio.2010.08.005.

Liu, M., & Hambleton, S. (2013). Laying the foundation for a taxonomic review of *Puccinia coronata s.l.* in a phylogenetic context. *Mycological Progress*, *12*(1), 63–89. https://doi.org/10.1007/s11557-012-0814-1.

Liu, J.-J., Sturrock, R. N., Sniezko, R. A., Williams, H., Benton, R., & Zamany, A. (2015). Transcriptome analysis of the white pine blister rust pathogen *Cronartium ribicola*: De novo assembly, expression profiling, and identification of candidate effectors. *BMC Genomics*, *16*(1), 1–16. https://doi.org/10.1186/s12864-015-1861-1.

Lo Presti, L., Lanver, D., Schweizer, G., Tanaka, S., Liang, L., Tollot, M., et al. (2015). Fungal effectors and plant susceptibility. *Annual Review of Plant Biology*, *66*(1), 513–545. https://doi.org/10.1146/annurev-arplant-043014-114623.

Loehrer, M., Vogel, A., Huettel, B., Reinhardt, R., Benes, V., Duplessis, S., et al. (2014). On the current status of *Phakopsora pachyrhizi* genome sequencing. *Frontiers in Plant Science*, *5*, 377. Retrieved from https://doi.org/10.3389/fpls.2014.00377.

Lorrain, C., Hecker, A., & Duplessis, S. (2015). Effector-mining in the poplar rust fungus *Melampsora larici-populina* secretome. *Frontiers in Plant Science*, *6*, 1051. https://doi.org/10.3389/fpls.2015.01051.

Maier, W., Begerow, D., Weiss, M., & Oberwinkler, F. (2003). Phylogeny of the rust fungi: An approach using nuclear large subunit ribosomal DNA sequences. *Canadian Journal of Botany*, *81*(1), 12–23.

Marthey, S., Aguileta, G., Rodolphe, F., Gendrault, A., Giraud, T., Fournier, E., et al. (2008). FUNYBASE: A FUNgal phYlogenomic dataBASE. *BMC Bioinformatics*, *9*(1), 456.

Martin, F., Aerts, A., Ahren, D., Brun, A., Danchin, E. G., Duchaussoy, F., et al. (2008). The genome of *Laccaria bicolor* provides insights into mycorrhizal symbiosis. *Nature*, *452*(7183), 88–92. https://doi.org/10.1038/nature06556.

McDowell, J. M. (2011). Genomes of obligate plant pathogens reveal adaptations for obligate parasitism. *Proceedings of the National Academy of Sciences of the United States of America*, *108*(22), 8921–8922. https://doi.org/10.1073/pnas.1105802108.

McTaggart, A. R., & Aime, M. C. (2012). ITS, problematic to amplify a rust barcode. *Inoculum*, *63*, 31.

McTaggart, A. R., Doungsa-ard, C., Geering, A. D. W., Aime, M. C., & Shivas, R. G. (2015). A co-evolutionary relationship exists between *Endoraecium* (Pucciniales) and its *Acacia* hosts in Australia. *Persoonia*, *35*, 50–62. https://doi.org/10.3767/003158515X687588.

McTaggart, A. R., Geering, A. D. W., & Shivas, R. G. (2014). The rusts on Goodeniaceae and Stylidiaceae. *Mycological Progress*, *13*(4), 1017–1025. https://doi.org/10.1007/s11557-014-0989-8.

McTaggart, A. R., Shivas, R. G., Doungsa-ard, C., Weese, T. L., Beasley, D. R., Hall, B. H., et al. (2016). Identification of rust fungi (Pucciniales) on species of *Allium* in Australia. *Australasian Plant Pathology*, *45*, 581–592. https://doi.org/10.1007/s13313-016-0445-0.

McTaggart, A. R., Shivas, R. G., van der Nest, M. A., Roux, J., Wingfield, B. D., & Wingfield, M. J. (2016). Host jumps shaped the diversity of extant rust fungi (Pucciniales). *New Phytologist*, *209*, 1149–1158. https://doi.org/10.1111/nph.13686.

Miller, M. E., Zhang, Y., Omidvar, V., Sperschneider, J., Schwessinger, B., Raley, C., et al. (2017). *De novo* assembly and phasing of dikaryotic genomes from two isolates of *Puccinia coronata* f. sp. *avenae*, the causal agent of oat crown rust. *bioRxiv*. https://doi.org/10.1101/179226.

Minnis, D., McTaggart, A. R., Rossman, A., & Aime, M. C. (2012). Taxonomy of mayapple rust: The genus *Allodus* resurrected. *Mycologia*, *104*(4), 942–950. https://doi.org/10.3852/11-350.

Mondo, S. J., Dannebaum, R. O., Kuo, R. C., Louie, K. B., Bewick, A. J., LaButti, K., et al. (2017). Widespread adenine N6-methylation of active genes in fungi. *Nature Genetics*, *49*(6), 964–968. https://doi.org/10.1038/ng.3859.

Morin, E., Kohler, A., Baker, A. R., Foulongne-Oriol, M., Lombard, V., Nagye, L. G., et al. (2012). Genome sequence of the button mushroom *Agaricus bisporus* reveals mechanisms governing adaptation to a humic-rich ecological niche. *Proceedings of the National Academy of Sciences of the United States of America*, *109*(43), 17501–17506. https://doi.org/10.1073/pnas.1206847109.

Nemri, A., Saunders, D., Anderson, C., Upadhyaya, N., Win, J., Lawrence, G., et al. (2014). The genome sequence and effector complement of the flax rust pathogen *Melampsora lini*. *Frontiers in Plant Science*, *5*, 98. https://doi.org/10.3389/fpls.2014.00098.

Novick, R. S. (2008). *Phylogeny, taxonomy, and life cycle evolution in the cedar rust fungi (*Gymnosporangium*)*. Ph.D. Dissertation, New Haven, CT: Yale University. 138 pp.

Padamsee, M., Kumar, T. K. A., Riley, R., Binder, M., Boyd, A., Calvo, A. M., et al. (2012). The genome of the xerotolerant mold *Wallemia sebi* reveals adaptations to osmotic stress and suggests cryptic sexual reproduction. *Fungal Genetics and Biology*, *49*, 217–226. https://doi.org/10.1016/j.fgb.2012.01.007.

Parra, G., Bradnam, K., & Korf, I. (2007). CEGMA: A pipeline to accurately annotate core genes in eukaryotic genomes. *Bioinformatics*, *23*(9), 1061–1067. https://doi.org/10.1093/bioinformatics/btm071.

Pedersen, C., van Themaat, E. V. L., McGuffin, L. J., Abbott, J. C., Burgis, T. A., Barton, G., et al. (2012). Structure and evolution of barley powdery mildew effector candidates. *BMC Genomics*, *13*(1), 694. https://doi.org/10.1186/1471-2164-13-694.

Perlin, M. H., Amselem, J., Fontanillas, E., Toh, S. S., Chen, Z., Goldberg, J., et al. (2015). Sex and parasites: Genomic and transcriptomic analysis of *Microbotryum lychnidis-dioicae*, the biotrophic and plant-castrating anther smut fungus. *BMC Genomics*, *16*, 461. https://doi.org/10.1186/s12864-015-1660-8.

Pernaci, M., De Mita, S., Andrieux, A., Pétrowski, J., Halkett, F., Duplessis, S., et al. (2014). Genome-wide patterns of segregation and linkage disequilibrium: The construction of a linkage genetic map of the poplar rust fungus *Melampsora larici-populina*. *Frontiers in Plant Science*, *5*, 454. https://doi.org/10.3389/fpls.2014.00454.

Persoons, A., Morin, E., Delaruelle, C., Payen, T., Halkett, F., Frey, P., et al. (2014). Patterns of genomic variation in the poplar rust fungus *Melampsora larici-populina* identify pathogenesis-related factors. *Frontiers in Plant Science, 5,* 450. https://doi.org/10.3389/fpls.2014.00450.

Petre, B., Joly, D., & Duplessis, S. (2014). Effector proteins of rust fungi. *Frontiers in Plant Science, 5,* 1–7.

Pfunder, M., Schurch, S., & Roy, B. A. (2001). Sequence variation and geographic distribution of pseudoflower-forming rust fungi (*Uromyces pisi s. lat.*) on *Euphorbia cyparissias*. *Mycological Research, 105*(1), 57–66.

Proost, S., Van Bel, M., Vaneechoutte, D., Van de Peer, Y., Inzé, D., Mueller-Roeber, B., et al. (2015). PLAZA 3.0: An access point for plant comparative genomics. *Nucleic Acids Research, 43*(D1), D974–D981. https://doi.org/10.1093/nar/gku986.

Raffaele, S., & Kamoun, S. (2012). Genome evolution in filamentous plant pathogens: Why bigger can be better. *Nature Reviews Microbiology, 10*(6), 417–430. https://doi.org/10.1038/nrmicro2790.

Ramos, A. P., Tavares, S., Tavares, D., Silva Mdo, C., Loureiro, J., & Talhinhas, P. (2015). Flow cytometry reveals that the rust fungus, *Uromyces bidentis* (Pucciniales), possesses the largest fungal genome reported—2489 Mbp. *Molecular Plant Pathology, 16*(9), 1006–1010. https://doi.org/10.1111/mpp.12255.

Riley, R., Salamov, A. A., Brown, D. W., Nagy, L. G., Floudas, D., Held, B. W., et al. (2014). Extensive sampling of basidiomycete genomes demonstrates inadequacy of the white-rot/brown-rot paradigm for wood decay fungi. *Proceedings of the National Academy of Sciences of the United States of America, 111*(27), 9923–9928. https://doi.org/10.1073/pnas.1400592111.

Rochi, L., Diéguez, M. J., Burguener, G., Darino, M. A., Pergolesi, M. F., Ingala, L. R., et al. (2016). Characterization and comparative analysis of the genome of *Puccinia sorghi* Schwein, the causal agent of maize common rust. *Fungal Genetics and Biology* https://doi.org/10.1016/j.fgb.2016.10.001.

Ronquist, F., & Huelsenbeck, J. P. (2003). MrBayes 3: Bayesian phylogenetic inference under mixed models. *Bioinformatics, 19*(12), 1572–1574. https://doi.org/10.1093/bioinformatics/btg180.

Rush, T. (2012). *Searching for alternative hosts and determining the variation in the internal transcribed spacer—ITS—region of* Phakopsora pachyrhizi *and the implications for currently used molecular diagnostic assays.* M.S. thesis. Baton Rouge, LA: Louisiana State University, 65 pp.

Rutter, W. B., Salcedo, A., Akhunova, A., He, F., Wang, S., Liang, H., et al. (2017). Divergent and convergent modes of interaction between wheat and *Puccinia graminis* f. sp. *tritici* isolates revealed by the comparative gene co-expression network and genome analyses. *BMC Genomics, 18*(1), 291. https://doi.org/10.1186/s12864-017-3678-6.

Sacristán, S., Vigouroux, M., Pedersen, C., Skamnioti, P., Thordal-Christensen, H., Micali, C., et al. (2009). Coevolution between a family of parasite virulence effectors and a class of LINE-1 retrotransposons. *PLoS One, 4*(10), e7463. https://doi.org/10.1371/journal.pone.0007463.

Saunders, D. G. O., Win, J., Cano, L. M., Szabo, L. J., Kamoun, S., & Raffaele, S. (2012). Using hierarchical clustering of secreted protein families to classify and rank candidate effectors of rust fungi. *PLoS One7,* e29847. https://doi.org/10.1371/journal.pone.0029847.

Savile, D. B. O. (1971). Coevolution of the rust fungi and their hosts. *The Quarterly Review of Biology, 46*(3), 211–218. https://doi.org/10.2307/2822510.

Schirawski, J., Mannhaupt, G., Munch, K., Brefort, T., Schipper, K., Doehlemann, G., et al. (2010). Pathogenicity determinants in smut fungi revealed by genome comparison. *Science, 330*(6010), 1546–1548. https://doi.org/10.1126/science.1195330.

Scholler, M., & Aime, M. C. (2006). On some rust fungi (Uredinales) collected in an *Acacia koa–Metrosideros polymorpha* woodland, Mauna Loa Road, Big Island, Hawaii. *Mycoscience*, *47*(3), 159–165. https://doi.org/10.1007/s10267-006-0286-8.

Schulze-Lefert, P., & Panstruga, R. (2011). A molecular evolutionary concept connecting nonhost resistance, pathogen host range, and pathogen speciation. *Trends in Plant Science*, *16*(3), 117–125. https://doi.org/10.1016/j.tplants.2011.01.001.

Shattock, R. C., & Preece, T. F. (2000). Tranzschel revisited: Modern studies of the relatedness of different rust fungi confirm his law. *Mycologist*, *14*(3), 113–117. https://doi.org/10.1016/S0269-915X(00)80086-5.

Simão, F. A., Waterhouse, R. M., Ioannidis, P., Kriventseva, E. V., & Zdobnov, E. M. (2015). BUSCO: Assessing genome assembly and annotation completeness with single-copy orthologs. *Bioinformatics*, *31*(19), 3210–3212. https://doi.org/10.1093/bioinformatics/btv351.

Simillion, C., Janssens, K., Sterck, L., & Van de Peer, Y. (2008). i-ADHoRe 2.0: An improved tool to detect degenerated genomic homology using genomic profiles. *Bioinformatics*, *24*(1), 127–128. https://doi.org/10.1093/bioinformatics/btm449.

Sjamsuridzal, W., Nishida, H., Ogawa, H., Kakishima, M., & Sugiyama, J. (1999). Phylogenetic positions of rust fungi parasitic on ferns: Evidence from 18S rDNA sequence analysis. *Mycoscience*, *40*(1), 21–27. https://doi.org/10.1007/bf02465669.

Spanu, P. D. (2011). The genomics of obligate (and nonobligate) biotrophs. *Annual Review of Phytopathology*, *50*(1), 91–109. https://doi.org/10.1146/annurev-phyto-081211-173024.

Spanu, P. D., Abbott, J. C., Amselem, J., Burgis, T. A., Soanes, D. M., Stüber, K., et al. (2010). Genome expansion and gene loss in powdery mildew fungi reveal tradeoffs in extreme parasitism. *Science*, *330*(6010), 1543.

Sperschneider, J., Taylor, J., Dodds, P. N., & Duplessis, S. (2017). Computational methods for predicting effectors in rust pathogens. In S. Periyannan (Ed.), *Wheat rust disease—Methods and protocols* (pp. 73–84). New York, NY: Springer Science+Business Media LLC.

Stamatakis, A. (2014). RAxML Version 8: A tool for phylogenetic analysis and post-analysis of large phylogenies. *Bioinformatics*, *30*, 1312–1313. https://doi.org/10.1093/bioinformatics/btu033.

Stone, C. L., Buitrago, M. L. P., Boore, J. L., & Frederick, R. D. (2010). Analysis of the complete mitochondrial genome sequences of the soybean rust pathogens *Phakopsora pachyrhizi* and *P. meibomiae*. *Mycologia*, *102*(4), 887–897. https://doi.org/10.3852/09-198.

Sydow, P., & Sydow, H. (1915). Monographia Uredinearum seu Specierum Omnium ad hunc usque Diem Descriptio et Adumbratio Systematica (Vol. 3). Lipsiae: Fratres Borntraeger.

Talhinhas, P., Azinheira, H. G., Vieira, B., Loureiro, A., Tavares, S., Batista, D., et al. (2014). Overview of the functional virulent genome of the coffee leaf rust pathogen *Hemileia vastatrix* with an emphasis on early stages of infection. *Frontiers in Plant Science*, *5*, 88. https://doi.org/10.3389/fpls.2014.00088.

Tan, M.-K., Collins, D., Chen, Z., Englezou, A., & Wilkins, M. R. (2014). A brief overview of the size and composition of the myrtle rust genome and its taxonomic status. *Mycology*, *5*(2), 52–63.

Tan, G., Muffato, M., Ledergerber, C., Herrero, J., Goldman, N., Gil, M., et al. (2015). Current methods for automated filtering of multiple sequence alignments frequently worsen single-gene phylogenetic inference. *Systematic Biology*, *64*, 778–791.

Tavares, S., Ramos, A. P., Pires, A. S., Azinheira, H. G., Caldeirinha, P., Link, T., et al. (2014). Genome size analyses of Pucciniales reveal the largest fungal genomes. *Frontiers in Plant Science*, *5*, 422.

Tisserant, E., Malbreil, M., Kuo, A., Kohler, A., Symeonidi, A., Balestrini, R., et al. (2013). Genome of an arbuscular mycorrhizal fungus provides insight into the oldest plant symbiosis. *Proceedings of the National Academy of Sciences of the United States of America*, *110*(50), 20117–20122. https://doi.org/10.1073/pnas.1313452110.

Toome, M., Kuo, A., Henrissat, B., Lipzen, A., Tritt, A., Yoshinaga, Y., et al. (2014). Draft genome sequence of a rare smut relative, *Tilletiaria anomala* UBC 951. *Genome Announcements*, *2*, e00539–14.

Toome, M., Ohm, R. A., Riley, R. W., James, T. Y., Lazarus, K. L., Henrissant, B., et al. (2014). Genome sequencing provides insight into the reproductive biology, nutritional mode, and ploidy of the fern pathogen *Mixia osmundae*. *New Phytologist*, *202*, 554–564. https://doi.org/10.1111/nph.12653.

Toome-Heller, M. (2016). Latest developments in the research of rust fungi and their allies (Pucciniomycotina). In D.-W. Li (Ed.), *Biology of microfungi* (pp. 147–168). Cham: Springer International Publishing.

Tranzschel, W. (1904). Über die Möglichkeit die Biologie wirtswechselnder Rostpilze auf Grund morphologischer Merkmale vorauszusehen. *Arbeiten der Kaiserlichen St Petersburger Naturforschenden Gesellschaft*, *35*, 311–313.

Tremblay, A., Hosseini, P., Li, S., Alkharouf, N. W., & Matthews, B. F. (2012). Identification of genes expressed by *Phakopsora pachyrhizi*, the pathogen causing soybean rust, at a late stage of infection of susceptible soybean leaves. *Plant Pathology*, *61*(4), 773–786. https://doi.org/10.1111/j.1365-3059.2011.02550.x.

Tremblay, A., Hosseini, P., Li, S., Alkharouf, N. W., & Matthews, B. F. (2013). Analysis of *Phakopsora pachyrhizi* transcript abundance in critical pathways at four time-points during infection of a susceptible soybean cultivar using deep sequencing. *BMC Genomics*, *14*(1), 614. https://doi.org/10.1186/1471-2164-14-614.

Tuskan, G. A., Difazio, S., Jansson, S., Bohlmann, J., Grigoriev, I., Hellsten, U., et al. (2006). The genome of black cottonwood, *Populus trichocarpa* (Torr. & Gray). *Science*, *313*(5793), 1596–1604. https://doi.org/10.1126/science.1128691.

Upadhyaya, N. M., Garnica, D. P., Karaoglu, H., Sperschneider, J., Nemri, A., Xu, B., et al. (2015). Comparative genomics of Australian isolates of the wheat stem rust pathogen *Puccinia graminis* f. sp. *tritici* reveals extensive polymorphism in candidate effector genes. *Frontiers in Plant Science*, *5*, 759. https://doi.org/10.3389/fpls.2014.00759.

van der Merwe, M., Ericson, L., Walker, J., Thrall, P. H., & Burdon, J. J. (2007). Evolutionary relationships among species of *Puccinia* and *Uromyces* (Pucciniaceae, Uredinales) inferred from partial protein coding gene phylogenies. *Mycological Research*, *111*, 163–175. https://doi.org/10.1016/j.mycres.2006.09.015.

van der Merwe, M. M., Walker, J., Ericson, L., & Burdon, J. J. (2008). Coevolution with higher taxonomic host groups within the *Puccinia/Uromyces* rust lineage obscured by host jumps. *Mycological Research*, *112*, 1387–1408. https://doi.org/10.1016/j.mycres.2008.06.027.

Vialle, A., Feau, N., Allaire, M., Didukh, M., Martin, F., Moncalvo, J.-M., et al. (2009). Evaluation of mitochondrial genes as DNA barcode for Basidiomycota. *Molecular Ecology Resources*, *9*, 99–113. https://doi.org/10.1111/j.1755-0998.2009.02637.x.

Virtudazo, E., Nakamura, H., & Kakishima, M. (2001). Ribosomal DNA-ITS sequence polymorphism in the sugarcane rust, *Puccinia kuehnii*. *Mycoscience*, *42*(5), 447–453. https://doi.org/10.1007/bf02464341.

Wingfield, B. D., Ericson, L., Szaro, T., & Burdon, J. J. (2004). Phylogenetic patterns in the Uredinales. *Australasian Plant Pathology*, *33*(3), 327–335. https://doi.org/10.1071/ap04020.

Wood, A. R., & Morris, M. J. (2007). Impact of the gall-forming rust fungus *Uromycladium tepperianum* on the invasive tree *Acacia saligna* in South Africa: 15 Years of monitoring. *Biological Control*, *41*(1), 68–77.

Yun, H. Y., Minnis, A. M., Kim, Y. H., Castlebury, L. A., & Aime, M. C. (2011). The rust genus *Frommeëlla* revisited: A later synonym of *Phragmidium* after all. *Mycologia*, *103*(6), 1451–1463.
Zambino, P. J., Kubelik, A. R., & Szabo, L. J. (2000). Gene action and linkage of avirulence genes to DNA markers in the rust fungus *Puccinia graminis*. *Phytopathology*, *90*(8), 819–826. https://doi.org/10.1094/PHYTO.2000.90.8.819.
Zhang, S., Skerker, J. M., Rutter, C. D., Maurer, M. J., Arkin, A. P., & Rao, C. V. (2016). Engineering *Rhodosporidium toruloides* for increased lipid production. *Biotechnology and Bioengineering*, *113*(5), 1056–1066. https://doi.org/10.1002/bit.25864.
Zheng, W., Huang, L., Huang, J., Wang, X., Chen, X., Zhao, J., et al. (2013). High genome heterozygosity and endemic genetic recombination in the wheat stripe rust fungus. *Nature Communications*, *4*, 2673. https://doi.org/10.1038/ncomms3673.

FURTHER READING

Carvalho, C. R., Fernandes, R. C., Carvalho, G. M. A., Barreto, R. W., & Evans, H. C. (2011). Cryptosexuality and the genetic diversity paradox in coffee rust, *Hemileia vastatrix*. *PLoS One*, *6*(11), e26387. https://doi.org/10.1371/journal.pone.0026387.
Townsend, J. P., & Lopez-Giraldez, F. (2010). Optimal selection of gene and ingroup taxon sampling for resolving phylogenetic relationships. *Systematic Biology*, *59*(4), 446–457. https://doi.org/10.1093/sysbio/syq025.

CHAPTER EIGHT

Advances in Fungal Phylogenomics and Their Impact on Fungal Systematics

Ning Zhang[1], Jing Luo, Debashish Bhattacharya
Rutgers University, New Brunswick, NJ, United States
[1]Corresponding author: e-mail address: ningz@rutgers.edu

Contents

Abstract

In the past decade, advances in next-generation sequencing technologies and bioinformatic pipelines for phylogenomic analysis have led to remarkable progress in fungal systematics and taxonomy. A number of long-standing questions have been addressed using comparative analysis of genome sequence data, resulting in robust multigene phylogenies. These have added to, and often surpassed traditional morphology or single-gene phylogenetic methods. In this chapter, we provide a brief history of fungal systematics and highlight some examples to demonstrate the impact of phylogenomics on this field. We conclude by discussing some of the challenges and promises in fungal biology posed by the ongoing genomics revolution.

Advances in Genetics, Volume 100
ISSN 0065-2660
https://doi.org/10.1016/bs.adgen.2017.09.004

1. A BRIEF HISTORY OF FUNGAL SYSTEMATICS

The kingdom Fungi is one of the most diverse and important eukaryotic kingdoms. It has been estimated that there are 1.5 million to 5.1 million species of Fungi on Earth but only less than 10% of these (approximately 135,000 species) have been identified and described (Blackwell, 2011; Hawksworth, 1991; Hibbett et al., 2016). Fungi are unicellular or multicellular heterotrophs with chitinous cell walls. They reproduce sexually (meiotically) or asexually (mitotically) by various types of spores. Many fungal species have more than one propagation method. Traditionally fungi were classified into five phyla mainly based on morphological characteristics: Chytridiomycota, the zoosporic fungi, are characterized by flagellated motile spores; Zygomycota that include bread molds, *Rhizopus*, *Mucor*, etc., are characterized by the thick-walled resting zygospores; Glomeromycota, the root symbiotic fungi that form arbuscular mycorrhizae with plant; Ascomycota consist of yeasts, *Aspergillus*, *Penicillium*, etc., produce meiotic ascospores borne internally in sac-like structures called asci; and Basidiomycota, the mushrooms, rusts, smuts, etc., produce elaborate fruiting bodies with club-shaped basidia that bear external meiotic basidiospores. Recently, fungal systematics have undergone significant revision due to the advances in molecular, microscopic, and other technologies, and we will give a brief review of the history and highlight the impact of phylogenomics on fungal systematics and taxonomy in this chapter.

1.1 Traditional Fungal Systematics Relying on Morphological or Other Phenotypic Characters (1729–1864)

Fungal systematics dates back to the 17th century after the development of the compound microscope by van Leeuwenhoek. Pier Antonio Micheli's publication Nova Plantarum Genera in 1729 is considered to be the pioneer work in this field (Alexopoulos, 1962). In traditional taxonomy, fungi were classified mainly based on their observable morphological and other phenotypic characters. There are numerous monographs and other literature that classify fungi exclusively based on morphology, for example, Dematiaceous Hyphomycetes (Ellis, 1971), A Reevaluation of the Bitunicate Ascomycetes with Keys to Families and Genera (von Arx & Müller, 1975), More Dematiaceous Hyphomycetes (Ellis, 1976), Genera of Hyphomycetes (Carmichael, Kendrick, Conners, & Sigler, 1980), The

Coelomycetes (Sutton, 1980), Prodromus to Class Loculoascomycetes (Barr, 1987), Illustrated Genera of Imperfect Fungi (Barnett & Hunter, 1997), and Coelomycetous Anamorphs with Appendage-Bearing Conidia (Nag Raj, 1993).

With the development of electronic microscopy techniques in recent years, fungal ultrastructures, such as the spindle pole body, cell wall, septa, and septal pores, have been used in fungal systematics (Guarro, Gené, & Stchigel, 1999). Information on the ultrastructures shed light on fungal evolution and contributed to the study of fungal systematics (Beckett, Heath, & Mclaughlin, 1974; Bracker, 1967; Kimbrough, 1994; Lü & Mclaughlin, 1991; Lutzoni et al., 2004).

1.2 Fungal Systematics Using Physiological and Biochemical Characters (1865–1989)

Although observable morphology and other phenotypic characters are the foundation of fungal taxonomy, some characters are unstable and provide limited systematic information. Other informative characters are needed for more reliable fungal classification. As early as the 19th century, color reactions associated with application of chemical stains were used to differentiate lichens and other ascomycetes (Nylander, 1866; Rolland, 1887). Chemical test, also called spot test, is a standard identification protocol in lichenology because the morphologically defined species generally have same or similar chemical components yielding constant color reactions in lichens (Elix, 1992). With advances in technology, more physiological and biochemical methods came into use and promoted fungal systematics. These included growth temperature and rate tests and nutrient utilization pattern analysis (Guarro et al., 1999) that were used in the classification and identification of certain culturable yeasts and filamentous fungi (de Hoog & Gerrits van den Ende, 1992; de Hoog et al., 1994; Paterson & Bridge, 1994; Rath, Carr, & Graham, 1995), such as *Aureobasidium* (de Hoog & Yurlova, 1994), *Penicillium* (Bridge et al., 1989), *Fusarium* (Wasfy, Bridge, & Brayford, 1987), *Phoma* (Monte, Bridge, & Sutton, 1990), and *Rhizoctonia* (Mordue, Currah, & Bridge, 1989). Secondary metabolites were found to be systematically informative and served as a component of integrative taxonomy. For example, chemotaxonomic characterization was used in the systematics of Xylariaceae and *Penicillium* (Frisvad & Filtenborg, 1990; Whalley & Edwards, 1995). Other characters, such as ubiquinone compounds, cellular fatty acids, cell wall composition, and protein patterns including isoenzymes and allozymes, were also found to be valuable in this regard. All physiological and biochemical characters were then used as important complementary information for fungal systematics and phylogeny, which helped traditional morphology-based fungal taxonomy step into the era of integrative fungal systematics.

1.3 Fungal Phylogeny Based on One or a Few Genes (1990–2005)

With the advent of the polymerase chain reaction (PCR) (Saiki et al., 1988) and Sanger sequencing (Sanger, Nicklen, & Coulson, 1977), since the early 1990s molecular phylogeny has continually been applied to fungal systematics and became the most powerful and popular systematic tool. Therefore, the characters used for fungal classification have evolved from morphology and development, to physiology and biochemistry, and more recently to protein and DNA sequences. Compared to the traditional tools in fungal systematics, molecular phylogeny has been shown to be inherently superior because phenotypic characters are often prone to convergent evolution, reduction, or loss. Numerous molecular studies based on one or a few genes have been carried out to investigate the phylogenetic relationships among fungal taxa at all taxonomic levels (Schoch et al., 2009). The best studied and the most commonly used locus in fungal phylogeny is the ribosomal RNA genes (rDNA) because: (1) rDNA is a multiple copy gene that is easily isolated and cloned, and (2) different regions of rDNA evolve at differing rates, making it useful for taxon resolving lineages at different taxonomic levels (Bruns, White, & Taylor, 1991; Hibbett, 1992). Internal transcribed spacers (ITS), intergenic spacer, large subunit (LSU), and small subunit (SSU) of rDNA have been well studied since the dawn of fungal molecular phylogenetics (White, Bruns, Lee, & Taylor, 1990). Generally, SSU and LSU rDNA genes were applied to higher taxonomic level delineation, whereas ITS was used for species or genus resolution. Because of its high levels of species discrimination rate and PCR success rate, the ITS of rDNA was assigned as a universal fungal DNA barcode (Schoch et al., 2012). DNA barcoding is the use of short standardized segments of the genome for accurate and rapid species identification. A challenge for fungal phylogenetics is data curation. Currently, there is a large amount of erroneous sequence data in GenBank and other public databases that were generated from misidentified or contaminated samples and may result in unreliable phylogeny and wrong classification. To address this problem, NCBI's fungal RefSeq curators and other mycologists have started curating a set of standard fungal ITS sequences derived from type materials (Schoch et al., 2014).

Apart from rDNA, other genes, in particular, single copy and housekeeping protein genes, represent the overwhelming majority of the cellular genome and are essential to biological functions, providing effective markers to track organismal evolution. These genes, such as actin (*ACT*) (Helgason, Watson, & Young, 2003), calmodulin (*CAL*) (Hong, Go, Shin, Frisvad, & Samson, 2005), histone (*H3*, *H4*) (Glass & Donaldson, 1995), glycerol-3-phosphate dehydrogenase (*GAPDH*) (Guerber, Liu, Correll, & Johnston, 2003; Templeton, Rikkerink, Solon, & Crowhurst, 1992), DNA replication licensing factor (*MCM7*) (Schmitt et al., 2009), RNA polymerase II largest subunit (*RPB1*) (Matheny, Liu, Ammirati, & Hall, 2002), RNA polymerase II 2nd largest subunit (*RPB2*) (Liu, Whelen, & Hall, 1999), and translation elongation factor 1 (*TEF1*) (Rehner & Buckley, 2005), also play important roles in fungal systematics. Fungal phylogenetics has also used mitochondrial genes such as the ATPase subunit 6 (*ATP6*) (Kretzer & Bruns, 1999), cytochrome *c* oxidase I (*COI*) (Seifert et al., 2007), and mitochondrial LSU and SSU of rDNA (mtLSU and mtSSU, respectively) (Seif et al., 2005; Zoller, Scheidegger, & Sperisen, 1999).

At the onset of fungal molecular phylogenetic studies, only single-gene sequences were used to build the phylogeny (Liu et al., 1999; Tehler, Farris, Lipscomb, & Källersjö, 2000). Soon after, multilocus (usually less than six loci) phylogenies appeared in the literature (Blackwell, Hibbett, Taylor, & Spatafora, 2006; Spatafora et al., 2006; Zhang et al., 2006). The latter had relatively higher resolution power than single-gene trees. In the Assembling the Fungal Tree of Life (AFTOL) project, LSU and SSU of rDNA as well as several protein coding genes, including *RPB1*, *RPB2*, *TEF1*, mtSSU, and *ATP6*, were chosen to explore the fungal phylogeny (http://aftol.org/about.php), which was used as basis for the classification of the kingdom Fungi (Hibbett et al., 2007; James et al., 2006). The majority of the recently published fungal monographs also have relied on multilocus phylogenies, such as the monographs on *Cladosporium* (Bensch, Braun, Groenewald, & Crous, 2012; Bensch et al., 2015), *Phoma* (Chen, Jiang, Zhang, Cai, & Crous, 2015), *Alternaria* (Woudenberg et al., 2015), Massarineae (Tanaka et al., 2015), and families of Sordariomycetes (Maharachchikumbura et al., 2016).

1.4 Fungal Phylogenomic Analysis (2006–Present)

Due to insufficient phylogenetic information and gene-specific noise, single or a few loci (multilocus) often generate incongruent phylogenies resulting in many poorly supported (by bootstrap analysis) nodes (Ebersberger et al., 2012; Fitzpatrick, Logue, Stajich, & Butler, 2006). It was estimated that at least 20 unlinked genes or 8000 randomly selected orthologous nucleotides on a genomic scale are required to reconstruct a robust systematic framework (Rokas, Williams, King, & Carroll, 2003). Phylogenomic analyses using large numbers of genes from various independent evolving regions throughout the whole genome or transcriptome will maximize the informativeness, reduce the stochastic error, and thus improve phylogenetic accuracy. Genomic data are the basis of this approach.

The first fungal genome (*Saccharomyces cerevisiae*) was published in 1996 (Goffeau et al., 1996). *Schizosaccharomyces pombe* was sequenced in 2002 (Wood et al., 2002) and *Neurospora crassa* in 2003 (Galagan et al., 2003). Subsequently, the first Basidiomycota genome of the white-rot *Phanerochaete chrysosporium* was reported in 2004 (Martinez et al., 2004). Advances in next-generation sequencing (NGS) technology have driven dramatic progress in fungal genome sequencing. NGS is rapid and high-throughput and represented by a number of different sequencing platforms including Roche 454, AB SOLiD, Illumina GA/HiSeq System, and PacBio RS (Goodwin, McPherson, & McCombie, 2016). Unlike Sanger sequencing with separate DNA synthesis and detection steps, NGS technology is superior with respect to massively parallel analysis with high-throughput and reduced cost, which makes large-scale genomic work faster and more affordable than ever. The number of available fungal genomes has exponentially increased, especially in recent years, with the efforts of genome consortia such as the Broad Institute Fungal Genome Initiatives, the 1000 Fungal Genomes Project at the Joint Genome Institute (Grigoriev et al., 2011), the TIGR and Genoscope sequencing projects. The 1000 Fungal Genome project aims to have at least two species genomes sequenced from each of the ~500 recognized families of Fungi (http://genome.jgi.doe.gov/programs/fungi/1000fungalgenomes.jsf). To date, approximate 800 fungal genomes have been sequenced and released, of which 50% have been published (http://genome.jgi.doe.gov/programs/fungi/index.jsf). Most released genomes were from Eurotiomycetes (14%), Dothideomycetes (14%), Sordariomycetes (14%), and Agariocomycetes (25%) (Fig. 1). These efforts provided necessary data and enabled fungal phylogenomics.

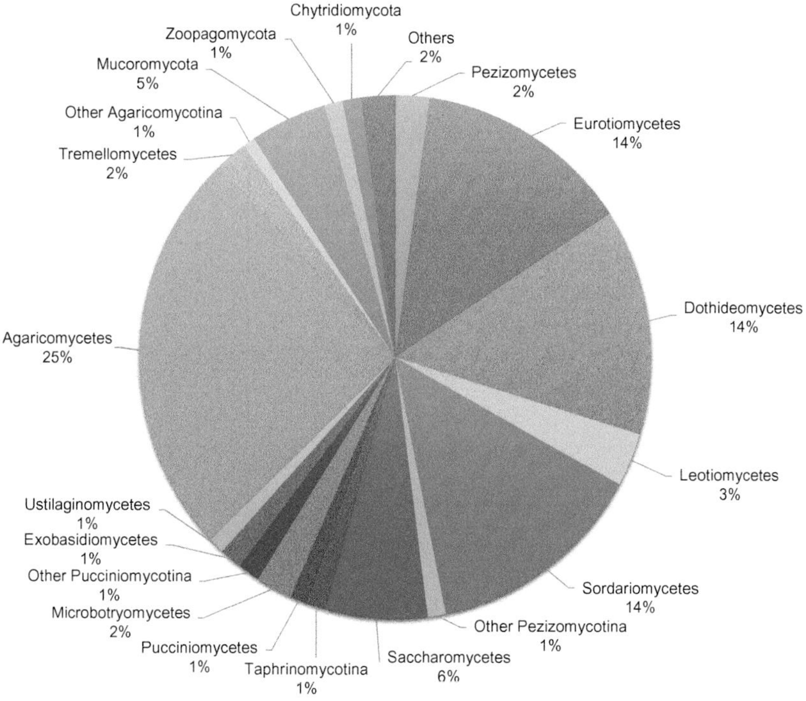

Fig. 1 Percentage of publicly available fungal genomes at the class or higher taxonomic levels. Taxa belonging to Ascomycota and Basidiomycota are illustrated at the class level. Other taxa are shown at the phylum or subphylum level. *Data are derived from the JGI, http://genome.jgi.doe.gov/programs/fungi/index.jsf.*

2. IMPACT OF PHYLOGENOMIC STUDIES ON FUNGAL SYSTEMATICS

Phylogenetic studies have regularly updated our understanding of fungal evolution and led to significant revision in fungal systematics. Evolution of the fungal phylogeny is shown in Fig. 2. The first phylogeny (Fig. 2A) represents traditional systematics of fungi based on morphology, physiological, and biochemical characters (Alexopoulos, 1962). In this system, fungi were placed in a division in the kingdom Plantae, Mycota, which included two subdivisions and 10 classes. In 2000, based on the SSU rDNA phylogeny (Tehler et al., 2000) (Fig. 2B), fungi were defined as a monophyletic group and divided into nine subgroups, with Oomycetes and Myxomycetes excluded. In 2007, a more comprehensive phylogeny of the kingdom Fungi

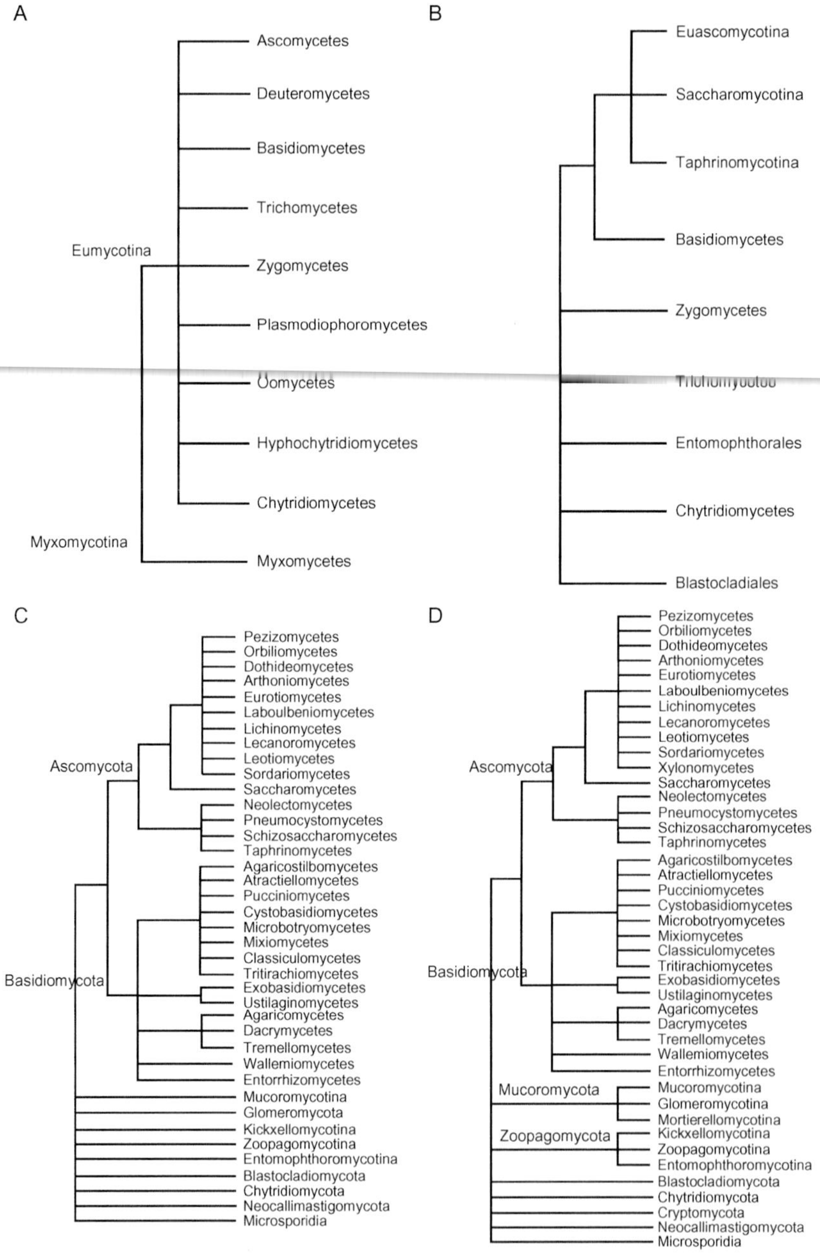

Fig. 2 (A) Fungal phylogeny based on morphological, physiological, and biochemical characters (Alexopoulos, 1962). (B) Fungal phylogeny based on SSU rDNA (Tehler et al., 2000). (C) Fungal phylogeny based on multiple genes (Hibbett et al., 2007). (D) Fungal Phylogeny based on genomic data (Spatafora et al., 2016).

was reconstructed based on multilocus DNA sequences. Ascomycota and Basidiomycota were classified as the subkingdom Dikarya, and additional nine major fungal groups were recognized with dramatic revisions of traditional Chytridiomycota and Zygomycota (Hibbett et al., 2007) (Fig. 2C). The current system was further developed with the help of phylogenomics. The phylum Mucoromycota was circumscribed to accommodate Glomeromycotina, Mortierellomycotina, and Mucoromycotina, and Zoopagomycota was established to accommodate Entomophthoromycotina, Kickxellomycotina, and Zoopagomycotina (Spatafora et al., 2016) (Fig. 2D).

2.1 Backbone Phylogeny of Fungi

Early phylogenomic studies focused on the backbone of the fungal tree of life or higher taxonomic level relationships, which are difficult to resolve using single or multilocus phylogeny. A robust phylogeny including 42 taxa of Ascomycota, Basidiomycota, and zygomycetes was reconstructed via whole genome analyses by Fitzpatrick et al. (2006), which suggested monophyletic Pezizomycotina and Saccharomycotina, and a close relationship between Leotiomycetes and Sordariomycetes. Robbertse, Reeves, Schoch, and Spatafora (2006) inferred a phylogenomic tree of 18 Ascomycota taxa, which resolved Pezizomycotina and Saccharomycotina, the early diverging Taphrinomycotina, and the sister relationship between Leotiomycetes and Sordariomycetes. Liu, Leigh, et al. (2009) and Liu, Steenkamp, et al. (2009) recovered the monophyly of Taphrinomycotina in Ascomycota from 54 fungal species and provided significant support for the paraphyly of the traditional Zygomycota. Wang, Xu, Gao, and Hao (2009) published another fungal phylogeny based on 82 taxa in 5 phyla, most of which are Ascomycota. This study confirmed the placement of Microsporidia in Fungi and reinforced a monophyletic Dikarya that included sister taxa Ascomycota and Basidiomycota proposed via multigene analyses by Hibbett et al. (2007). By using 99 fungal genomes and 109 fungal expressed sequenced tag set, Ebersberger et al. (2012) constructed a kingdom-wide fungal phylogeny, including the early branching lineages, Ascomycota and Basidiomycota. This study revealed Ustilaginomycotina as sister taxa to Agaricomycotina, Dothideomycetes as sister of Eurotiomycetes, and the alliance of Blastocladiomycota with Chytridiomycota.

2.2 Phylogeny of Polyporales

In addition to the backbone, phylogenomic analysis also was used to resolve lower taxonomic level relationships, for example, the order Polyporales (Binder et al., 2013). The Polyporales is a diverse order in Agaricomycetes of Basidiomycota that includes 1801 species (Kirk, Cannon, Minter, & Stalpers, 2008). The taxonomy in Polyporales did not reflect the evolutionary history because many genera and families were found to be polyphyletic or paraphyletic based on single or multilocus phylogenies (Binder et al., 2005; Miettinen, Larsson, Sjökvist, & Larsson, 2012). Binder et al. (2013) used 10 Polyporales taxa with available genome data to generate phylogenomic trees based on 356 genes and 71 genes, respectively. They were compared with each other, and with a six-gene tree of 373 Polyporales and related taxa. In the 356-gene tree, four well-supported major clades, i.e., antrodia, gelatoporia, core polyporoid, and phlebioid were recognized, and all the internal braches received full support (100%). The 71-gene tree showed an identical topology but with lower branch supports for *Gelatoporia subvermispora* and the antrodia clade (89%), and *Rhodonia placenta* and *Wolfiporia cocos* (92%). These four clades were also found from the six-gene tree but with even weaker or no support, e.g., those for the antrodia clade and gelatoporia clade (no support), the antrodia clade (71%), the phlebioid clade (96%). A denser sampling of larger and identical gene sets across the genomic scale was therefore recommended to advance Polyporales phylogeny and systematics. Although only limited taxa were included, the phylogenomic analysis was considered to have great potential of producing better resolved and highly supported phylogenies.

2.3 Phylogeny and Taxonomy of Magnaporthales

Magnaporthales is an order in Sordariomycetes of Ascomycota with about 200 species. Magnaporthales contains important pathogens of cereals and grasses, e.g., the rice blast fungus *Pyricularia oryzae* (syn. *Magnaporthe oryzae*), the take-all pathogen of cereals *Gaeumannomyces graminis*, and the summer patch pathogen of turfgrass *Magnaporthiopsis poae*. Due to a lack of convincing morphological and developmental characters, these fungi have historically been placed in various orders, including Diaporthales (Krause & Webster, 1972), Dothideales (von Arx & Müller, 1975), Sordariales (Conway & Barr, 1977; Shearer, 1989), Phyllachorales (Barr, 1977), Polystigmatales (Hawksworth, Sutton, & Ainsworth, 1983), and Amphisphaeriales (Eriksson, 1984). Based on rDNA phylogenies, a new order Magnaporthales was proposed to accommodate these fungi (Thongkantha et al., 2009). However, consensus regarding their phylogenetic affinities was not reached. In an SSU phylogeny, Magnaporthales formed a sister clade with Ophiostomatales (Zhang & Blackwell, 2001), whereas an LSU tree suggested a close relationship of Magnaporthales with Sordariales, Chaetosphaeriales, and Boliniales (Huhndorf, Greif, Mugambi, & Miller, 2008). In another SSU rDNA tree, Magnaporthales formed a sister clade to Diaporthales and Ophiostomatales (Thongkantha et al., 2009). In the same paper, however, the LSU rDNA tree grouped these fungi with Chaetosphaeriales and Sordariaceae. A four-gene phylogeny found the rice blast fungus to be close to Diaporthales; however, no other Magnaporthales taxa were included in the analysis (Zhang et al., 2006).

Phylogenetic relationships within the Magnaporthales were also long unclear. Spore morphology was considered as the most important criterion to differentiate genera in this group. For example, *Gaeumannomyces* species had filiform (needle-like) ascospores, while *Magnaporthe* had fusiform (spindle-shaped) and three septate ascospores with center pigmented cells (Cannon, 1994). Perithecial wall structure was also used for classifying these fungi, such as *Muraerita* (Huhndorf et al., 2008). However, these genus concepts based on morphology were not supported by the molecular phylogeny. Based on SSU, LSU, ITS, *MCM7*, *RPB1*, and *TEF1*, genera *Gaeumannomyces*, *Magnaporthe*, *Harpophora*, and *Pyricularia* were revised, and anamorphic and ecological characters were suggested more informative than teleomorphic characteristics in defining monophyletic genera (Luo, Walsh, & Zhang, 2015; Luo & Zhang, 2013; Zhang, Zhao, & Shen, 2011). By using LSU and *RPB1*, three families were established for these fungi (Klaubauf et al., 2014).

A recent phylogenomic study was conducted to uncover the evolutionary history of Magnaporthales (Luo, Qiu, et al., 2015). To address the phylogenetic position of Magnaporthales in Sordariomycetes, 6 Magnaporthales species and 15 non-Magnaporthales species representing the major lineages of Pezizomycotina with two Saccharomycetes as outgroup were included in the analysis. A robust phylogeny based on 226 genes revealed that Magnaporthales is monophyletic and forms a sister group to the order Ophiostomatales. The close relationship between these two orders is supported by several overlooked phenotypic characters (Luo, Qiu, et al., 2015; Luo, Walsh, et al., 2015). To study the evolutionary relationships within Magnaporthales, 24 Magnaporthales and five outgroup taxa were included in the phylogenomic analysis. The resulting tree supported three major clades corresponding to three families, Magnaporthaceae (root clade), Pyriculariaceae (blast clade), and Ophioceraceae (wood clade) (Klaubauf et al., 2014). These results indicated that morphology-based genera often are polyphyletic, whereas biological and ecological characteristics, such as the hosts and substrates and infection structures correspond better to the true evolutionary history of Magnaporthales fungi. These phylogenomic studies shed light on the evolution of pathogenicity of these economically important species and provide a foundation for future research in various fields.

2.4 Phylogeny of Zygomycetes

Compared to Basidiomycota and Ascomycota, relatively few phylogenomic studies have been done for other fungal phyla. The zygomycetes include 1065 known species (Kirk et al., 2008) that are saprobes or parasites from mostly terrestrial habitats. Many zygomycetes taxa circumscribed using morphology were found to be nonmonophyletic (Hibbett et al., 2007). A comprehensive phylogenomic analysis of these fungi was recently done by Spatafora et al. (2016). A total of 46 taxa including 25 zygomycetes fungi based on 192 protein-coding loci were used to resolve their relationships at the phylum-level. Based on the phylogeny, a new classification was proposed (Spatafora et al., 2016) (Fig. 2D). The paraphyletic Zygomycota was revised, and a new system was established to accommodate 2 phyla, 6 subphyla, 4 classes, and 16 orders. Zoopagomycota is composed of Entomophthoromycotina, Kickxellomycotina, and Zoopagomycotina. This is the earliest diverging lineage of zygomycetes and comprises parasites and pathogens of small animals and other fungi. Mucoromycota is a later diverging lineage that is a sister clade to Dikarya. The proposed Mucoromycota includes Glomeromycotina, Mortierellomycotina, and Mucoromycotina that are generally arbuscular mycorrhizal (AM) fungi, root endophytes, and plant material decomposers, respectively.

3. CHALLENGES FACING FUNGAL PHYLOGENOMICS

Despite advances in phylogenomics, many fungal groups are still woefully undersampled. The AM fungi, for example, cannot grow on media, which limits the nucleic acid quantity and quality required for phylogenetic analyses. Moreover, they do not have any single-cell stage with only one nucleus, and their large spores contain thousands of nuclei, making genome analysis challenging. The current taxonomy of the AM fungi still relies primarily on spore morphology with guidance provided by rDNA gene analysis

(Schüßler & Walker, 2010). *Rhizophagus irregularis* is the only Glomeromycotina species that has a published genome (Tisserant et al., 2013). In a recent phylogenomic study, *R. irregularis* showed a sister relationship to Mortierellomycotina and was put into Mucoromycota (Spatafora et al., 2016). However, the current phylogeny within Glomeromycotina is based on the SSU rDNA gene and is far from being robust. A major obstacle for molecular phylogenetic study of the AM fungi is obtaining high quality, uncontaminated DNA for genome sequencing. Single-cell genome sequencing provides a potential option for these unculturable symbiotic organisms (Gawad, Koh, & Quake, 2016). However, the relatively large size of the *R. irregularis* genome (153 Mbp, 28,232 genes; Tisserant et al., 2013), if common among other AM fungi, may make it difficult to successfully assemble these single cell data. Another option is to develop bioinformatic tools to screen for AM genomes, or large genomic contigs from these taxa when analyzing fungus-enriched environmental metagenomes (Vogel & Moran, 2013).

Although most sequenced genomes are members of the Ascomycota and Basidiomycota (Fig. 1), some groups within these two phyla have received less attention and have little or no genome data available because of their relatively less economic importance or difficulty in obtaining pure culture for genomic sequencing. For example, Lecanoromycetes is the largest and most varied class of Ascomycota with 14,199 known species (Kirk et al., 2008); however, only four species have sequenced genomes (Fig. 3). Another example is the Orbiliomycetes, with only two released genomes out of the 288 known species. Furthermore, there are many fungal classes that have no genome data published yet, such as Arthoniomycetes, Geoglossomycetes, and Lichinomycetes in the Ascomycota, and Agaricostilbomycetes, Atractiellomycetes, Classiculomycetes, Cryptomycocolacomycetes, Cystobasidiomycetes, and Entorrhizomycetes in the Basidiomycota. In addition to known fungal taxa, undescribed environmental fungi are another important resource that can significantly fill gaps, and perhaps alter the structure of the fungal tree of life. With advances in high-throughput sequencing and bioinformatics, fungal systematics will be enriched by including genomes from a wide range of symbiotic and environmental samples in future analyses.

A final note of caution needs to be made about horizontal gene transfer (HGT), the bane of multigene phylogenetics. It has long been known that when unaccounted for, HGT can mislead phylogenies by introducing reticulate relationships among taxa (Chan, Bhattacharya, & Reyes-Prieto, 2012; Soanes & Richards, 2014). An example is a recent study we conducted that showed the grass pathogen *Magnaporthiopsis incrustans* (Magnaporthales) and

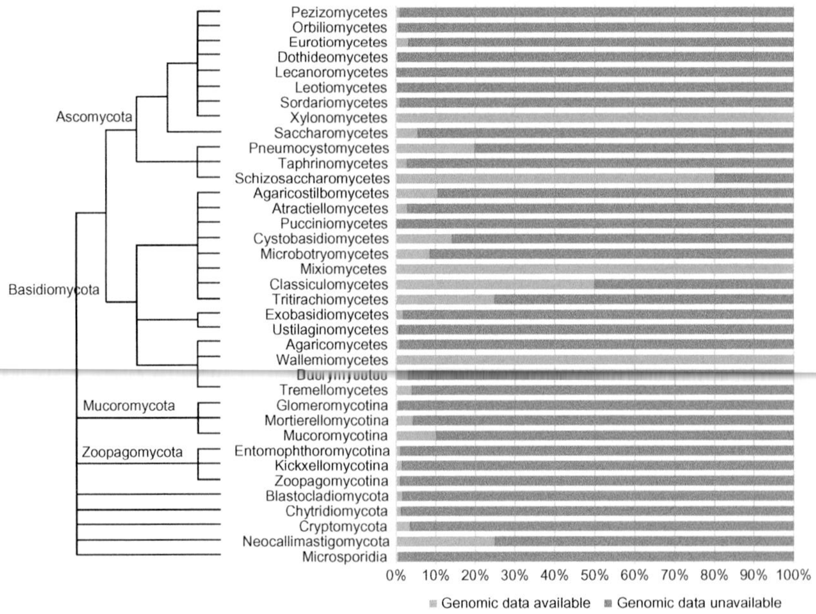

Fig. 3 Relative proportion of fungal taxa with released genomes among published taxa at the class or higher levels. *These inferences are derived from the JGI database, http://genome.jgi.doe.gov/programs/fungi/index.jsf, and the known fungal species data are from the Dictionary of the Fungi (Kirk, P. M., Cannon, P. F., Minter, D. W., & Stalpers, J. A. (2008).* Ainsworth & Bisby's Dictionary of the Fungi. *Wallingford, CT: CABI).*

members of the genus *Colletotrichum* to share at least 90 genes derived through HGT (Qiu, Cai, Luo, Bhattacharya, & Zhang, 2016). Many of these HGTs are physically linked (supporting the transfer event) and show twofold enrichment in carbohydrate-active enzymes (CAZymes) associated with plant cell wall degradation, and presumably underlie the origin of plant pathogenesis in these taxa. These results lead to two major conclusions: (1) it is critical to test for vertical inheritance of candidate genes in phylogenomic analyses, and (2) when present, HGTs can provide key insights into lineage evolution and how selection leads to the spread of valuable "genetic goods" (e.g., that enable a major lifestyle transition) across taxonomically diverse fungi.

4. GENERAL CONCLUSIONS

Molecular phylogenetics opened a new era in fungal systematics by providing the capacity to reconstruct a more robust fungal phylogeny, independent of morphology. The utilization of genome sequence data further improved phylogenetic resolution and linked species phylogeny to gene functions. To date, about 800 fungal genomes (0.6% of known fungal species) are publicly available, which forms the foundation for future systematic research. The development of next-generation high-throughput sequencing greatly accelerated access to genomic data and will soon make routine the application of this rich information to fungal taxonomic and systematic studies. These efforts promise to result in a robust tree of life for kingdom Fungi that will stand the test of time.

ACKNOWLEDGMENTS

The research was partially supported by Grants from the National Science Foundation of the United States (DEB 1145174 and 1452971).

REFERENCES

Alexopoulos, C. J. (1962). *Introductory mycology*. New York: John Wiley & Sons.

Barnett, H. L., & Hunter, B. B. (1997). *Illustrated genera of imperfect fungi*. St. Paul, MN, USA: American Phytopathological Society (APS Press).

Barr, M. E. (1977). *Magnaporthe*, *Telimenella*, and *Hyponectria* (Physosporellaceae). *Mycologia*, *69*, 952–966.

Barr, M. E. (1987). *Prodromus to class Loculoascomycetes*. MA, USA: Department of Botany, University of Massachusetts.

Beckett, A., Heath, I. B., & Mclaughlin, D. J. (1974). *An atlas of fungal ultrastructure*. London, UK: Longman Group.

Bensch, K., Braun, U., Groenewald, J. Z., & Crous, P. W. (2012). The genus *Cladosporium*. *Studies in Mycology*, *72*, 1–401.

Bensch, K., Groenewald, J. Z., Braun, U., Dijksterhuis, J., de Jesús Yáñez-Morales, M., & Crous, P. W. (2015). Common but different: The expanding realm of *Cladosporium*. *Studies in Mycology*, *82*, 23–74.

Binder, M., Hibbett, D. S., Larsson, K. H., Larsson, E., Langer, E., & Langer, G. (2005). The phylogenetic distribution of resupinate forms across the major clades of mushroom-forming fungi (Homobasidiomycetes). *Systematics and Biodiversity*, *3*, 113–157.

Binder, M., Justo, A., Riley, R., Salamov, A., Lopez-Giraldez, F., Sjokvist, E., et al. (2013). Phylogenetic and phylogenomic overview of the Polyporales. *Mycologia*, *105*, 1350–1373.

Blackwell, M. (2011). The fungi: 1, 2, 3 ... 5.1 million species? *American Journal of Botany*, *98*, 426–438.

Blackwell, M., Hibbett, D. S., Taylor, J. W., & Spatafora, J. W. (2006). Research coordination networks: A phylogeny for kingdom Fungi (Deep Hypha). *Mycologia*, *98*, 829–837.

Bracker, C. E. (1967). Ultrastructure of fungi. *Annual Review of Phytopathology*, *5*, 343–374.

Bridge, P. D., Hawksworth, D. L., Kozakiewicz, Z., Onions, A. H., Paterson, R. R., Sackin, M. J., et al. (1989). A reappraisal of the terverticillate penicillia using biochemical, physiological and morphological features. I. Numerical taxonomy. *Journal of General Microbiology*, *135*, 2941–2966.

Bruns, T. D., White, T. J., & Taylor, J. W. (1991). Fungal molecular systematics. *Annual Review of Ecology and Systematics*, *22*, 525–564.

Cannon, P. F. (1994). The newly recognized family Magnaporthaceae and its interrelationships. *Systema Ascomycetum*, *13*, 25–42.

Carmichael, J. W., Kendrick, W. B., Conners, J. L., & Sigler, L. (1980). *Genera of hyphomycetes*. Edmonton, Alberta, Canada: University of Alberta Press.

Chan, C. X., Bhattacharya, D., & Reyes-Prieto, A. (2012). Endosymbiotic and horizontal gene transfer in microbial eukaryotes: Impacts on cell evolution and the tree of life. *Mobile Genetic Elements*, *2*, 101–105.

Chen, Q., Jiang, J. R., Zhang, G. Z., Cai, L., & Crous, P. W. (2015). Resolving the Phoma enigma. *Studies in Mycology*, *82*, 137–217.

Conway, K., & Barr, M. (1977). Classification of *Ophioceras dolichostomum*. *Mycotaxon*, *5*(2), 376–380.

de Hoog, G. S., & Gerrits van den Ende, A. H. G. (1992). Nutritional pattern and ecophysiology of *Hortaea werneckii*, agent of human tinea nigra. *Antonie Van Leeuwenhoek*, *62*, 321–329.

de Hoog, G. S., Marvin-Sikkema, F. D., Lahpoor, G. A., Gottschall, J. C., Prins, R. A., & Gueho, E. (1994). Ecology and physiology of the emerging opportunistic fungi *Pseudallescheria boydii* and *Scedosporium prolificans*. *Mycoses*, *37*, 71–78.

de Hoog, G. S., & Yurlova, N. A. (1994). Conidiogenesis, nutritional physiology and taxonomy of *Aureobasidium* and *Hormonema*. *Antonie Van Leeuwenhoek*, *65*, 41–54.

Ebersberger, I., de Matos Simoes, R., Kupczok, A., Gube, M., Kothe, E., Voigt, K., et al. (2012). A consistent phylogenetic backbone for the fungi. *Molecular Biology and Evolution*, *29*, 1319–1334.

Elix, J. A. (1992). Lichen chemistry. *Flora of Australia*, *54*, 23–29.

Ellis, M. B. (1971). *Dematiaceous hyphomycetes*. Kew, UK: Commonwealth Mycological Institute.

Ellis, M. B. (1976). *More dematiaceous hyphomycetes*. Kew, UK: Commonwealth Mycological Institute.

Eriksson, O. (1984). *Outline of the ascomycetes—1984*. Umea, Sweden: University of Umeå.

Fitzpatrick, D. A., Logue, M. E., Stajich, J. E., & Butler, G. (2006). A fungal phylogeny based on 42 complete genomes derived from supertree and combined gene analysis. *BMC Evolutionary Biology*, *6*, 99.

Frisvad, J. C., & Filtenborg, O. (1990). Secondary metabolites as consistent criteria in *Penicillium* taxonomy and a synoptic key to *Penicillium* subgenus *Penicillium*. In R. A. Samson & J. I. Pitt (Eds.), *Modern concepts in penicillium and aspergillus classification* (pp. 373–384). Boston, MA: Springer.

Galagan, J. E., Calvo, S. E., Borkovich, K. A., Selker, E. U., Read, N. D., Jaffe, D., et al. (2003). The genome sequence of the filamentous fungus *Neurospora crassa*. *Nature*, *422*, 859–868.

Gawad, C., Koh, W., & Quake, S. R. (2016). Single-cell genome sequencing: Current state of the science. *Nature Reviews Genetics*, *17*, 175–188.

Glass, N. L., & Donaldson, G. C. (1995). Development of primer sets designed for use with the PCR to amplify conserved genes from filamentous ascomycetes. *Applied and Environmental Microbiology*, *61*, 1323–1330.

Goffeau, A., Barrell, B. G., Bussey, H., Davis, R. W., Dujon, B., Feldmann, H., et al. (1996). Life with 6000 genes. *Science*, *274*, 546–567.

Goodwin, S., McPherson, J. D., & McCombie, W. R. (2016). Coming of age: Ten years of next-generation sequencing technologies. *Nature Reviews Genetics*, *17*, 333–351.

Grigoriev, I. V., Cullen, D., Goodwin, S. B., Hibbett, D., Jeffries, T. W., Kubicek, C. P., et al. (2011). Fueling the future with fungal genomics. *Mycology*, *2*, 192–209.

Guarro, J., Gené, J., & Stchigel, A. M. (1999). Developments in fungal taxonomy. *Clinical Microbiology Reviews*, *12*, 454–500.

Guerber, J. C., Liu, B., Correll, J. C., & Johnston, P. R. (2003). Characterization of diversity in *Colletotrichum acutatum* sensu lato by sequence analysis of two gene introns, mtDNA and intron RFLPs, and mating compatibility. *Mycologia, 95*, 872–895.

Hawksworth, D. L. (1991). The fungal dimension of biodiversity: Magnitude, significance, and conservation. *Mycological Research, 95*, 641–655.

Hawksworth, D. L., Sutton, B. C., & Ainsworth, G. C. (1983). *Ainsworth & Bisby's dictionary of the fungi (including the Lichens)*. Kew, UK: Commonwealth Mycological Institute.

Helgason, T., Watson, I. J., & Young, J. P. W. (2003). Phylogeny of the glomerales and diversisporales (Fungi: Glomeromycota) from actin and elongation factor 1-alpha sequences. *FEMS Microbiology Letters, 229*, 127–132.

Hibbett, D. (1992). Ribosomal RNA and fungal systematics. *Transactions of the Mycological Society of Japan, 33*, 533–556.

Hibbett, D., Abarenkov, K., Koljalg, U., Opik, M., Chai, B., Cole, J., et al. (2016). Sequence-based classification and identification of Fungi. *Mycologia, 108*, 1049–1068.

Hibbett, D. S., Binder, M., Bischoff, J. F., Blackwell, M., Cannon, P. F., Eriksson, O. E., et al. (2007). A higher-level phylogenetic classification of the Fungi. *Mycological Research, 111*, 509–547.

Hong, S. B., Go, S. J., Shin, H. D., Frisvad, J. C., & Samson, R. A. (2005). Polyphasic taxonomy of *Aspergillus fumigatus* and related species. *Mycologia, 97*, 1316–1329.

Huhndorf, S. M., Greif, M., Mugambi, G. K., & Miller, A. N. (2008). Two new genera in the Magnaporthaceae, a new addition to *Ceratosphaeria* and two new species of *Lentomitella*. *Mycologia, 100*, 940–955.

James, T. Y., Kauff, F., Schoch, C. L., Matheny, P. B., Hofstetter, V., Cox, C. J., et al. (2006). Reconstructing the early evolution of Fungi using a six-gene phylogeny. *Nature, 443*, 818–822.

Kimbrough, J. W. (1994). Septal ultrastructure and ascomycete systematics. In D. Hawksworth (Ed.), *Ascomycete systematics: Problems and perspectives in the nineties* (pp. 127–141). New York: Plenum Press.

Kirk, P. M., Cannon, P. F., Minter, D. W., & Stalpers, J. A. (2008). *Ainsworth & Bisby's dictionary of the fungi*. Wallingford, CT: CABI.

Klaubauf, S., Tharreau, D., Fournier, E., Groenewald, J. Z., Crous, P. W., de Vries, R. P., et al. (2014). Resolving the polyphyletic nature of Pyricularia (Pyriculariaceae). *Studies in Mycology, 79*, 85–120.

Krause, R. A., & Webster, R. K. (1972). The morphology, taxonomy, and sexuality of the rice stem rot fungus, *Magnaporthe salvinii* (*Leptosphaeria salvinii*). *Mycologia, 64*, 103–114.

Kretzer, A. M., & Bruns, T. D. (1999). Use of atp6 in fungal phylogenetics: An example from the boletales. *Molecular Phylogenetics and Evolution, 13*, 483–492.

Liu, Y., Leigh, J. W., Brinkmann, H., Cushion, M. T., Rodriguez-Ezpeleta, N., Philippe, H., et al. (2009). Phylogenomic analyses support the monophyly of Taphrinomycotina, including *Schizosaccharomyces* fission yeasts. *Molecular Biology and Evolution, 26*, 27–34.

Liu, Y., Steenkamp, E. T., Brinkmann, H., Forget, L., Philippe, H., & Lang, B. F. (2009). Phylogenomic analyses predict sistergroup relationship of nucleariids and fungi and paraphyly of zygomycetes with significant support. *BMC Evolutionary Biology, 9*, 272.

Liu, Y. J., Whelen, S., & Hall, B. D. (1999). Phylogenetic relationships among ascomycetes: Evidence from an RNA polymerase II subunit. *Molecular Biology and Evolution, 16*, 1799–1808.

Lü, H. S., & Mclaughlin, D. J. (1991). Ultrastructure of the septal pore apparatus and early septum initiation in *Auricularia auricula-judae*. *Mycologia, 83*, 322–334.

Luo, J., Qiu, H., Cai, G., Wagner, N. E., Bhattacharya, D., & Zhang, N. (2015). Phylogenomic analysis uncovers the evolutionary history of nutrition and infection mode in rice blast fungus and other Magnaporthales. *Scientific Reports, 5*, 9448.

Luo, J., Walsh, E., & Zhang, N. (2015). Toward monophyletic generic concepts in Magnaporthales: Species with Harpophora asexual states. *Mycologia, 107*, 641–646.

Luo, J., & Zhang, N. (2013). *Magnaporthiopsis*, a new genus in Magnaporthaceae (Ascomycota). *Mycologia, 105*, 1019–1029.

Lutzoni, F., Kauff, F., Cox, C. J., McLaughlin, D., Celio, G., Dentinger, B., et al. (2004). Assembling the fungal tree of life: Progress, classification, and evolution of subcellular traits. *American Journal of Botany, 91*, 1446–1480.

Maharachchikumbura, S. S. N., Hyde, K. D., Jones, E. B. G., McKenzie, E. H. C., Bhat, J. D., Dayarathne, M. C., et al. (2016). Families of sordariomycetes. *Fungal Diversity, 79*, 1–317.

Martinez, D., Larrondo, L. F., Putnam, N., Gelpke, M. D., Huang, K., Chapman, J., et al. (2004). Genome sequence of the lignocellulose degrading fungus *Phanerochaete chrysosporium* strain RP78. *Nature Biotechnology, 22*, 695–700.

Matheny, P. B., Liu, Y. J., Ammirati, J. F., & Hall, B. D. (2002). Using RPB1 sequences to improve phylogenetic inference among mushrooms (Inocybe, Agaricales). *American Journal of Botany, 89*, 688–698.

Miettinen, O., Larsson, E., Sjökvist, E., & Larsson, K.-H. (2012). Comprehensive taxon sampling reveals unaccounted diversity and morphological plasticity in a group of dimitic polypores (Polyporales, Basidiomycota). *Cladistics, 28*, 251–270.

Monte, E., Bridge, P. D., & Sutton, B. C. (1990). Physiological and biochemical studies in Coelomycetes. *Phoma. Studies in Mycology, 32*, 21–28.

Mordue, J. E. M., Currah, R. S., & Bridge, P. D. (1989). An integrated approach to *Rhizoctonia* taxonomy: Cultural, biochemical and numerical techniques. *Mycological Research, 92*, 78–90.

Nag Raj, T. R. (1993). *Coelomycetous anamorphs with appendage bearing conidia.* Waterloo, Ontario: Mycologue Publications.

Nylander, W. (1866). Hypochlorite of lime and hydrate of potash, two new criteria in the study of Lichens. *Botanical Journal of the Linnean Society, 9*, 358–365.

Paterson, R. R. M., & Bridge, P. D. (1994). *Biochemical techniques for filamentous fungi.* Wallingford, UK: CAB International.

Qiu, H., Cai, G., Luo, J., Bhattacharya, D., & Zhang, N. (2016). Extensive horizontal gene transfers between plant pathogenic fungi. *BMC Biology, 14*, 41.

Rath, A. C., Carr, C. J., & Graham, B. R. (1995). Characterization of *Metarhizium anisopliae* strains by carbohydrate utilization (API50CH). *Journal of Invertebrate Pathology, 65*, 152–161.

Rehner, S. A., & Buckley, E. (2005). A *Beauveria* phylogeny inferred from nuclear ITS and EF1-alpha sequences: Evidence for cryptic diversification and links to *Cordyceps* teleomorphs. *Mycologia, 97*, 84–98.

Robbertse, B., Reeves, J. B., Schoch, C. L., & Spatafora, J. W. (2006). A phylogenomic analysis of the Ascomycota. *Fungal Genetics and Biology, 43*, 715–725.

Rokas, A., Williams, B. L., King, N., & Carroll, S. B. (2003). Genome-scale approaches to resolving incongruence in molecular phylogenies. *Nature, 425*, 798–804.

Rolland, L. (1887). De la coloration en bleu développée par l'iode sur divers champignons. *Bulletin de la Société Mycologique de France, 3*, 134–137.

Saiki, R. K., Gelfand, D. H., Stoffel, S., Scharf, S. J., Higuchi, R., Horn, G. T., et al. (1988). Primer-directed enzymatic amplification of DNA with a thermostable DNA polymerase. *Science, 239*, 487–491.

Sanger, F., Nicklen, S., & Coulson, A. R. (1977). DNA sequencing with chain-terminating inhibitors. *Proceedings of the National Academy of Sciences of the United States of America, 74*, 5463–5467.

Schmitt, I., Crespo, A., Divakar, P. K., Fankhauser, J. D., Herman-Sackett, E., Kalb, K., et al. (2009). New primers for promising single-copy genes in fungal phylogenetics and systematics. *Persoonia, 23*, 35–40.

Schoch, C. L., Robbertse, B., Robert, V., Vu, D., Cardinali, G., Irinyi, L., et al. (2014). Finding needles in haysacks: Linking scientific names, reference specimens and molecular data for Fungi. *Database*, *2014*, 1–21.

Schoch, C. L., Seifert, K. A., Huhndorf, S., Robert, V., Spouge, J. L., Levesque, C. A., et al. (2012). Nuclear ribosomal internal transcribed spacer (ITS) region as a universal DNA barcode marker for Fungi. *Proceedings of the National Academy of Sciences of the United States of America*, *109*, 6241–6246.

Schoch, C. L., Sung, G. H., Lopez-Giraldez, F., Townsend, J. P., Miadlikowska, J., Hofstetter, V., et al. (2009). The Ascomycota tree of life: A phylum-wide phylogeny clarifies the origin and evolution of fundamental reproductive and ecological traits. *Systematic Biology*, *58*, 224–239.

Schüßler, A., & Walker, C. (2010). *The glomeromycota: A species list with new families and new genera*. The Royal Botanic Garden Kew: Botanische Staatssammlung Munich, and Oregon State University.

Seif, E., Leigh, J., Liu, Y., Roewer, I., Forget, L., & Lang, B. F. (2005). Comparative mitochondrial genomics in zygomycetes: Bacteria-like RNase P RNAs, mobile elements and a close source of the group I intron invasion in angiosperms. *Nucleic Acids Research*, *33*, 734–744.

Seifert, K. A., Samson, R. A., Dewaard, J. R., Houbraken, J., Levesque, C. A., Moncalvo, J. M., et al. (2007). Prospects for fungus identification using CO1 DNA barcodes, with *Penicillium* as a test case. *Proceedings of the National Academy of Sciences of the United States of America*, *104*, 3901–3906.

Shearer, C. A. (1989). *Pseudohalonectria* (Lasiosphaeriaceae), an antagonistic genus from wood in freshwater. *Canadian Journal of Botany*, *67*, 1944–1955.

Soanes, D., & Richards, T. A. (2014). Horizontal gene transfer in eukaryotic plant pathogens. *Annual Review of Phytopathology*, *52*, 583–614.

Spatafora, J. W., Chang, Y., Benny, G. L., Lazarus, K., Smith, M. E., Berbee, M. L., et al. (2016). A phylum-level phylogenetic classification of zygomycete fungi based on genome-scale data. *Mycologia*, *108*, 1028–1046.

Spatafora, J. W., Sung, G. H., Johnson, D., Hesse, C., O'Rourke, B., Serdani, M., et al. (2006). A five-gene phylogeny of Pezizomycotina. *Mycologia*, *98*, 1018–1028.

Sutton, B. C. (1980). *The coelomycetes: Fungi imperfecti with pycnidia acervuli and stromata*. Kew, UK: Commonwealth Mycological Institute.

Tanaka, K., Hirayama, K., Yonezawa, H., Sato, G., Toriyabe, A., Kudo, H., et al. (2015). Revision of the *Massarineae* (Pleosporales, Dothideomycetes). *Studies in Mycology*, *82*, 75–136.

Tehler, A., Farris, J. S., Lipscomb, D. L., & Källersjö, M. (2000). Phylogenetic analyses of fungi based on large rDNA data sets. *Mycologia*, *92*, 459–474.

Templeton, M. D., Rikkerink, E. H. A., Solon, S. L., & Crowhurst, R. N. (1992). Cloning and molecular characterization of the glyceraldehyde-3-phosphate dehydrogenase-encoding gene and cDNA from the plant pathogenic fungus *Glomerella cingulata*. *Gene*, *122*, 225–230.

Thongkantha, S., Jeewon, R., Vijaykrishna, D., Lumyong, S., McKenzie, E. H. C., & Hyde, K. D. (2009). Molecular phylogeny of Magnaporthaceae (Sordariomycetes) with a new species *Ophioceras chiangdaoense* from *Dracaena loureiroi* in Thailand. *Fungal Diversity*, *34*, 157–173.

Tisserant, E., Malbreil, M., Kuo, A., Kohler, A., Symeonidi, A., Balestrini, R., et al. (2013). Genome of an arbuscular mycorrhizal fungus provides insight into the oldest plant symbiosis. *Proceedings of the National Academy of Sciences of the United States of America*, *110*, 20117–20122.

Vogel, K. J., & Moran, N. A. (2013). Functional and evolutionary analysis of the genome of an obligate fungal symbiont. *Genome Biology and Evolution*, *5*, 891–904.

von Arx, J. A., & Müller, E. (1975). A re-evaluation of the bitunicate Ascomycetes with keys to families and genera. *Studies in Mycology*, *9*, 136–142.

Wang, H., Xu, Z., Gao, L., & Hao, B. (2009). A fungal phylogeny based on 82 complete genomes using the composition vector method. *BMC Evolutionary Biology*, *9*, 195.

Wasfy, E. H., Bridge, P. D., & Brayford, D. (1987). Preliminary studies on the use of biochemical and physiological tests for the characterization of *Fusarium* isolates. *Mycopathologia*, *99*, 9–13.

Whalley, A. J. S., & Edwards, R. L. (1995). Secondary metabolites and systematic arrangement within the Xylariaceae. *Canadian Journal of Botany*, *73*, 802–810.

White, T., Bruns, T., Lee, S., & Taylor, J. (1990). Amplification and direct sequencing of fungal ribosomal RNA genes for phylogenetics. In M. Innis, et al. (Ed.), *PCR protocols: A guide to methods and applications* (pp. 315–322). New York: Academic Press.

Wood, V., Gwilliam, R., Rajandream, M. A., Lyne, M., Lyne, R., Stewart, A., et al. (2002). The genome sequence of *Schizosaccharomyces pombe*. *Nature*, *415*, 871–880.

Woudenberg, J. H. C., Seidl, M. F., Groenewald, J. Z., de Vries, M., Stielow, J. B., Thomma, B. P. H. J., et al. (2015). *Alternaria* section *Alternaria*: Species, formae speciales or pathotypes? *Studies in Mycology*, *82*, 1–21.

Zhang, N., & Blackwell, M. (2001). Molecular phylogeny of dogwood anthracnose fungus (*Discula destructiva*) and the Diaporthales. *Mycologia*, *93*, 355–365.

Zhang, N., Castlebury, L. A., Miller, A. N., Huhndorf, S. M., Schoch, C. L., Seifert, K. A., et al. (2006). An overview of the systematics of the Sordariomycetes based on a four-gene phylogeny. *Mycologia*, *98*, 1076–1087.

Zhang, N., Zhao, S., & Shen, Q. (2011). A six-gene phylogeny reveals the evolution of mode of infection in the rice blast fungus and allied species. *Mycologia*, *103*, 1267–1276.

Zoller, S., Scheidegger, C., & Sperisen, C. (1999). PCR primers for the amplification of mitochondrial small subunit ribosomal DNA of lichen-forming ascomycetes. *The Lichenologist*, *31*, 511–516.